AF352925

Assessment of Different Contaminants in Freshwater

Assessment of Different Contaminants in Freshwater

Origin, Fate, and Ecological Impact

Editors

Diana M. P. Galassi
Tiziana Di Lorenzo
Grant Hose

MDPI • Basel • Beijing • Wuhan • Barcelona • Belgrade • Manchester • Tokyo • Cluj • Tianjin

Editors
Diana M. P. Galassi
University of L'Aquila
Italy

Grant Hose
Macquarie University
Australia

Tiziana Di Lorenzo
Research Institute on Terrestrial Ecosystems
CNR—National Research Council of Italy
Italy

Editorial Office
MDPI
St. Alban-Anlage 66
4052 Basel, Switzerland

This is a reprint of articles from the Special Issue published online in the open access journal *Water* (ISSN 2073-4441) (available at: https://www.mdpi.com/journal/water/special_issues/freshwater_assessment).

For citation purposes, cite each article independently as indicated on the article page online and as indicated below:

LastName, A.A.; LastName, B.B.; LastName, C.C. Article Title. *Journal Name* **Year**, *Article Number*, Page Range.

ISBN 978-3-03943-000-0 (Hbk)
ISBN 978-3-03943-001-7 (PDF)

Cover image courtesy of Diana Maria Paola Galassi.

Contents

About the Editors

Diana M. P. Galassi is a zoologist and environmental biologist. Her main fields of research include the systematics, ecology, and biogeography of the Copepoda Cyclopoida, Harpacticoida, Calanoida, and Isopoda Microparasellidae (Crustacea); the study of the main distribution drivers for freshwater meiofauna in both surface waters and aquatic groundwater-dependent ecosystems in both natural and polluted conditions; the assessment of ecological and biogeographic factors that affect the spatial distribution of groundwater invertebrates at different spatial scales; and the role of groundwater invertebrates in groundwater biodiversity assessment and conservation. Her research is predominantly multidisciplinary, by coupling hydrochemistry, hydrogeology, statistical ecology, geostatistics, and biology of groundwater.

Tiziana Di Lorenzo is an aquatic ecologist and ecotoxicologist with a background in groundwater ecology. She carries out field studies aimed at evaluating groundwater ecosystem conditions with particular reference to porous aquifers subject to anthropic stress, including climate change. Her research deals mainly with freshwater meiofauna with an emphasis on groundwater crustaceans. She has recently focused on physiology studies with particular reference to respiration as a proxy of energy consumption of groundwater organisms. She explores the behavior of groundwater crustaceans with the aim of assessing behavioral changes under chemical stress. At her laboratories, Tiziana Di Lorenzo is supplied with an 80-μL and 2 mL multiplate microrespirometer, specifically designed for respirometric studies with small-sized invertebrates.

Grant Hose is an aquatic ecologist and ecotoxicologist. His research examines the response of groundwater and surface water ecosystems, invertebrate and microbial communities to environmental change, and develops tools for assessing change in ecosystem health and condition. He leads a diverse and enthusiastic research group who has undertaken field surveys and laboratory manipulative experiments to identify correlative and causal links to community change. His team has expertise in invertebrate taxonomy and using environmental DNA (eDNA) to characterize the composition and function of aquatic ecosystems. His current research focuses on the ecology of groundwater ecosystems and assessment of ecosystem health, ecological risk assessments and ecotoxicology for groundwater biota, and the roles of stygofauna in providing ecosystem services.

Preface to "Assessment of Different Contaminants in Freshwater"

Freshwater ecosystems cover over 15% of the world's surface and provide pivotal ecosystem services that sustain human society. However, fast-growing anthropogenic activities have deleterious impacts on these ecosystems. In this Special Issue, we collect 10 studies that encompass five different factors of freshwater contamination: landfill leaks, nutrients, heavy metals, emerging organic contaminants, and marble slurry. Using different approaches, the studies detailed the direct and indirect effects that these contaminants have on a range of freshwater organisms, from bacteria to vertebrates. Although the papers here focused on specific case studies, they also exemplified common issues, such as expanding in groundwaters, hyporheic zones, streams, lakes, and ponds around the world. Every issue presented in this Special Issue is in dire need of being continuously discussed among scientists, end-users, and policymakers. To this end, the Special Issue presents a new free software suite to analyze ecological risk and conservation priority of freshwater ecosystems. The software can support local authorities in the preparation of management plans for freshwater basins pursuant in the water directives in Europe.

Diana M. P. Galassi, Tiziana Di Lorenzo, Grant Hose
Editors

Editorial

Assessment of Different Contaminants in Freshwater: Origin, Fate and Ecological Impact

Tiziana Di Lorenzo [1,*], Grant C. Hose [2] and Diana M.P. Galassi [3]

[1] Research Institute on Terrestrial Ecosystems CNR—National Research Council of Italy, Via Madonna del Piano 10, 50019 Florence, Italy

[2] Department of Biological Sciences, Macquarie University, Sydney, NSW 2109, Australia; grant.hose@mq.edu.au

[3] Department of Life, Health and Environmental Sciences, University of L'Aquila, Via Vetoio 1, 67100 L'Aquila, Italy; dianamariapaola.galassi@univaq.it

* Correspondence: tiziana.dilorenzo@cnr.it

Received: 1 June 2020; Accepted: 23 June 2020; Published: 24 June 2020

Abstract: Freshwater ecosystems cover over 15% of the world's surface and provide ecosystem services that are pivotal in sustaining human society. However, fast-growing anthropogenic activities have deleterious impacts on these ecosystems. In this Special Issue, we collect ten studies encompassing five different factors of freshwater contamination: landfill leaks, nutrients, heavy metals, emerging organic contaminants and marble slurry. Using different approaches, the studies detailed the direct and indirect effects that these contaminants have on a range of freshwater organisms, from bacteria to vertebrates. Although the papers covered here focused on specific case studies, they exemplify common issues that are expanding in groundwaters, hyporheic zones, streams, lakes and ponds around the world. All the aspects of these issues are in dire need of being continuously discussed among scientists, end-users and policy-makers. To this end, the Special Issue presents a new free software suite for the analysis of the ecological risk and conservation priority of freshwater ecosystems. The software can support local authorities in the preparation of management plans for freshwater basins pursuant to the Water Directives in Europe.

Keywords: heavy metals; EOCs; nitrate; landfill; marble slurry; neonicotinoids; microplastics; software

---

1. Introduction

Freshwater sustains human society with some 93,000 km^3 of water stored in lakes and rivers and much more in groundwater or as ice [1], covering about 15% of the world's surface [2]. From nanograms of pharmaceuticals to large pieces of plastics, freshwater ecosystems are a sink for anthropogenic contamination, which is predicted to become more severe in the coming decades as a result of population growth and urbanization. Contaminants can enter freshwater directly, through countless anthropogenic activities ranging from legal and illegal discharges from factories to imperfect water treatment plants, landfill leachates, mines, quarries and agricultural runoff. Mining activities are the major contributors of heavy metals [3], while emerging organic compounds (EOCs), including pesticides, pharmaceuticals and personal care products, are mainly released by wastewater treatment plants and agricultural practices [4]. Long-lasting contamination with nitrates, nitrites, ionized ammonia and pesticides in groundwater is a common consequence of intensive agriculture [5,6]. Modern high-tech methodologies, such as nanofiltration, electrodeposition, and sequestration, can only help in stemming the tide of aquatic pollution without solving the problem completely [7]. Strict obedience to environmental laws and regulations is the only way of bringing freshwater pollution down to its barest minimum [8].

As a result of contamination, biodiversity loss continues to occur in freshwater ecosystems at a rate double that in other ecosystems [9]. Rivers and lakes sustain nearly 10% of all described animal species, including 40% of the world's fish diversity and a third of all vertebrate species [10], while groundwater hosts at least 5000 species of invertebrates, of which 70% are represented by crustaceans [11,12]. These numbers translate into a great diversity of life forms, consisting of prokaryotes and microscopic single-cell eukaryotes through to meiofauna and macrofauna and up to vertebrates. Freshwater organisms are subjected directly to the toxic actions of pollutants. Their responses to contaminants are varied, and the most drastic responses are represented by death or migration, potentially leading to local extinction. Other responses may include reductions in reproductive capacity and morphological and physiological changes. Changes at the community, population and individual levels not only constitute direct evidence of the ecological impact that a contaminant can have in freshwater ecosystems but can also serve as bioindicators to track the source and fate of contamination.

This Special Issue collects ten studies on the response of freshwater ecosystems to anthropogenic pollution. The papers detail the direct and indirect effects that pollutants have on a range of freshwater aquatic organisms, from microbes to vertebrates. The overview covers a variety of contaminants, from nutrients and heavy metals to emerging contaminants and microplastics, and from landfill leaks to marble quarry slurries. The studies involve different types of fresh water ecosystems, from groundwater to mountain karst ponds, hyporheic zones, stream waters and lakes. The papers of the Special Issue are hereafter presented with reference to the types of water bodies to which they refer. A final paragraph is dedicated to a new software tool for the assessment of the ecological risk and the conservation priority of freshwater ecosystems.

2. Overview

2.1. Groundwater

Changes in groundwater quality, such those induced by municipal solid waste disposal in landfills, can directly affect the functional properties of aquatic microbial communities. Melita et al. [13] explored the structural alterations, as well as the functional responses, of groundwater microbial communities to a gradient of geochemical alterations induced by a municipal solid waste landfill in central Italy. The microbial communities were characterized by flow cytometry and the Biolog EcoPlates^{TM} assay, with the aim of highlighting biogeochemical cycling patterns along the gradient. The microbial communities from the pristine and mildly-altered groundwaters showed a low affinity for most of the carbon sources. By contrast, the microbial community thrived in the altered groundwater conditions defined by high concentrations of organic matter and fecal bacteria. The profiles of the carbon substrate use highlighted the efficient metabolization of a large number of organic compounds, including those with complex molecular structures.

Nitrate is likely to be the most common pollutant in groundwater and is such a serious pollutant that the EU dedicated the subject of a specific directive to it. In alluvial aquifers of agricultural areas, the ion can reach concentrations up to hundreds of milligrams per liter. In aquifers that are well oxygenated and have a low organic content, denitrification processes (i.e., the reduction of nitrate) do not take place or take place at such low rates that they do not allow the recovery of the groundwater body. Nitrate contamination can, therefore, persist for years [14]. Di Lorenzo et al. [15] analyzed the effects of long-term nitrate contamination on the groundwater fauna of an alluvial aquifer in central Italy. The study revealed that structural traits of the biological assemblages, such as the ratios of juveniles to adults and of males to females, as well as the relationships between abundances and biomasses, provide indicators of the alteration of the communities in a more efficient way than classical taxonomy-based analyses, which are focused on species richness and abundances only.

Both taxonomy-based and trait-based approaches, although providing important information, are not able to identify which is the lethal concentration of a pollutant for a given species. To obtain this information, it is necessary to carry out ecotoxicological tests; that is to say, experiments in controlled laboratory conditions. Unfortunately, ecotoxicological studies with the so-called stygobiotic species (i.e., species that carry out their entire life cycles in groundwater) are extremely few due to the low abundances with which these animals are collected and the difficulty of breeding them in the laboratory [16,17]. Hose et al. [18] provided an important contribution to the field of ecotoxicology with stygofauna, through testing the effects of some heavy metals—such as As, Cr and Zn—on an undescribed species of syncarid (Malacostraca: Syncarida: Bathynellidae). The study revealed the high sensitivity, both in terms of survival rates and bioaccumulation, of syncarids to metals, with particular reference to As. Zinc was the least toxic metal for the syncarids, which showed significant metal bioaccumulation at exposure concentrations higher than 1 mg of Zn/L.

If the studies on the toxic effects of chemical pollutants on stygobiotic species are few, the analysis of the impact of non-chemical stressors, such as marble slurries, on stygobiotic communities are, to date, non-existent. This Special Issue presents the first preliminary study on this topic. Piccini et al. [19] characterized, from a sedimentological and chemical point of view, the fine marble powder that is produced at quarries during the extraction and squaring of marble blocks by means of chain and diamond-wire saws. The study was conducted on an iconic area of marble quarry activity, the Apuan Alps in central Italy, a karst area exploited for marble since the time of the ancient Romans. The results of the study showed that, during intense rain events, the marble slurry—which, in the Apuan Alps, is called *marmettola*—is washed into the karst springs, making the water temporarily unsuitable for domestic uses. A very preliminary investigation suggests that the *marmettola*, consisting of calcite granules with a diameter between 8 and 32 µm, could fill the voids of the Apuan Alps karst system, making the voids unavailable for some stygobiotic taxa. The results need to be supported by further studies; however, stygobiotic amphipods are, in fact, unable to penetrate fine-grained sediment with a diameter <60 µm, as already observed by Korbel et al. [20].

2.2. Streams and Hyporheic Zones

Insecticides, especially neonicotinoids, are among the most commonly used and detected EOCs in stream waters worldwide [21]. Hunn et al. [22] investigated the effects of the neonicotinoid imidacloprid on the mayfly *Deleatidium* spp. (Leptophlebiidae), collected from the Silver Stream, a fourth-order stream near Dunedin in New Zealand. Imidacloprid exposure had severe lethal and sublethal effects on *Deleatidium* spp., with significant impairment at concentrations as low as 1 µg/L of imidacloprid. Starvation worsened the effect of the neonicotinoid on the mayfly species, affecting their ability to swim or right themselves and leading to immobility.

The contamination of freshwaters with EOCs is predicted to become more pronounced in the coming decades [4]. The hyporheic zone (i.e., the zone where water is exchanged between the open channel and the saturated permeable streambed sediments) is particularly susceptible to EOC contamination due to the long residence times of these toxicants in the pore-spaces and, consequently, long exposure periods for organisms dwelling in the hyporheic zone. Peralta-Maraver et al. [23] reported the effect of EOC pollution on the composition and abundance–body mass (N–M) relationships of the hyporheic communities (prokaryotes, single cell eukaryotes, meiofauna and macrofauna) of thirty streams in the UK. Emerging organic compounds negatively affected the hyporheic assemblages, inducing significant changes in the N–M coefficients, with an increase in the biomasses and abundances of the organisms chiefly associated with pollutant-tolerant groups such as Asellidae and Oligochaeta.

Under a scenario of pollution and a concomitant increase in tolerant, large-sized taxa, bioaccumulation is expected to be more acute. Labuschagne et al. [24] tested the potential for the bioaccumulation of platinum group elements (PGEs) using artificial and natural bivalves. Platinum group elements are pollutants of emerging concern almost everywhere but particularly in areas where PGEs are mined. The study was undertaken along a contamination gradient in the Hex River, South Africa and compared the accumulation of metals in a passive sampling device made up of artificial mussels with the uptake by transplanted individuals of the freshwater clam *Corbicula fluminali africana*. Both bioindicators were efficient in highlighting the presence of As, Cd, Co, Cr, Ni, Pb, Pt, V and

Zn, but the uptake pattern for Pt, Cr and Ni—between the artificial mussels and natural clams—was different. The results showed that a combination of the two bioindicators, rather than the use of only one of the two, provided a more holistic assessment of PGE contamination in the Hex River.

2.3. Lakes and Karst Mountain Ponds

Morphological deformities of chironomids are known to provide information about contamination by heavy metals, pesticides and, in general, substances that act as endocrine disruptors [25]. Goretti et al. [26] examined the incidence of mentum deformities in populations of *Chironomus plumosus* (Diptera: Chironomidae) in Lake Trasimeno, central Italy. The 26-month study showed that <10% of the *C. plumous* population had a mentum deformity, with a higher incidence during spring and in the littoral populations. The most common type of deformity found was the "round/filed teeth" type. Compared to the results of the years 2000–2010, the incidence of *C. plumosus* mentum deformities has decreased significantly. This result is suggestive of a recovery of the chemical conditions of Lake Trasimeno in the last decade.

The study of Iannella et al. [27] focused on karst mountain ponds, which are high-altitude ecosystems usually exploited as watering points for livestock. The ponds may show high ammonium concentrations and the significant depletion of dissolved oxygen due to eutrophication induced by cattle excrement. The study, which examined the karst ponds of the Apennines in central Italy, showed that the enrichment of the ponds with organic nutrients due to the manure may have a significant effect on the diet of the Italian crested newt *Triturus carnifex*, a "Near Threatened" species included in the Italian IUCN red list. In particular, the study reported an impoverishment of the *T. carnifex* diet in the most nutrient-enriched ponds, probably due to a low diversity of the benthic invertebrate assemblages. Alarmingly, the study also reported the first evidence of the occurrence of microplastics in the stomach contents of *T. carnifex*.

2.4. AQUALIFE Software

Finally, Strona et al. [28] presented the AQUALIFE software, a user-friendly online interface (available at http://app.aqualifeproject.eu), that can be used to assess the ecological and hydrological-hydromorphological risks to aquatic ecosystems dependent on groundwater (e.g., springs, hyporheic zones, streams and rivers, aquifers etc.) along with their conservation priority. The Ecological Risk suite of the AQUALIFE software computes the probability (from 0 to ∞) that a toxicant may cause the impairment of a biological aquatic community at a measured environmental concentration. The Hydrological-Hydromorphological Risk suite identifies and scores the risks posed by human-induced alterations to the hydrological connectivity of springs and hyporheic zones. Finally, the Conservation Priority Index suite determines the conservation priority of aquatic ecosystems based on selected species traits such as distribution, endemicity, degree of groundwater dependence, ecological affinity to groundwater, ecological niche breadth, thermal tolerance, microhabitat preferences, trophic role, life span, frequency of occurrence, abundance, evolutionary origin and phylogenetic rarity. With its multi-metric facet, the AQUALIFE software provides a novel tool to assist policy-makers and local authorities in managing groundwater-dependent ecosystems.

3. Conclusions

In this Special Issue, we had the pleasure of collecting ten studies that each increased the level of knowledge regarding the origin, fate and ecological impact of contaminants on freshwater ecosystems. The studies covered various geographical areas, from Europe to South Africa to Australia and New Zealand, and numerous types of aquatic environments, from groundwater to karst mountain ponds, rivers, lakes and the areas of transition between surface water and groundwater. The stressors examined in these ten studies ranged from heavy metals to emerging organic pollutants, and also included non-chemical contaminants, such as the dust produced at marble quarries. The approaches used were ecotoxicological, taxonomy-based and trait-based, and touched upon many of the components of the freshwater communities, from bacteria to vertebrates. We believe that the collation of these papers may contribute to and stimulate further interest in some critical issues of vulnerability in freshwater ecosystems.

Author Contributions: D.M.P.G. conceived the special issue and all authors were in charge of overall direction and planning. T.D.L. took lead in writing the manuscript with input from all authors. G.C.H. provided critical feedback. All authors have read and agreed to the published version of the manuscript.

Funding: This research received no external funding.

Acknowledgments: The Guest Editors (DMPG, GCH and TDL) thank both the research community for offering and contributing a wide range of valuable papers, and the publisher MDPI for allocating resources and support towards this Special Issue.

Conflicts of Interest: The author declares no conflict of interest.

References

1. Bunn, S.E. Grand Challenge for the Future of Freshwater Ecosystems. *Front. Environ. Sci.* **2016**, *4*, 257. [CrossRef]
2. Bassem, S.M. Water pollution and aquatic biodiversity. *Biodivers. Int. J.* **2020**, *4*, 10–16.
3. Erasmus, J.; Malherbe, W.; Zimmermann, S.; Lorenz, A.; Nachev, M.; Wepener, V.; Sures, B.; Smit, N.J. Metal accumulation in riverine macroinvertebrates from a platinum mining region. *Sci. Total. Environ.* **2020**, *703*, 134738. [CrossRef]
4. Pal, A.; Gin, K.Y.-H.; Lin, A.Y.-C.; Reinhard, M. Impacts of emerging organic contaminants on freshwater resources: Review of recent occurrences, sources, fate and effects. *Sci. Total. Environ.* **2010**, *408*, 6062–6069. [CrossRef] [PubMed]
5. Boy-Roura, M.; Nolan, B.T.; Menció, A.; Mas-Pla, J. Regression model for aquifer vulnerability assessment of nitrate pollution in the Osona region (NE Spain). *J. Hydrol.* **2013**, *505*, 150–162. [CrossRef]
6. Lewis, K.; Tzilivakis, J.; Warner, D.; Green, A. An international database for pesticide risk assessments and management. *Hum. Ecol. Risk Assess.* **2016**, *22*, 1–15. [CrossRef]
7. Vörösmarty, C.J.; Pahl-Wostl, C.; Bunn, S.E.; Lawford, R. Global water, the anthropocene and the transformation of a science. *Curr. Opin. Environ. Sustain.* **2013**, *5*, 539–550. [CrossRef]
8. UN-United Nations. Report of the Inter-Agency and Expert Group on Sustainable Development Goal Indicators. In Proceedings of the 47th Session of the United Nations Statistical Commission, New York, NY, USA, 8–11 March 2016.
9. MEA-Millennium Ecosystem Assessment. *Ecosystems and Human Well-being: Synthesis*; Island Press: Washington, DC, USA, 2005.
10. Balian, E.V.; Segers, H.; Martens, K.; Leveque, C. The Freshwater Animal Diversity Assessment: An overview of the results. *Freshw. Anim. Diversity Assess.* **2008**, *198*, 627–637. [CrossRef]
11. Guzik, M.T.; Austin, A.D.; Cooper, S.J.; Harvey, M.S.; Humphreys, W.F.; Bradford, T.; Tomlinson, M. Is the Australian subterranean fauna uniquely diverse? *Invertebr. Syst.* **2010**, *24*, 407–418. [CrossRef]
12. Por, F.D.; Botosaneanu, L. Stygofauna Mundi, a Faunistic, Distributional, and Ecological Synthesis of the World Fauna Inhabiting Subterranean Waters (Including the Marine Interstitial). *J. Crustac. Boil.* **1987**, *7*, 203. [CrossRef]
13. Melita, M.; Amalfitano, S.; Preziosi, E.; Ghergo, S.; Frollini, E.; Parrone, D.; Zoppini, A. Physiological Profiling and Functional Diversity of Groundwater Microbial Communities in a Municipal Solid Waste Landfill Area. *Water* **2019**, *11*, 2624. [CrossRef]
14. Di Lorenzo, T.; Fiasca, B.; Tabilio, A.D.C.; Murolo, A.; Di Cicco, M.; Galassi, D.M.P. The weighted Groundwater Health Index (wGHI) by Korbel and Hose (2017) in European groundwater bodies in nitrate vulnerable zones. *Ecol. Indic.* **2020**, *116*, 106525. [CrossRef]
15. Di Lorenzo, T.; Murolo, A.; Fiasca, B.; Di Camillo, A.T.; Di Cicco, M.; Lombardo, P. Potential of A Trait-Based Approach in the Characterization of An N-Contaminated Alluvial Aquifer. *Water* **2019**, *11*, 2553. [CrossRef]
16. Di Lorenzo, T.; Di Marzio, W.D.; Fiasca, B.; Galassi, D.M.P.; Korbel, K.; Iepure, S.; Pereira, J.L.; Reboleira, A.S.P.; Schmidt, S.; Hose, G.C. Recommendations for ecotoxicity testing with stygobiotic species in the framework of groundwater environmental risk assessment. *Sci. Total. Environ.* **2019**, *681*, 292–304. [CrossRef] [PubMed]

17. Castaño-Sánchez, A.; Hose, G.C.; Reboleira, A.S.P. Ecotoxicological effects of anthropogenic stressors in subterranean organisms: A review. *Chemosphere* **2020**, *244*, 125422. [CrossRef] [PubMed]

18. Hose, G.C.; Symington, K.; Lategan, M.J.; Siegele, R. The Toxicity and Uptake of As, Cr and Zn in a Stygobitic Syncarid (Syncarida: Bathynellidae). *Water* **2019**, *11*, 2508. [CrossRef]

19. Piccini, L.; Di Lorenzo, T.; Costagliola, P.; Lombardo, P. Marble Slurry's Impact on Groundwater: The Case Study of the Apuan Alps Karst Aquifers. *Water* **2019**, *11*, 2462. [CrossRef]

20. Korbel, K.L.; Stephenson, S.; Hose, G.C. Sediment size influences habitat selection and use by groundwater macrofauna and meiofauna. *Aquat. Sci.* **2019**, *81*, 39. [CrossRef]

21. Sánchez-Bayo, F.; Hyne, R.V. Detection and analysis of neonicotinoids in river waters–Development of a passive sampler for three commonly used insecticides. *Chemosphere* **2014**, *99*, 143–151. [CrossRef] [PubMed]

22. Hunn, J.; Macaulay, S.; Matthaei, C.D. Food Shortage Amplifies Negative Sublethal Impacts of Low-Level Exposure to the Neonicotinoid Insecticide Imidacloprid on Stream Mayfly Nymphs. *Water* **2019**, *11*, 2142. [CrossRef]

23. Peralta-Maraver, I.; Posselt, M.; Perkins, D.; Robertson, A. Mapping Micro-Pollutants and Their Impacts on the Size Structure of Streambed Communities. *Water* **2019**, *11*, 2610. [CrossRef]

24. Labuschagne, M.; Wepener, V.; Nachev, M.; Zimmermann, S.; Sures, B.; Smit, N.J. The Application of Artificial Mussels in Conjunction with Transplanted Bivalves to Assess Elemental Exposure in a Platinum Mining Area. *Water* **2019**, *12*, 32. [CrossRef]

25. Deliberalli, W.; Cansian, R.L.; Pereira, A.A.M.; Loureiro, R.C.; Hepp, L.U.; Restello, R.M. The effects of heavy metals on the incidence of morphological deformities in Chironomidae (Diptera). *Zoologia* **2018**, *35*, 1–7. [CrossRef]

26. Goretti, E.; Pallottini, M.; Pagliarini, S.; Catasti, M.; La Porta, G.; Selvaggi, R.; Gaino, E.; Di Giulio, A.M.; Ali, A. Use of Larval Morphological Deformities in Chironomus plumosus (Chironomidae: Diptera) as an Indicator of Freshwater Environmental Contamination (Lake Trasimeno, Italy). *Water* **2019**, *12*, 1. [CrossRef]

27. Iannella, M.; Console, G.; D'Alessandro, P.; Cerasoli, F.; Mantoni, C.; Ruggieri, F.; Di Donato, F.; Biondi, M. Preliminary Analysis of the Diet of Triturus carnifex and Pollution in Mountain Karst Ponds in Central Apennines. *Water* **2019**, *12*, 44. [CrossRef]

28. Strona, G.; Fattorini, S.; Fiasca, B.; Di Lorenzo, T.; Di Cicco, M.; Lorenzetti, W.; Boccacci, F.; Lombardo, P. AQUALIFE Software: A New Tool for a Standardized Ecological Assessment of Groundwater Dependent Ecosystems. *Water* **2019**, *11*, 2574. [CrossRef]

Article

Physiological Profiling and Functional Diversity of Groundwater Microbial Communities in a Municipal Solid Waste Landfill Area

Marco Melita *, Stefano Amalfitano, Elisabetta Preziosi, Stefano Ghergo, Eleonora Frollini, Daniele Parrone and Annamaria Zoppini *

Water Research Institute, National Research Council (IRSA-CNR), 00015 Monterotondo, Rome, Italy; amalfitano@irsa.cnr.it (S.A.); preziosi@irsa.cnr.it (E.P.); ghergo@irsa.cnr.it (S.G.); frollini@irsa.cnr.it (E.F.); parrone@irsa.cnr.it (D.P.)

* Correspondence: melita@irsa.cnr.it (M.M.); zoppini@irsa.cnr.it (A.Z.); Tel.: +39-06-9067-2753 (M.M.); +39-06-9067-2792 (A.Z.)

Received: 14 November 2019; Accepted: 7 December 2019; Published: 12 December 2019

Abstract: The disposal of municipal solid wastes in landfills represents a major threat for aquifer environments at the global scale. The aim of this study was to explore how groundwater geochemical characteristics can influence the microbial community functioning and the potential degradation patterns of selected organic substrates in response to different levels of landfill-induced alterations. Groundwaters collected from a landfill area were monitored by assessing major physical-chemical parameters and the microbiological contamination levels (total coliforms and fecal indicators—Colilert-18). The aquatic microbial community was further characterized by flow cytometry and Biolog EcoPlates^{TM} assay. Three groundwater conditions (i.e., pristine, mixed, and altered) were identified according to their distinct geochemical profiles. The altered groundwaters showed relatively higher values of organic matter concentration and total cell counts, along with the presence of fecal indicator bacteria, in comparison to samples from pristine and mixed conditions. The kinetic profiles of the Biolog substrate degradation showed that the microbial community thriving in altered conditions was relatively more efficient in metabolizing a larger number of organic substrates, including those with complex molecular structures. We concluded that the assessment of physiological profiling and functional diversity at the microbial community level could represent a supportive tool to understand the potential consequences of the organic contamination of impacted aquifers, thus complementing the current strategies for groundwater management.

Keywords: Biolog EcoPlates^{TM}; flow cytometry; microbial community; metabolic fingerprint; groundwater quality; hydrogeochemistry

1. Introduction

Groundwater is the second largest freshwater deposit representing 14% of all inland waters and one of the most important drinking water resource [1,2]. Several factors may impair groundwater environment including drought, excessive withdrawals, and pollutant contamination derived by anthropogenic activities [3–5]. Awareness of the importance of groundwater resources led the European Commission to define legislative tools for ensuring the groundwater ecosystem quality and fostering the related research efforts [6]. The environmental policy started to consider aquifers as a living ecosystem to be preserved [7]. To date, the anthropogenic impact on groundwater biological communities is largely disregarded, although a wide range of organisms are adapted to live under the limiting conditions imposed by darkness and low trophic resources [8,9]. Among those, prokaryotic microorganisms are the most relevant in terms of biomass [10]. The environmental factors that influence the dynamics of aquatic microbial communities are increasingly attracting the research interest, given the key role played by microorganisms in the ecosystem while processing and cycling organic carbon and nutrients [11,12]. Changes in water physical-chemical characteristics were found to affect the structural and functional properties of the microbial communities [13–15].

Among the global threats for the aquifer environment, the municipal solid waste disposal in landfills is of major concern for either human or environmental health, since pollutants can promptly migrate into groundwater by the leaching of percolate [11,16,17]. To counteract and prevent "significant adverse environmental effects" derived from landfill practices, a detailed characterization of groundwater quality is required and based on representative hydrogeological conceptual models (European Landfill Directive; 1999/31/EC) [18]. However, despite the good chemical status is determined by compliance with water quality standards, the ecological approach to evaluate the anthropogenic impact on groundwaters has been poorly considered [19].

Different approaches are suitable for monitoring and predicting the effects of anthropogenic groundwater alterations, offering an insight into the relationships occurring between geochemical parameters and the microbial functional properties [13,20,21]. The microbial phylogenetic changes, observed so far in many literature studies [22–24], were not likely exhaustive to explain the microbial functional responses to the anthropic impact. A fundamental understanding of the microbially-driven processes and the following ecosystem responses to anthropogenic stress factors is still needed.

Starting from the outcomes of one-year monitoring activities in a municipal solid waste landfill in Central Italy [21], we aimed to explore the functional properties of groundwater microbial communities and their responses to different levels of landfill-induced geochemical alterations. The hypothesis is that changes in groundwater quality will affect the functioning of the resident microbial communities by inducing preferential metabolic pathways. This approach could complement the current geochemical assessments to better elucidate the effects of the anthropic activities on the aquifer environment.

2. Materials and Methods

2.1. Field Sampling and Major Water Characteristics

The study was carried out at a groundwater-monitoring network located within the area of a municipal solid non-hazardous waste landfill facility in Central Italy (Figure 1). The landfill, duly equipped with liners and leachate collection systems, is operating since 2002, following the Landfill Directive 1999/31/EC. Between 2003 and 2018, it received 2.9 million tons of waste in 4 different basins, deriving mainly from the mechanical-biological treatment of municipal waste (65%), and non-hazardous waste of other origin. The yearly amount varied from 170,000 tons to 2,700,000 tons. A water table aquifer (depth 5–15 m) was found only in the northern part of the facility. A confined aquifer (depth 5–27 m) underlies the entire landfill area, with groundwater flowing from North to South through lacustrine sediments formed by clays, silts, and sandy layers [21].

During two sampling campaigns (October 2016 and May 2017), 22 water samples were collected from 12 piezometers (Figure 1), by following specific guidelines for groundwater sampling in industrial areas [25]. The landfill is composed of four different basins, three of which are currently in the "after-care" phase and only the fourth basin was active during sampling. Basin 4 (840,000 m^3) is located between Pz07 and Pz01. The distance of Pz07 from the waste load area is approximately 50 m. Pz01 is at about 50 m from the southern end of the basin 4.

Major physical-chemical water characterization was assessed on-site by measuring electrical conductivity, oxidation-reduction potential (ORP), temperature, pH, dissolved oxygen with a flow cell equipped with AQUAREAD probes. The concentration of NO_3, SO_4, F, Cl (Ionic Chromatography—Dionex DX-120), HCO_3 (HCl titration), Si, Mg, Ca, Na, K (Inductively Coupled Plasma—Optical Emission Spectrometry—Perkin Elmer P400), DOC (Shimadzu TOC-5000A analyser) were successively measured in laboratory [21].

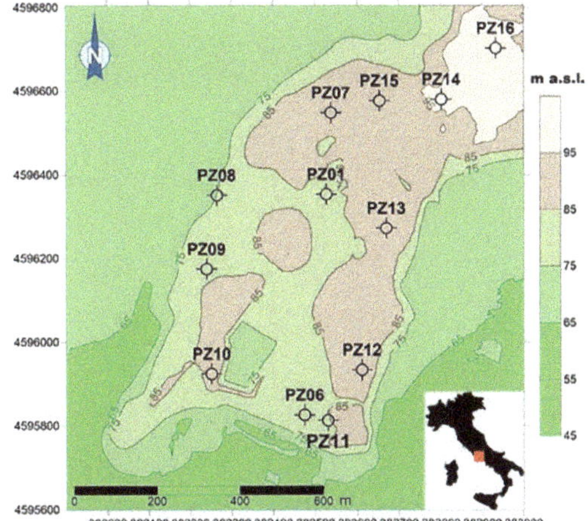

Figure 1. Location of the waste landfill facility area (Central Italy), with the indication of sampling sites.

The occurrence of total coliforms (TC) and fecal indicator bacteria (*E. coli*) was estimated using the Colilert-18 test kit (IDEXX Laboratories Inc., Westbrook, Maine, USA) [26]. Duplicate samples were collected in 100-mL sterile polystyrene bottles and analyzed within 6 h. The positive wells for the activities on ONPG (yellow) and MUG (fluorescent) substrata were counted and referred to the Quanti-Tray/2000 Most Probable Number (MPN) table to quantify the total coliforms and *E. coli*, respectively.

2.2. Microbial Community Characterization by Flow Cytometry

Subsamples for flow cytometric analyses were collected in 2-mL Eppendorf safe-lock tubes, fixed with formaldehyde (2%, final concentration) and analyzed within 24 h. The microbial community characterization was performed by the Flow Cytometer A50-micro (Apogee Flow System, Hertfordshire, UK) equipped with a solid-state laser set at 20 mV and tuned to an excitation wavelength of 488 nm. The total prokaryotic cell counts (TCC) were measured by staining with SYBR Green I (1:10,000 dilution; Molecular Probes, Invitrogen) for 10 min in the dark at room temperature. The light scattering signals (forward and side scatters), the green fluorescence (530/30 nm) and red fluorescence (>610 nm) were acquired with a threshold sets on the green channel. Samples were run at low flow rates to keep the number of events below 1000 event s^{-1}.

The data were analyzed using the Apogee Histogram Software (v2.05). The TCC was determined by their signatures in a plot of the side scatter vs. the green fluorescence. The intensity of green fluorescence emitted by SYBR positive cells allowed for the discrimination among cell groups exhibiting two different nucleic acid content (cells with low NA content—LNA; cells with high NA content—HNA) [27]. This method was successfully applied for prokaryotic cell counting in different groundwater settings [15,21].

2.3. Metabolic Potential, Functional Diversity and Community Level Physiological Profile

The Biolog EcoPlates^{TM} assay (Biolog, Inc., Hayward, California, USA) was used to detect the microbial degradative activity of 31 organic carbon sources, contained in triplicate in 96 micro-well plates (plus three control wells) and labeled with the respiration-sensitive tetrazolium dye. Within 6 h from sampling, 150 µL of sampled water was introduced in each well. The plates were incubated at 20 °C in the dark. The analysis of the color development for each substrate along the incubation time provided information on the degradation rate of each substance. The development of formazan was detected through the measurement of the absorbance at 590 nm (VICTOR™ X3 Multilabel plate reader.

PerkinElmer Inc., Waltham, Maine, USA). Data acquisition was performed at the starting time (time 0) and every 24 h, over a period of 15 days. The optical density (OD 590 nm, expressed as absorbance units) of each substrate was subtracted by the corresponding initial OD value (t = 0) to eliminate the intrinsic absorbance signal [28]. Negative values were set to 0.

The mean degradative activity on all 31 substrates was intended as the microbial metabolic potential, expressed as Average Well Color Development (AWCD), the average of adjusted absorbance of the whole microplate [29]:

$$AWCD = \frac{1}{31} \sum (OD_{590})$$ (1)

The AWCD kinetic profiles were calculated over the entire incubation time, based on the density-dependent logistic growth equation:

$$y = AWCD_t = \frac{K}{1 + e^{-\mu(t-s)}}$$ (2)

where:

- K = the asymptote/carrying capacity
- μ = exponential rate of AWCD change
- t = time (h)
- s = the time when $y = K/2$

These parameters were utilized to compare the microbial functional properties in different samples [30]. Data elaboration were performed in the interval between 0 and 168 h of incubation, representing the exponential phase before the stationary phase. In particular, the 24 h endpoint represented the prompt microbial responsiveness in substrate degradation, while the 168-h endpoint represented the maximum potential degradation activity.

Following Weber et al. [31], the functional diversity was related to the rates at which the Biolog substrates were utilized by the aquatic microbial community. Herein, the functional diversity was estimated in terms of equitability (E), taking into account both Shannon index (H) and substrate richness (i.e., number of positive wells in the microplates—S):

$$E = \frac{H}{\ln S}$$ (3)

where:

- $H = -\sum p_i (\ln p_i)$
- p_i = the ratio of the activity of a particular substrate to the sums of activities of all substrates
- S = number of positive wells in the microplates

The standardization of initial density inoculum was carried out by normalization of optical density (OD_{590}) of each substrate by the corresponding Average Well Color Development (AWCD) value [28,32,33]. To simplify the interpretation of the results, the 31 substrates utilized by the Biolog test, were grouped into six categories of chemical compounds as follows: carbohydrates, polymers, carboxylic acids, phenolic compounds, amino acids and amines [34]. The Area Under the Curve (AUC) describing the variation of OD with time was calculated for each well using a trapezoidal approximation [35]. All values, hereafter reported both for classes of compounds and single substrates, were expressed as AUC and utilized for the univariate analysis.

2.4. Statistical Analysis

Geometric means were used to characterize physical, chemical and microbial characteristics of the groundwater groups. In the case of *E. coli* < 1 MPN/100 mL, we assumed 0.5 MPN/100 mL as lowest value.

The multivariate analysis of similarities (ANOSIM) was used to test the difference among the sample groups by considering all major physical, chemical, and microbial parameters. A multi-group similarity percentage test (SIMPER), through the Bray-Curtis similarity measure (multiplied by 100), was run to identify the percentage contribution of selected data to the average dissimilarity.

The non-parametric Kruskal-Wallis univariate test was performed on all parameters in order to identify statistical differences between the groundwater groups.

A non-metric multi-dimensional scaling ordination plot (nMDS), based on the Gower dissimilarity matrix, was used to graphically visualize hydrogeochemical and microbiological variation patterns [36]. The values of the degradation activity performed on the different categories of organic compounds were incorporated in a nMDS analysis with a vector-fitting procedure, in which the length of the arrow is proportional to the correlation between the nMDS-axes and each variable. This method allowed determining which degraded class of compounds were significantly correlated with the nMDS ordination. In particular, we focused on those substrates that contributed for 50% of the overall SIMPER dissimilarities between the groundwater groups.

A non-parametric multivariate analysis of variance (NPMANOVA) was performed on AWCD-normalized Biolog data (i.e., on either substrate classes or single substrates) to highlight differences in the community-level physiological profiles (CLPP) between the identified groundwater groups.

3. Results

3.1. Groundwater Physical, Chemical and Microbial Characteristics

The major groundwater physical, chemical and microbial characterization (Tables S1 and S2) allowed identifying three statistically distinct groundwater groups (ANOSIM, $p < 0.05$), herein defined as "pristine", "mixed", and "altered".

The pristine groundwater group included ten samples from six piezometers (Pz6-2016, Pz6-2017, Pz12-2016, Pz12-2017, Pz13-2017, Pz14-2016, Pz14-2017, Pz15-2016, Pz15-2017, and Pz16-2017) (Figure 1). These samples were anoxic groundwaters characterized by the lowest mean values of ORP and DO, the lowest mean EC along with the lowest mean concentration of SO_4 and Ca (Table 1).

Table 1. Physical and chemical characteristics of different groundwater groups presented as geometric means. The median was used in case of negative ORP values. Data are shown in order of contribution to the variability among groups obtained through the SIMPER test. EC = electrical conductivity; ORP = redox potential; T = temperature; DO = dissolved oxygen; DOC = Dissolved Organic Carbon). a, b, c indicate statistical differences among groups (Kruskal-Wallis, $p < 0.05$).

Parameter	Unit	Pristine	Mixed	Altered
EC	µS/cm	766.4 [a]	915.6	1452.7
ORP	mV	−119.3 [a]	74.3 [b]	−40.8 [c]
T	°C	17.8	19.1	17.3
pH		7.2 [a]	7.0 [b]	6.7 [c]
DO	mg/L	0.2 [a]	3.1 [b]	0.4 [a]
NO_3	mg/L	0.9 [a]	18.2 [b]	0.8 [a]
SO_4	mg/L	11.7 [a]	45.9 [b]	232.5 [c]
K	mg/L	10.5 [a]	2.1 [b]	5.1 [ab]
DOC	mg/L	2.2 [a]	0.8 [b]	2.0 [a]
F	mg/L	0.4 [a]	0.2 [b]	0.2 [ab]
Cl	mg/L	7.9 [a]	19.5 [b]	11.3 [a]
Si	mg/L	16.4 [a]	7.5 [b]	19.7 [a]
Mg	mg/L	23.1 [b]	22.0 [b]	45.7 [a]
Ca	mg/L	94.1 [b]	135.4 [c]	238.3 [a]
Na	mg/L	9.9	8.2	8.8
HCO_3	mg/L	419.8 [b]	424.5 [b]	632.4 [a]

The mixed groundwater group included six samples from 3 piezometers, where the water table aquifer is probably mixed with the lower one (Pz7-2016, Pz7-2017, Pz8-2016, Pz8-2017, Pz9-2016, and Pz9-2017). These samples were more oxidized groundwaters and showed the highest mean values of ORP, DO, NO_3 and Cl, along with the lowest mean DOC and Si concentrations.

The altered groundwater group, including six samples from four piezometers (Pz1-2016, Pz10-2016, Pz10-2017, Pz11-2016, Pz11-2017, Pz13-2016), was characterized by anoxic, slightly acidic groundwaters with the highest EC, SO_4, Mg, Ca and HCO_3 concentrations.

Fecal contamination was detected only in the altered groundwater samples in the 2016 campaign. Total prokaryotic cell counts (TCC = $7.93 \times 10^4 - 1.69 \times 10^7$ cells/mL) showed a large variability among the groundwater groups, with relatively high values in altered conditions, along with significant high percentages of LNA cells (Table 2).

Table 2. Microbial community characteristics (geometric means) of the three groundwater groups. In case of negative *E. coli*, half of detection limit was used for mathematical calculation (i.e., 0.5 MPN/100 mL). TCC = Total Cell Counts; HNA cells = cells with high nucleic acid content; LNA cells = cells with low nucleic acid content. a, b, c indicate statistical differences among groups (Kruskal-Wallis, $p < 0.05$).

Parameter	Unit	Pristine	Mixed	Altered
Total Cells Count	10^5 cells/mL	4.4	3.1	6.6
HNA Cells	% of TCC	52.8 [b]	52.6 [b]	27.1 [a]
LNA Cells	% of TCC	41.6 [b]	46.4 [b]	65.5 [a]
Total Coliforms	MPN/100 mL	157.8	103.0	292.6
E. coli	MPN/100 mL	<1	<1	3.3

The different contribution of the macro-chemical elements and microbial characteristics to the identified groundwater groups was graphically summarized in an ordination plot (Figure 2). The multivariate approach allowed for the testing of inter-group dissimilarity, along with highlighting the variation patterns of the groundwater major characteristics.

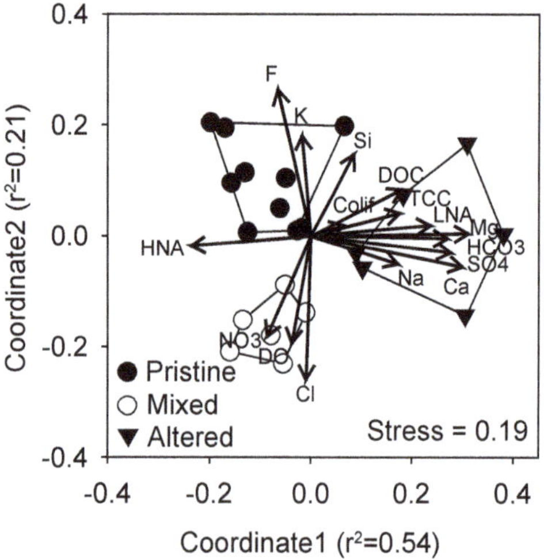

Figure 2. NMDS ordination plot of sampled waters. Vectors represent the major geochemical and microbial community characteristics that concurred to the identification of the three different groundwater groups. Stress value indicates the significant concordance between the distance among samples in the NMDS plot and the actual Gower distance among samples.

3.2. Kinetic Profiles of Metabolic Potential and Functional Diversity

The kinetic parameters, calculated from the sigmoidal curves of the AWCD values over time, highlighted the overall differences in the metabolic responses of microbial communities from the three groundwater groups (Figure 3a and Table 3). The microbial potential activity reached a similar theoretical maximum K value in all groundwater groups. The microbial community from the pristine groundwater differed significantly, showing the highest exponential rate of AWCD change (μ) (Table 3). In altered conditions, the microbial community showed $AWCD_{24h}$ values two-fold higher than that developed under pristine and mixed conditions and the lowest time to reach K/2, hence indicating a higher affinity for the available substrates.

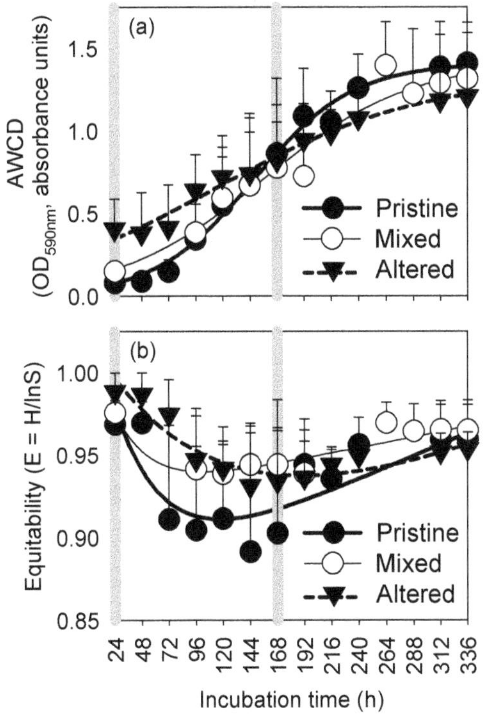

Figure 3. Kinetic profiles of (**a**) metabolic potential (Average Color Well Development—AWCD) and (**b**) functional diversity (substrate utilization equitability—E) (Equation (3)) of the identified groundwater groups. Gray lines indicate 24 h and 168 h incubation time.

Table 3. Kinetic parameters of the fitted logistic growth equations (Equation (2)). K = asymptote, s = time to reach K/2, μ = exponential rate of AWCD change, r^2 = correlation coefficient.

Parameter	Unit	Pristine	Mixed	Altered
K	OD$_{590nm}$	1.42	1.41	1.33
s	h	157.1	147	117.1
μ	h^{-1}	0.029	0.023	0.021
r^2		0.98	0.99	0.99

After 24 h of incubation, the functional microbial diversity, indicated by the equitability values, of microbial communities from pristine and mixed sample groups showed similar values (E = 0.93 and 0.92, respectively), which were significantly lower than those found in altered conditions (E = 0.96).

The equitability under pristine conditions reached the lowest values with respect to altered and mixed conditions, indicating a heterogeneous distribution in substrate degradation (Figure 3b and Table 4). The equitability showed lower values at 168 h (E = 0.903, 0.944, and 0.934, respectively for pristine, mixed groundwaters) than at 24 h.

Table 4. Utilization patterns of organic substrates in different groundwater groups after 24 h and 168 h of incubation. Values are given as percentage of the total OD_{590}.

24 h

Classes of Substrates	Substrates	Pristine										Mixed							Altered				
Carbohydrates	β-methyl-D-glucoside	4.7	6.4	4.8	4.9	4.6	5.1	5.3	3.8	5.6	4.4	4.8	5.6	3.3	3.4	3.6	4.8	3.3	3.1	4.9	3.5	3.5	3.5
	D-xylose	4.1	1.5	3.2	2.9	4.0	1.6	0.0	5.7	1.5	3.0	3.6	3.6	1.9	3.7	4.5	3.4	3.9	3.1	3.6	3.7	3.3	3.6
	i-erythritol	7.0	4.2	5.2	5.7	6.3	6.6	8.1	5.6	7.0	4.6	5.8	5.3	6.7	4.9	6.0	5.9	4.1	3.7	4.5	4.0	3.7	6.5
	D-mannitol	5.5	7.2	5.1	5.5	5.0	5.2	7.1	4.4	8.6	4.9	5.6	6.8	5.7	3.6	5.3	5.3	3.9	2.9	4.1	3.4	3.6	5.2
	N-acetyl-D-glucosamine	4.9	6.9	5.5	5.4	5.0	4.4	7.3	4.0	6.1	6.2	5.3	4.6	3.8	4.5	4.3	4.1	3.8	2.9	3.3	3.6	3.8	5.0
	D-cellobiose	5.0	2.9	5.1	5.6	6.6	3.7	6.2	4.8	5.9	5.3	5.5	3.2	1.7	4.6	5.0	5.7	3.5	4.7	4.2	3.6	3.7	4.7
	α-D-lactose	6.0	3.3	5.0	6.1	5.2	5.5	1.7	4.7	6.7	5.5	5.2	7.7	1.8	4.1	5.2	6.2	3.2	5.0	4.0	2.9	4.1	4.4
	glucose-1-phosphate	2.7	1.9	1.5	2.7	1.9	0.0	7.3	2.3	0.0	2.0	2.5	1.1	2.9	2.5	1.4	1.8	2.5	2.5	2.0	2.2	3.0	1.6
	D,L-α-glycerol phosphate	2.0	2.8	0.0	0.6	3.1	0.0	0.0	0.9	2.8	1.1	1.5	2.0	2.7	1.8	1.9	1.7	2.5	1.6	2.1	2.5	2.9	0.5
Polymers	tween40	5.7	5.1	5.7	4.1	6.7	3.0	3.3	4.5	4.6	7.6	4.6	4.0	5.3	4.5	7.8	4.2	4.4	5.7	3.6	3.3	4.0	6.4
	tween80	0.0	2.4	3.4	3.6	0.0	2.4	0.1	2.6	0.0	1.6	0.0	0.0	1.7	3.1	2.9	3.4	3.7	3.3	3.0	3.3	3.8	3.3
	α-cyclodextrin	2.3	1.7	2.3	2.3	1.7	1.1	3.1	2.4	2.7	2.4	2.0	2.5	3.7	2.2	1.3	3.0	2.3	2.8	3.1	3.6	2.9	1.7
	glycogen	5.0	6.3	3.8	2.6	2.8	3.8	4.4	3.6	4.2	4.4	3.9	5.5	2.7	2.8	3.3	3.9	3.0	3.4	3.0	3.3	3.7	3.9
Carboxylic Acids	D-galactonic acid γ-lactone	4.5	2.5	4.6	3.0	3.7	3.9	4.3	2.9	2.4	3.4	4.3	2.8	4.4	2.6	4.2	3.5	3.7	3.8	3.4	3.5	2.5	4.3
	D-galacturonic acid	4.6	2.8	0.7	2.4	2.8	2.8	2.2	3.1	2.8	0.0	2.6	3.8	3.5	3.1	3.0	3.5	1.9	2.9	3.2	3.0	3.1	1.9
	γ-hydroxybutyric acid	4.1	1.8	4.3	5.3	3.9	11	1.0	2.7	9.3	7.6	5.0	2.6	2.7	3.8	2.7	3.8	4.1	3.9	2.6	3.7	3.9	5.1
	D-glucosaminic acid	4.3	2.7	3.1	4.2	3.8	3.2	7.0	3.2	5.8	4.4	4.1	4.1	4.5	3.3	3.7	3.6	2.5	2.4	3.1	3.1	3.2	3.0
	itaconic acid	2.6	2.0	2.9	2.6	2.7	2.1	2.7	2.4	3.2	2.2	2.6	3.4	1.9	2.2	2.0	2.2	3.5	3.1	2.6	2.8	2.4	2.4
	α-ketobutyric acid	0.0	2.7	3.7	3.0	0.0	5.9	0.8	3.2	0.0	2.6	4.6	0.0	2.3	2.9	1.5	1.2	2.0	1.5	2.1	2.8	3.1	2.7
	D-malic acid	0.0	2.1	2.8	1.6	3.0	0.0	0.5	2.3	1.7	1.5	0.0	3.4	3.7	3.5	3.3	1.8	2.6	3.8	3.0	3.4	3.6	1.8
	pyruvic acid methyl ester	0.0	1.7	2.9	0.0	0.0	0.0	0.0	0.7	0.0	0.8	0.0	0.0	2.8	1.0	0.5	0.0	2.6	2.2	2.0	3.4	3.6	0.4
Phenolic Compounds	2-hydroxy benzoic acid	0.0	1.9	2.1	0.0	2.2	0.0	0.0	2.2	0.0	0.0	0.0	2.2	3.3	3.5	3.5	5.0	3.2	2.5	4.4	3.3	3.4	1.7
	4-hydroxy benzoic acid	0.0	1.5	3.2	1.0	0.0	1.2	0.2	3.6	0.0	1.4	0.0	0.0	1.2	1.6	0.8	0.0	3.2	1.9	2.2	2.9	2.7	2.8
Aminoacids	L-arginine	4.1	3.6	3.0	4.1	3.8	4.4	6.4	3.7	2.9	5.9	3.6	4.6	4.1	3.2	3.9	3.7	2.8	3.7	3.5	3.5	3.3	3.9
	L-asparagine	3.8	3.8	2.8	3.0	3.4	3.6	3.1	2.7	1.2	2.8	3.5	3.7	4.6	4.1	3.1	2.5	3.8	4.2	3.7	3.1	3.0	3.5
	L-phenylalanine	3.0	2.3	1.6	2.3	2.0	2.4	1.4	2.8	1.9	1.8	2.4	2.0	2.9	2.5	1.5	1.4	3.3	2.1	2.0	2.9	1.3	2.0
	L-serine	3.4	3.8	2.3	3.1	3.5	3.9	5.0	2.7	2.6	2.5	3.6	4.2	3.5	4.2	3.1	3.3	3.4	3.6	3.2	3.3	3.3	2.9
	L-threonine	3.8	2.8	4.7	4.3	4.2	4.0	4.3	3.9	3.7	4.6	4.1	1.5	3.9	4.2	3.7	4.3	4.1	4.1	3.5	3.7	3.6	4.2
	glycyl-L-glutamic acid	3.6	2.6	1.8	2.8	2.2	0.0	1.5	2.3	2.6	2.4	3.0	3.3	3.1	3.4	2.1	2.9	2.3	3.2	3.0	2.7	2.3	2.0
Amines	phenylethyl-amine	0.0	2.7	0.0	2.5	2.3	5.7	1.2	3.9	2.3	0.9	3.3	3.0	2.1	2.9	2.2	1.2	3.8	3.0	4.2	3.1	2.7	2.5
	putrescine	3.2	4.0	2.7	2.8	3.4	3.8	3.6	2.6	2.1	2.3	2.8	3.5	1.5	2.2	2.7	2.7	3.1	3.4	2.9	3.0	3.1	2.6

168 h

Classes of substrates	Substrates	Pristine										Mixed							Altered				
Carbohydrates	β-methyl-D-glucoside	4.8	4.4	8.8	7.6	1.3	2.6	5.8	2.1	11	5.8	3.4	5.7	3.8	2.8	1.5	5.6	4.8	5.5	4.5	5.3	2.6	8.2
	D-xylose	3.7	3.8	0.3	0.5	3.4	0.3	0.0	1.9	0.1	0.2	1.2	0.6	0.9	1.3	0.9	1.5	1.2	1.2	2.7	1.8	2.3	0.7
	i-erythritol	1.7	1.5	1.2	1.7	1.4	1.1	2.9	2.5	0.9	2.4	2.7	2.4	1.7	2.9	2.1	4.2	2.7	1.6	3.4	2.5	2.8	2.9
	D-mannitol	5.0	5.1	8.2	9.2	2.1	3.1	7.6	3.7	5.3	10	6.7	3.4	6.0	5.6	6.1	5.3	5.6	5.3	3.7	4.7	5.9	5.7
	N-acetyl-D-glucosamine	4.0	4.9	6.1	8.1	3.0	6.8	7.4	7.4	1.4	6.0	4.3	7.0	5.1	4.9	4.2	4.0	5.0	5.4	3.0	3.0	7.2	5.3
	D-cellobiose	5.7	4.7	5.5	3.1	4.6	2.0	3.6	3.2	5.0	7.4	4.2	2.2	3.0	4.0	2.3	4.9	4.9	7.5	4.6	6.2	2.6	6.2
	α-D-lactose	4.3	3.8	2.1	2.9	3.0	6.9	2.2	5.0	1.9	4.2	3.8	2.5	1.8	2.4	1.8	4.9	2.3	2.5	3.5	3.0	2.9	2.1
	glucose-1-phosphate	2.2	2.9	2.1	4.6	0.4	3.7	4.0	2.3	0.3	2.0	2.1	0.5	2.8	2.4	0.8	1.6	3.3	2.8	1.6	3.8	4.5	3.8
	D,L-α-glycerol phosphate	1.0	2.0	0.3	1.6	1.7	1.3	1.2	1.7	1.4	0.7	1.0	1.1	1.4	1.2	1.6	1.6	2.4	1.3	2.1	2.2	2.0	1.4
Polymers	tween40	3.2	3.7	3.9	5.8	2.9	6.0	4.0	4.1	7.3	4.0	4.9	8.3	3.0	4.9	7.6	6.1	3.7	5.1	2.7	4.0	4.7	3.3
	tween80	4.4	3.8	4.1	2.4	0.0	8.7	6.7	1.6	7.4	2.5	4.4	5.9	4.1	4.5	8.6	2.8	5.1	3.1	3.4	4.8	5.1	4.6
	α-cyclodextrin	4.5	3.6	2.3	1.5	1.0	3.7	3.2	4.1	5.4	2.6	3.2	3.8	3.9	3.8	3.2	1.8	1.0	2.1	3.2	2.7	3.2	2.5
	glycogen	4.7	4.3	11	3.2	0.4	1.9	2.4	2.8	1.4	7.1	3.0	4.5	4.7	3.0	3.8	3.6	4.8	6.8	2.6	4.2	2.6	10
Carboxylic Acids	D-galactonic acid γ-lactone	1.9	2.9	1.0	1.3	2.3	1.0	2.6	0.8	0.5	6.1	2.8	3.2	2.9	3.7	3.1	5.3	3.5	1.5	3.3	2.0	1.9	2.0
	D-galacturonic acid	3.4	3.0	6.4	2.6	8.2	1.5	3.0	3.4	5.0	0.3	1.6	2.9	3.6	3.2	2.6	2.5	2.8	1.4	2.6	2.0	2.2	3.9
	γ-hydroxybutyric acid	0.8	1.1	0.5	1.5	3.1	1.2	0.4	0.6	1.4	1.1	3.1	0.5	0.6	1.7	1.0	2.9	2.3	1.2	2.2	1.6	2.4	1.9
	D-glucosaminic acid	2.3	2.6	0.7	2.1	3.3	2.9	1.7	1.0	1.4	3.1	3.3	4.0	4.7	4.3	6.3	4.4	1.9	1.7	2.3	2.2	2.2	2.9
	itaconic acid	4.1	3.9	6.5	4.2	5.3	6.3	2.0	6.3	1.9	3.3	3.9	3.2	4.4	3.9	3.8	4.9	4.5	5.1	9.1	3.5	2.0	3.8
	α-ketobutyric acid	1.9	3.0	0.8	1.6	0.1	3.4	0.3	1.4	0.1	0.6	5.8	1.3	1.5	1.7	0.8	0.6	0.9	0.6	1.3	1.7	2.0	0.4
	D-malic acid	1.8	1.5	0.6	0.9	1.9	0.7	0.6	1.0	0.6	2.4	1.0	1.6	5.3	4.5	3.4	1.2	2.3	1.9	2.8	2.7	2.7	0.7
	pyruvic acid methyl ester	4.3	3.0	3.9	2.9	9.7	9.9	3.9	5.5	4.1	3.0	5.1	5.6	3.5	2.9	4.0	2.8	4.5	4.9	5.5	4.3	4.3	4.7
Phenolic Compounds	2-hydroxy benzoic acid	1.0	1.0	0.2	0.0	0.5	0.0	4.0	0.4	0.0	0.1	0.6	0.8	2.0	2.1	0.8	3.6	1.6	0.9	2.9	2.1	2.2	0.5
	4-hydroxy benzoic acid	3.3	4.4	2.1	3.0	6.8	4.4	5.4	6.4	0.9	4.7	3.8	3.9	5.3	3.4	4.4	1.9	4.1	6.1	2.0	3.8	1.7	4.9
Aminoacids	L-arginine	5.1	2.9	1.4	3.4	1.6	1.8	2.1	3.9	0.3	1.4	3.8	4.2	2.6	3.7	3.8	5.5	3.6	3.9	3.4	4.6	2.9	1.9
	L-asparagine	4.2	4.2	5.3	2.4	3.5	1.6	5.4	6.8	3.9	4.4	4.4	6.8	4.6	4.5	3.6	2.3	5.6	3.5	3.7	4.7	3.5	4.2
	L-phenylalanine	3.6	3.2	1.8	4.8	1.6	2.6	1.6	2.3	2.1	1.3	2.1	1.5	2.6	1.6	1.3	1.6	1.9	1.6	1.2	2.7	1.3	1.5
	L-serine	3.6	4.1	5.7	2.0	3.2	1.6	3.0	2.3	0.6	8.0	2.7	2.9	3.6	4.2	1.9	2.9	3.9	4.3	3.7	3.6	6.9	1.1
	L-threonine	1.5	2.0	1.3	2.3	3.7	0.9	2.0	2.9	0.6	0.9	2.3	2.1	2.1	2.6	1.4	2.9	2.6	2.2	3.2	2.4	2.6	2.0
	glycyl-L-glutamic acid	1.6	1.9	1.8	3.9	1.6	1.1	0.7	1.4	1.5	0.5	2.0	1.6	2.1	2.2	2.2	1.6	1.6	3.0	2.4	2.2	1.8	1.1
Amines	phenylethyl-amine	3.8	2.8	0.1	2.2	1.3	7.4	6.1	5.2	0.3	0.2	2.5	2.3	3.8	2.7	4.6	0.4	3.3	1.1	3.6	1.8	1.8	0.7
	putrescine	2.9	4.1	4.1	6.8	0.9	3.6	4.2	5.9	8.2	3.7	4.4	4.2	2.8	3.5	6.4	4.9	2.3	5.2	4.0	3.9	6.9	4.7

<1	1–2	2–3	3–4	4–5	5–6	6–7	7–8	8–9	9–10	>10

3.3. Community Level Physiological Profiles

The CLPP data were expressed as percentage contribution of the degradative activity on each substrate to the total. The data integration by nMDS representations summarizes how changes in the hydro-geochemical conditions corresponded to a shift in the utilization of organic substrates provided (Figure 4).

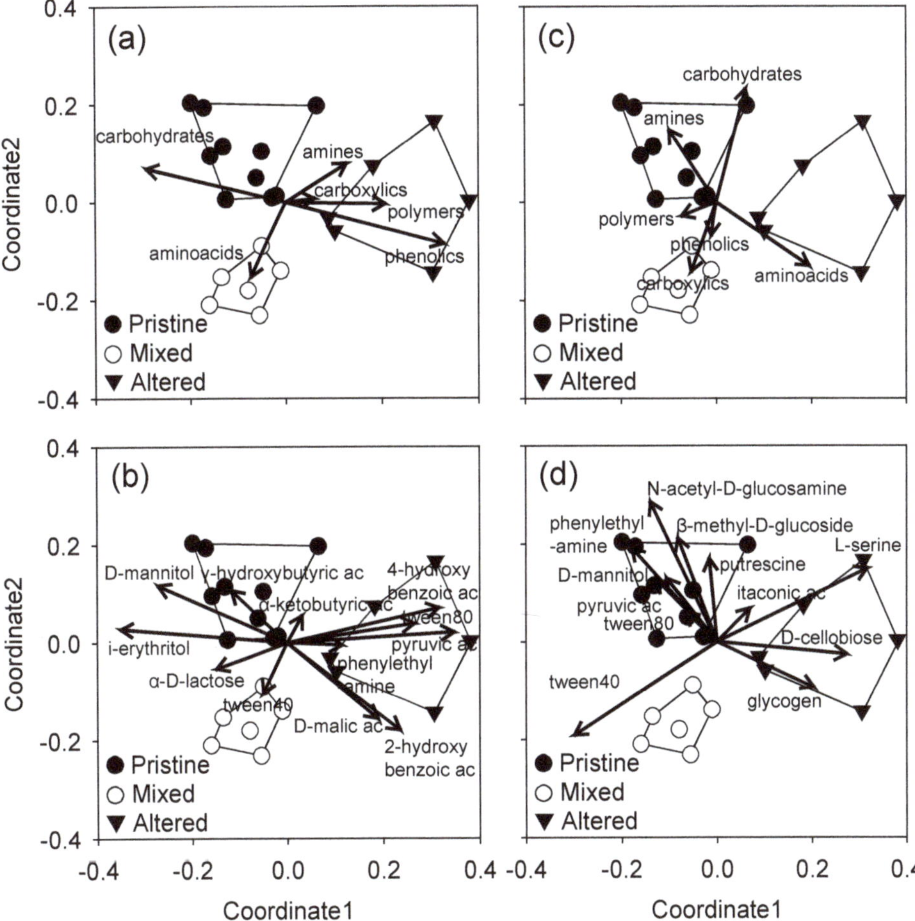

Figure 4. Variation patterns of the six class of compounds and of a selection of 12 substrates (i.e., those explaining 50% of dissimilarity among groups) at 24 h (**a**), (**c**) and 168 h (**b**), (**d**). Data were expressed as AUC values normalized for the AWCD. NMDS ordination plot of sampled waters is based on geochemical and microbial community characteristics (see Figure 2).

The utilization patterns of organic substrates in the early stage of incubation (CLPP$_{24h}$) showed that microbial communities from pristine conditions fed preferentially on carbohydrates (Table 4; Figure 4a), which contributed on average for 38.8% of the total absorbance. Inside this class of compounds, i-erythritol was the most degraded compound (6.0%), followed by D-mannitol and D-lactose, (5.9% and 5.5% respectively) (Figure 4b). Carbohydrates utilization decreased significantly to 33% at 168h incubation (NPMANOVA; $p < 0.01$) and shifted toward the utilization of other compounds such as β-methyl-D-glucoside and N-acetyl-D-glucosamine (Figure 4d). Carboxylic acids contributed to the $CLPP_{24h}$ for 22.0%, although only four out of eight compounds were utilized (Table 4). Most of the contribution was assigned to the γ-hydroxibutyrric acid (5.0%) (Figure 4b). The utilization of pyruvic acid methyl ester increased consistently at 168 h, passing from 0.6% to 5.0%. Aminoacids and polymers contributed for 18.9% and 12.9% of the total absorbance, respectively. The $CLPP_{168h}$ changed significantly due to the significant high utilization (NPMANOVA; $p < 0.05$) of polymers, phenolic compounds and amines (Figure 4c). In particular tween 40 (4.5%), phenylethyl-amine (2.9%) mostly drove this change (Figure 4d).

The $CLPP_{24h}$ in mixed conditions relied preferentially on carbohydrates (36.5%), mainly driven by i-erythritol (5.8%) and D-mannitol (5.4%) (Table 4), that declined after 168 h incubation (NPMANOVA; $p < 0.01$) (27.3%) (Table 4). The carboxylic acids contribution varied significantly in the $CLPP_{24h}$ and $CLPP_{168h}$ (22.1% and 25.1%, respectively) and relied mostly on D-glucosaminic acid (3.9% and 4.5%, respectively) (Table 4). Polymer utilization (13.0%) at 24 h was driven mainly by tween40 degradation (5.1%) (Figure 4a,b). As for pristine conditions, amines and phenolic compounds, contributed at lesser extent to the $CLPP_{24h}$ (5.0% and 3.5% respectively). Polymers and phenolic compounds increased significantly (NPMNOVA; $p < 0.05$) from $CLPP_{24h}$ to $CLPP_{168h}$ (17.9% and 5.4%, respectively) and mainly contributed, together with carboxylic acid degradation, to define the CLPP at 168h (Figure 4c).

In altered conditions, the $CLPP_{24h}$ was characterized by a wide utilization of the provided substrates (Table 4) and no statistical changes were recorded in the degradation patterns over the incubation time (Table 4). Carbohydrates (31.5%) were utilized preferentially, although at lesser extent with respect to the pristine and mixed conditions (NPMANOVA; $p < 0.05$). A relatively higher contribution of carboxylic acids (23.5%), polymers (14.1%), amines (6.2%) and phenolic compounds (5.7%) was also observed (Figure 4a). Compounds that mostly contributed to the $CLPP_{24h}$ were D-malic acid and pyruvic acid methyl ester (3.0% and 2.4%, respectively), tween-80 (3.4%), phenylethyl-amine (3.2%) and 2-hydroxy benzoic acid and 4-hydroxy benzoic acid (3.1% and 2.6% respectively) (Figure 4b). The relative $CLPP_{168h}$ was characterized for enhanced degradative skills toward specific substrata such as: D-cellobiose (5.4%), glycogen (5.2%), itaconic acid (4.7%), and L-serine (3.9%) (Figure 4d).

4. Discussion

In this study, we found that the landfill-induced geochemical alteration determined changes in the microbial functional diversity and metabolic physiological profile with an increased ability in the substrate utilization passing from pristine to altered conditions (i.e., altered > mixed > pristine). Although partly affected by the limitations of all experimental approaches performed in laboratory conditions, the Biolog assay and the kinetic profiles of substrate degradation allowed assessing the potential metabolic capacity of groundwater microbial communities, as a suitable information to better understand the effects of potential inputs of allochthons organic substrates on the aquifer. This methodological approach was consistently applied in several aquatic settings [37–40].

The Biolog assay relies on the different capacity of aquatic microorganisms to adapt to the spatial and temporal variations in quantity and quality of dissolved OM pool available in the aquatic environment [41,42]. Overall, the bioavailable OM fraction decrease by the transit from surface to groundwater, passing through several soil and sediment layers [43–45]. Therefore, groundwater ecosystems can show oligotrophic conditions and scarce productivity [10,46,47].

Microbial communities from the pristine and mixed groundwater showed the typical traits found in subsurface aquatic environments, with a relatively lower affinity for most of the Biolog substrates [10,48,49]. The heterogeneity in substrate utilization and the decreasing values of functional diversity over time suggested a relatively lower metabolic versatility to adapt to newly available substrates with respect to those in altered conditions. The physical-chemical characteristics in pristine conditions were compatible with the preferential utilization of carbohydrates by microbial communities, being their utilization associated to unbalanced nutrient conditions [50] and decreasing redox potential [36,39]. Carbohydrates represent an important energy-rich carbon source, and constitute storage molecules for the aquatic bacterial metabolism [51], being promptly and preferentially utilized through the catabolic heterotrophic pathway that involve oxidation of simple or complex carbohydrates and oxygen as a terminal electron acceptor [52,53]. According to the degradation patterns of C-rich substrates, the microbial communities in pristine and mixed condition were likely to be affected by C-limitation. Only after 168 h, microbial communities in pristine conditions developed the ability to degrade polymers (i.e., tween 80) and amines (phenylethyl-amine and putrescine). Polymers are complex carbon substrates, including compounds of synthetic origin. In particular, tween 40 and tween 80 are nonionic surfactants used as oil-in-water emulsifiers in pharmaceuticals, cosmetics, and cleaning compounds. The delay in utilizing tween 80 may reflect the metabolic plasticity [54] that modulates the microbial community functions, suggesting a higher sensitivity of pristine communities to contingent events of contamination by anthropogenic organic pollutants. The shift toward the utilization of amines in the pristine condition might indicate the presence of opportunistic classes of bacteria and a condition of nitrogen deprivation [37,55]. Unlike C-enriched substrates (i.e., carbohydrates and polymers), amines are enriched in nitrogen and utilized by bacteria to synthetize organic molecules such as amino acids and proteins [56].

In altered conditions, microbial communities were relatively more efficient in metabolizing most of all provided substrates, since the early stage of incubation. Although the degrading activity on Biolog carbon sources did not reflect the presence of these specific compounds in the groundwater, a higher homogeneity in substrates utilization and a relatively higher affinity with organic substrates, suggested a higher number of metabolic pathways, which may allow microbial communities to exploit several carbon sources [20,57,58]. The input of OMs of different origins and quality can induce the development of a well-structured community and rare taxa that contribute to enhance the OM biodegradation [42,59]. In addition, the occurrence of fecal indicator bacteria (i.e., *E. coli*) in altered conditions indicated the input of microorganisms originating from the surface environment, suggesting the co-presence of microorganisms with different functional traits other than those originally available in groundwater. In altered groundwater a higher prokaryotic cell abundance and a structural shift toward LNA cells was observed as similarly found for impacted groundwater [24,57,60]. Notably, LNA cells were found to be specifically composed by selected microbial taxa, which were also considered differently responsive to environmental changes than HNA cells [24,61].

In altered groundwater, the metabolic profile relied mainly on carboxylic acids, phenolic compounds, polymers, and amines. Carboxylic acids can be considered part of the labile pool of the organic matter in the aquatic environment, which is an important carbon source for the aquatic microorganisms [62,63]. These substrates are naturally occurring as products of the bacterial degradation and photochemical degradation of the organic matter in surface waters [64,65]. The prompt ability (24 h) in the utilization of complex polymers (tween 80) might indicate the presence of organic pollutants in altered groundwater, being this functional trait linked to polluted environment [55,63]. In this study, organic pollutants were detected at low concentration in some of the altered groundwater sites [25].

Interestingly, glycogen was one of the polymers utilized by microbial communities in altered conditions in the last stage of incubation. Glycogen, representing a source of readily available glucose for many organisms, is a highly branched polysaccharide that consists of glucose units linked in a linear chain. Glycogen is accumulated in bacterial cells during the stationary phase or during inorganic nutrient limitation, conditions that very likely occurred in the last stage of incubation [66]. Moreover, the presence in these groups with specific functional abilities might be deduced, being the degradation of glycogen performed by a set of extracellular enzymes (i.e., glycogen phosphorylases and glycosidases such as α-amylases) owned only by those bacteria that complete the degradative processes to carbohydrates [67]. The consumption of N-rich substrates (phenylethyl-amine and L-serine) across the incubation time of samples from altered conditions, was compatible with a strong nitrogen depletion of the microbial degradative processes given the relatively high activity occurring with respect to pristine and mixed conditions.

5. Conclusions

Changes in groundwater quality can directly affect the functional properties of the aquatic microbial communities with implications on the pattern of the biogeochemical cycling. The Biolog assay provided valuable information for tracking different levels of groundwater alterations. It is topical to understand the structural and functional diversity of the microbial community implied in the C-cycling and ecosystem services (i.e., nitrogen removal, pollutants degradation, and DOC assimilation). The microbial metabolic potential and the environmental factors shaping microbial community functioning are important to understand the potential drivers of biodegradation, thus helping to develop advanced strategies for groundwater management.

Supplementary Materials: The following are available online at http://www.mdpi.com/2073-4441/11/12/2624/s1, Table S1: Values of physical and chemical parameters measured in the sampling sites in two sampling campaigns, Table S2: Microbial parameters measured in the sampling sites in two sampling campaigns.

Author Contributions: All authors contributed to conceptualization, writing, and revising the present manuscript.

Funding: This work was financially supported by the Regione Lazio (Area Ciclo Integrato Rifiuti), Italy (Det. G9473 of 30/7/2015) and MAD s.r.l., Italy (contract 1471/2018).

Acknowledgments: We wish to thank Domenico Mastroianni and Francesca Falconi for major anions, cations and DOC analyses.

Conflicts of Interest: The authors declare no conflict of interest. The funders had no role in the design of the study; in the collection, analyses, or interpretation of data; in the writing of the manuscript, or in the decision to publish the results.

References

1. Downing, J.A.; Prairie, Y.T.; Cole, J.J.; Duarte, C.M.; Tranvik, L.J.; Striegl, R.G.; McDowell, W.H.; Kortelainen, P.; Caraco, N.F.; Melack, J.M.; et al. The global abundance and size distribution of lakes, ponds, and impoundments. *Limnol. Oceanogr.* **2006**, *51*, 2388–2397. [CrossRef]

2. Pham, H.V.; Torresan, S.; Critto, A.; Marcomini, A. Alteration of freshwater ecosystem services under global change–A review focusing on the Po River basin Italy and the Red River basin Vietnam. *Sci. Total Environ.* **2019**, *652*, 1347–1365. [CrossRef] [PubMed]

3. Mpelasoka, F.; Hennessy, K.; Jones, R.; Bates, B. Comparison of suitable drought indices for climate change impacts assessment over Australia towards resource management. *Int. J. Climatol.* **2008**, *28*, 1283–1292. [CrossRef]

4. Kløve, B.; Ala-Aho, P.; Bertrand, G.; Gurdak, J.J.; Kupfersberger, H.; Kværner, J.; Muotka, T.; Mykrä, H.; Preda, E.; Rossi, P.; et al. Climate change impacts on groundwater and dependent ecosystems. *J. Hydrol.* **2014**, *518*, 250–266. [CrossRef]

5. Gozlan, R.E.; Karimov, B.K.; Zadereev, E.; Kuznetsova, D.; Brucet, S. Status, trends, and future dynamics of freshwater ecosystems in Europe and Central Asia. *Inland Waters* **2019**, *9*, 78–94. [CrossRef]

6. EU Directive on the protection of groundwater against pollution and deterioration. European Parliament and Council Directive 2006 2006/118/EC of 12 December 2006. *Off. J. Eur. Union* **2006**, *372*, 19–31.

7. Griebler, C.; Stein, H.; Kellermann, C.; Berkhoff, S.; Brielmann, H.; Schmidt, S.; Selesi, D.; Steube, C.; Fuchs, A.; Hahn, H.J. Ecological assessment of groundwater ecosystems-Vision or illusion? *Ecol. Eng.* **2010**, *36*, 1174–1190. [CrossRef]

8. Di Lorenzo, T.; Galassi, D.M.P. Agricultural impact on Mediterranean alluvial aquifers: Do groundwater communities respond? *Fundam. Appl. Limnol. Arch. Hydrobiol.* **2013**, *182*, 271–282. [CrossRef]

9. Griebler, C.; Avramov, M. Groundwater ecosystem services: A review. *Freshw. Sci.* **2015**, *34*, 355–367. [CrossRef]

10. Griebler, C.; Lueders, T. Microbial biodiversity in groundwater ecosystems. *Freshw. Biol.* **2009**, *54*, 649–677. [CrossRef]

11. Moser, D.P.; Fredrickson, J.K.; Geist, D.R.; Arntzen, E.V.; Peacock, A.D.; Li, S.-M.W.; Spadoni, T.; McKinley, J.P. Biogeochemical Processes and Microbial Characteristics across Groundwater–Surface Water Boundaries of the Hanford Reach of the Columbia River. *Environ. Sci. Technol.* **2003**, *37*, 5127–5134. [CrossRef]

12. Yagi, J.M.; Neuhauser, E.F.; Ripp, J.A.; Mauro, D.M.; Madsen, E.L. Subsurface ecosystem resilience: Long-term attenuation of subsurface contaminants supports a dynamic microbial community. *ISME J.* **2010**, *4*, 131–143. [CrossRef] [PubMed]

13. Röling, W.F.M.; van Breukelen, B.M.; Braster, M.; Lin, B.; van Verseveld, H.W. Relationships between Microbial Community Structure and Hydrochemistry in a Landfill Leachate-Polluted Aquifer. *Appl. Environ. Microbiol.* **2001**, *67*, 4619–4629. [CrossRef] [PubMed]

14. D'Angeli, I.M.; Serrazanetti, D.I.; Montanari, C.; Vannini, L.; Gardini, F.; De Waele, J. Geochemistry and microbial diversity of cave waters in the gypsum karst aquifers of Emilia Romagna region, Italy. *Sci. Total Environ.* **2017**, *598*, 538–552. [CrossRef] [PubMed]

15. Amalfitano, S.; Del Bon, A.; Zoppini, A.; Ghergo, S.; Fazi, S.; Parrone, D.; Casella, P.; Stano, F.; Preziosi, E. Groundwater geochemistry and microbial community structure in the aquifer transition from volcanic to alluvial areas. *Water Res.* **2014**, *65*, 384–394. [CrossRef] [PubMed]

16. Röling, W.F.M.; Van Breukelen, B.M.; Braster, M.; Goeltom, M.T.; Groen, J.; Van Verseveld, H.W. Analysis of microbial communities in a landfill leachate polluted aquifer using a new method for anaerobic physiological profiling and 16S rDNA based fingerprinting. *Microb. Ecol.* **2000**, *40*, 177–188. [CrossRef] [PubMed]

17. Castañeda, S.S.; Sucgang, R.J.; Almoneda, R.V.; Mendoza, N.D.S.; David, C.P.C. Environmental isotopes and major ions for tracing leachate contamination from a municipal landfill in Metro Manila, Philippines. *J. Environ. Radioact.* **2012**, *110*, 30–37. [CrossRef] [PubMed]

18. EU Directive on Landfill of Waste. European Union Council Directive 1999/31/EC issued 26 April 1999. *Off. J. Eur. Communities* **1999**, *182*, 228–246.

19. Smith, R.J.; Jeffries, T.C.; Roudnew, B.; Fitch, A.J.; Seymour, J.R.; Delpin, M.W.; Newton, K.; Brown, M.H.; Mitchell, J.G. Metagenomic comparison of microbial communities inhabiting confined and unconfined aquifer ecosystems. *Environ. Microbiol.* **2012**, *14*, 240–253. [CrossRef]

20. Janniche, G.S.; Spliid, H.; Albrechtsen, H.-J. Microbial Community-Level Physiological Profiles (CLPP) and herbicide mineralization potential in groundwater affected by agricultural land use. *J. Contam. Hydrol.* **2012**, *140–141*, 45–55. [CrossRef]

21. Preziosi, E.; Frollini, E.; Zoppini, A.; Ghergo, S.; Melita, M.; Parrone, D.; Rossi, D.; Amalfitano, S. Disentangling natural and anthropogenic impacts on groundwater by hydrogeochemical, isotopic and microbiological data: Hints from a municipal solid waste landfill. *Waste Manag.* **2019**, *84*, 245–255. [CrossRef] [PubMed]

22. Barba, C.; Folch, A.; Gaju, N.; Sanchez-Vila, X.; Carrasquilla, M.; Grau-Martínez, A.; Martínez-Alonso, M. Microbial community changes induced by Managed Aquifer Recharge activities: Linking hydrogeological and biological processes. *Hydrol. Earth Syst. Sci.* **2019**, *23*, 139–154. [CrossRef]

23. Salo, M.J. Anthropogenically Driven Changes to Shallow Groundwater in Southeastern Wisconsin and Its Effects on the Aquifer Microbial Communities. *ProQuest Diss. Theses* **2019**, 1–178.

24. Song, Y.; Mao, G.; Gao, G.; Bartlam, M.; Wang, Y. Structural and Functional Changes of Groundwater Bacterial Community During Temperature and pH Disturbances. *Microb. Ecol.* **2019**, *78*, 428–445. [CrossRef] [PubMed]

25. Frollini, E.; Rossi, D.; Rainaldi, M.; Parrone, D.; Ghergo, S.; Preziosi, E. A proposal for groundwater sampling guidelines: Application to a case study in southern Latium. *Rend. Online Soc. Geol. Ital.* **2019**, *47*, 46–51. [CrossRef]

26. Eaton, A.D.; American Public Health Association; Water Environment Federation; American Water Works Associations. *Standard Methods for the Examination of Water and Wastewater*, 21st ed.; APHA-AWWA-WEF: Washington, DC, USA, 2005.

27. Amalfitano, S.; Fazi, S.; Ejarque, E.; Freixa, A.; Romaní, A.M.; Butturini, A. Deconvolution model to resolve cytometric microbial community patterns in flowing waters. *Cytom. Part A* **2018**, *93*, 194–200. [CrossRef] [PubMed]

28. Insam, H.; Goberna, M. *Molecular Microbial Ecology Manual*; Kowalchuk, G.A., de Bruijn, F.J., Head, I.M., Akkermans, A.D., van Elsas, J.D., Eds.; Springer: Dordrecht, The Netherlands, 2004; ISBN 978-1-4020-2176-3.

29. Garland, J.L.; Mills, A.L. Classification and Characterization of Heterotrophic Microbial Communities on the Basis of Patterns of Community-Level Sole-Carbon-Source Utilization. *Appl. Environ. Microbiol.* **1991**, *57*, 2351–2359.

30. Salomo, S.; Münch, C.; Röske, I. Evaluation of the metabolic diversity of microbial communities in four different filter layers of a constructed wetland with vertical flow by Biolog^{TM} analysis. *Water Res.* **2009**, *43*, 4569–4578. [CrossRef] [PubMed]

31. Weber, K.P.; Gehder, M.; Legge, R.L. Assessment of changes in the microbial community of constructed wetland mesocosms in response to acid mine drainage exposure. *Water Res.* **2008**, *42*, 180–188. [CrossRef]

32. Garland, J.L. Analytical approaches to the characterization of samples of microbial communities using patterns of potential C source utilization. *Soil Biol. Biochem.* **1996**, *28*, 213–221. [CrossRef]

33. Konopka, A.; Oliver, L.; Turco, R.F., Jr. The Use of Carbon Substrate Utilization Patterns in Environmental and Ecological Microbiology. *Microb. Ecol.* **1998**, *35*, 103–115. [CrossRef] [PubMed]

34. Insam, H. A New Set of Substrates Proposed for Community Characterization in Environmental Samples. In *Microbial Communities*; Insam, H., Rangger, A., Eds.; Springer: Berlin/Heidelberg, Germany, 1997; pp. 259–260. [CrossRef]

35. Guckert, J.B.; Carr, G.J.; Johnson, T.D.; Hamm, B.G.; Davidson, D.H.; Kumagai, Y. Community analysis by Biolog: Curve integration for statistical analysis of activated sludge microbial habitats. *J. Microbiol. Methods* **1996**, *27*, 183–197. [CrossRef]

36. Clarke, K.R.; Green, R.H. Statistical design and analysis for a biological effects study. *Mar. Ecol. Prog. Ser.* **1988**, *46*, 213–226. [CrossRef]

37. Tiquia, S.M. Metabolic diversity of the heterotrophic microorganisms and potential link to pollution of the Rouge River. *Environ. Pollut.* **2010**, *158*, 1435–1443. [CrossRef] [PubMed]

38. Taş, N.; Brandt, B.W.; Braster, M.; van Breukelen, B.M.; Röling, W.F.M. Subsurface landfill leachate contamination affects microbial metabolic potential and gene expression in the Banisveld aquifer. *FEMS Microbiol. Ecol.* **2018**, *94*, 1–12. [CrossRef] [PubMed]

39. Preston-Mafham, J.; Boddy, L.; Randerson, P.F. Analysis of microbial community functional diversity using sole-carbon-source utilisation profiles-A critique. *FEMS Microbiol. Ecol.* **2002**, *42*, 1–14.

40. Christian, B.W.; Lind, O.T. Multiple carbon substrate utilization by bacteria at the sediment–water interface: Seasonal patterns in a stratified eutrophic reservoir. *Hydrobiologia* **2007**, *586*, 43–56. [CrossRef]

41. Kothawala, D.N.; Stedmon, C.A.; Müller, R.A.; Weyhenmeyer, G.A.; Köhler, S.J.; Tranvik, L.J. Controls of dissolved organic matter quality: Evidence from a large-scale boreal lake survey. *Glob. Chang. Biol.* **2014**, *20*, 1101–1114. [CrossRef]

42. Logue, J.B.; Stedmon, C.A.; Kellerman, A.M.; Nielsen, N.J.; Andersson, A.F.; Laudon, H.; Lindström, E.S.; Kritzberg, E.S. Experimental insights into the importance of aquatic bacterial community composition to the degradation of dissolved organic matter. *ISME J.* **2016**, *10*, 533–545. [CrossRef]

43. Griebler, C.; Malard, F.; Lefébure, T. Current developments in groundwater ecology—from biodiversity to ecosystem function and services. *Curr. Opin. Biotechnol.* **2014**, *27*, 159–167. [CrossRef]

44. Shen, Y.; Chapelle, F.H.; Strom, E.W.; Benner, R. Origins and bioavailability of dissolved organic matter in groundwater. *Biogeochemistry* **2015**, *122*, 61–78. [CrossRef]

45. Meckenstock, R.U.; Elsner, M.; Griebler, C.; Lueders, T.; Stumpp, C.; Aamand, J.; Agathos, S.N.; Albrechtsen, H.-J.; Bastiaens, L.; Bjerg, P.L.; et al. Biodegradation: Updating the Concepts of Control for Microbial Cleanup in Contaminated Aquifers. *Environ. Sci. Technol.* **2015**, *49*, 7073–7081. [CrossRef] [PubMed]

46. Pretty, J.L.; Hildrew, A.G.; Trimmer, M. Nutrient dynamics in relation to surface–subsurface hydrological exchange in a groundwater fed chalk stream. *J. Hydrol.* **2006**, *330*, 84–100. [CrossRef]

47. Hofmann, R.; Griebler, C. DOM and bacterial growth efficiency in oligotrophic groundwater: Absence of priming and co-limitation by organic carbon and phosphorus. *Aquat. Microb. Ecol.* **2018**, *81*, 55–71. [CrossRef]

48. Gounot, A.-M. Microbial oxidation and reduction of manganese: Consequences in groundwater and applications. *FEMS Microbiol. Rev.* **1994**, *14*, 339–349. [CrossRef]

49. Hemme, C.L.; Tu, Q.; Shi, Z.; Qin, Y.; Gao, W.; Deng, Y.; Nostrand, J.D.; van Wu, L.; He, Z.; Chain, P.S.G.; et al. Comparative metagenomics reveals impact of contaminants on groundwater microbiomes. *Front. Microbiol.* **2015**, *6*, 1205. [CrossRef]

50. Zoppini, A.; Puddu, A.; Fazi, S.; Rosati, M.; Sist, P. Extracellular enzyme activity and dynamics of bacterial community in mucilaginous aggregates of the northern Adriatic Sea. *Sci. Total Environ.* **2005**, *353*, 270–286. [CrossRef]

51. Arnosti, C.; Bell, C.; Moorhead, D.L.; Sinsabaugh, R.L.; Steen, A.D.; Stromberger, M.; Wallenstein, M.; Weintraub, M.N. Extracellular enzymes in terrestrial, freshwater, and marine environments: Perspectives on system variability and common research needs. *Biogeochemistry* **2014**, *117*, 5–21. [CrossRef]

52. Keith, S.C.; Arnosti, C. Extracellular enzyme activity in a river-bay-shelf transect: Variations in polysaccharide hydrolysis rates with substrate and size class. *Aquat. Microb. Ecol.* **2001**, *24*, 243–253. [CrossRef]

53. Rosenstock, B.; Simon, M. Consumption of Dissolved Amino Acids and Carbohydrates by Limnetic Bacterioplankton According to Molecular Weight Fractions and Proportions Bound to Humic Matter. *Microb. Ecol.* **2003**, *45*, 433–443. [CrossRef]

54. Arnosti, C. Patterns of Microbially Driven Carbon Cycling in the Ocean: Links between Extracellular Enzymes and Microbial Communities. *Adv. Oceanogr.* **2014**, *2014*, 1–12. [CrossRef]

55. Oest, A.; Alsaffar, A.; Fenner, M.; Azzopardi, D.; Tiquia-Arashiro, S.M. Patterns of change in metabolic capabilities of sediment microbial communities in river and lake ecosystems. *Int. J. Microbiol.* **2018**, *2018*, 5–7. [CrossRef] [PubMed]

56. Hollibaugh, J.T.; Azam, F. Microbial degradation of dissolved proteins in seawater1. *Limnol. Oceanogr.* **1983**, *28*, 1104–1116. [CrossRef]

57. Cho, J.-C.; Kim, S.-J. Increase in Bacterial Community Diversity in Subsurface Aquifers Receiving Livestock Wastewater Input. *Appl. Environ. Microbiol.* **2000**, *66*, 956–965. [CrossRef]

58. Feris, K.P.; Hristova, K.; Gebreyesus, B.; Mackay, D.; Scow, K.M. A Shallow BTEX and MTBE Contaminated Aquifer Supports a Diverse Microbial Community. *Microb. Ecol.* **2004**, *48*, 589–600. [CrossRef]

59. Anantharaman, K.; Brown, C.T.; Hug, L.A.; Sharon, I.; Castelle, C.J.; Probst, A.J.; Thomas, B.C.; Singh, A.; Wilkins, M.J.; Karaoz, U.; et al. Thousands of microbial genomes shed light on interconnected biogeochemical processes in an aquifer system. *Nat. Commun.* **2016**, *7*, 13219. [CrossRef]

60. Gözdereliler, E. Groundwater Bacteria: Diversity, Activity and Physiology of Pesticide Degradation at Low Concentrations. PhD thesis, Technical University of Denmark, Lyngby, Denmark, 2012.

61. Proctor, C.R.; Besmer, M.D.; Langenegger, T.; Beck, K.; Walser, J.-C.; Ackermann, M.; Bürgmann, H.; Hammes, F. Phylogenetic clustering of small low nucleic acid-content bacteria across diverse freshwater ecosystems. *ISME J.* **2018**, *12*, 1344–1359. [CrossRef]

62. Obernosterer, I.; Benner, R. Competition between biological and photochemical processes in the mineralization of dissolved organic carbon. *Limnol. Oceanogr.* **2004**, *49*, 117–124. [CrossRef]

63. Sala, M.M.; Estrada, M. Seasonal changes in the functional diversity of bacterioplankton in contrasting coastal environments of the NW Mediterranean. *Aquat. Microb. Ecol.* **2006**, *44*, 1–9. [CrossRef]

64. Ding, H.; Sun, M.-Y. Biochemical degradation of algal fatty acids in oxic and anoxic sediment–seawater interface systems: Effects of structural association and relative roles of aerobic and anaerobic bacteria. *Mar. Chem.* **2005**, *93*, 1–19. [CrossRef]

65. Bertilsson, S.; Tranvik, L.J. Photochemical transformation of dissolved organic matter in lakes. *Limnol. Oceanogr.* **2000**, *45*, 753–762. [CrossRef]

66. Preiss, J.; Romeo, T. *Molecular Biology and Regulatory Aspects of Glycogen Biosynthesis in Bacteria*; Cohn, W.E., Moldave, K.B.T.-P., Eds.; Academic Press: New York, NY, USA, 1994; Volume 47, pp. 299–329.

67. Henrissat, B.; Deleury, E.; Coutinho, P.M. Glycogen metabolism loss: A common marker of parasitic behaviour in bacteria? *Trends Genet.* **2002**, *18*, 437–440. [CrossRef]

Article

Potential of A Trait-Based Approach in the Characterization of An N-Contaminated Alluvial Aquifer

Tiziana Di Lorenzo [1,*], Alessandro Murolo [2], Barbara Fiasca [2], Agostina Tabilio Di Camillo [2], Mattia Di Cicco [2] and Diana Maria Paola Galassi [2]

[1] Research Institute on Terrestrial Ecosystems (IRET-CNR), Via Madonna del Piano 10, 50019 Sesto Fiorentino, Firenze, Italy

[2] Department of Life, Health and Environmental Sciences, University of L'Aquila, Via Vetoio 1, Coppito, 67100 L'Aquila, Italy; alessandro.murolo92@gmail.com (A.M.); barbara.fiasca@univaq.it (B.F.); ago.tabilio@gmail.com (A.T.D.C.); diciccomattiauni@yahoo.it (M.D.C.); dianamariapaola.galassi@univaq.it (D.M.P.G.)

* Correspondence: tiziana.dilorenzo@cnr.it

Received: 25 October 2019; Accepted: 28 November 2019; Published: 3 December 2019

Abstract: Groundwater communities residing in contaminated aquifers have been investigated mainly through taxonomy-based approaches (i.e., analyzing taxonomic richness and abundances) while ecological traits have been rarely considered. The aim of this study was to assess whether a trait analysis adds value to the traditional taxonomy-based biomonitoring in N-contaminated aquifers. To this end, we monitored 40 bores in the Vomano alluvial aquifer (VO_GWB, Italy) for two years. The aquifer is a nitrate vulnerable zone according to the Water Framework Directive. The traditional taxonomy-based approach revealed an unexpectedly high biodiversity (38 taxa and 5725 individuals), dominated by crustaceans, comparable to that of other unpolluted alluvial aquifers worldwide. This result is in contrast with previous studies and calls into question the sensitivity of stygobiotic species to N-compounds. The trait analysis provided an added value to the study, unveiling signs of impairments of the groundwater community such as low juveniles-to-adults and males-to-females ratios and a crossover of biomasses and abundances curves suggestive of an intermediate alteration of the copepod assemblages.

Keywords: crustaceans; copepods; stygobiotic; traits; groundwater; nitrate; ammonium; nitrite; nitrogen; contamination

---

1. Introduction

The taxonomy-based approach, which is traditionally focused on the analysis of taxonomic richness and abundances, has long been used to characterize and assess the impairment of freshwater ecosystems [1]. Following this approach, biological samples are collected from impaired sites and the taxa are identified, enumerated, and then compared to those of pristine reference sites (e.g., [2]). Trait-based approaches have been developed over the years alternatively to the taxonomy-based one or, more often, in combination with it. Traits are measurable physiological, morphological, and ecological characteristics of organisms, populations, or communities, such as the body size, the trophic role, the duration of life cycle, the reproductive rate, and so forth [3]. The trait-based approach allows understanding ecosystems in ways often unattainable by studies that are exclusively focused on taxonomic diversity. In the first place, traits i) reflect the species (population or community) adaptation level to their environment [4] and ii) can be used as measures of functional diversity [5]. Secondly, trait-based approaches work well across geographies since many traits are shared by widely distributed taxa [3]. From this perspective, some traits can be even applied across ecoregions [6]. Other strengths are that they add diagnostic knowledge, do not require new sampling methodology, and can supplement taxonomic analyses, while weaknesses are redundancy, autocorrelation, and difficulty in the measurements [3]. One of the most useful and fascinating aspects of trait-based approaches is that they can reveal a substantive trait variation (e.g., metabolic changes) before the taxonomic composition of the community is impaired by the loss of some species [1]. Similarly, some traits may have an earlier diagnostic potential regarding ecosystem recovery [1]. However, the decision to prefer a trait-based approach over a taxonomy-based one must be taken with caution; rather, the best option is the combination of the two approaches, at least for freshwater ecosystems [1]. In fact, trait-based approaches may fail to diagnose impacts on imperiled species which are rare in terms of abundance, because of functional redundancy [1]. On the other hand, the diagnostic power of a taxonomy-based approach decreases if the community is not characterized at the species level, but rather described at the genus or even family resolution, as required in several biomonitoring programs [7,8].

Although trait-based approaches have an old tradition in freshwater ecology (reviewed in [1]), they have not been widely applied to groundwater ecosystems. Groundwater habitats, and alluvial aquifers in particular, are characterized by lack of light and by environmental conditions that are more stable compared to surface water environments [9]. The absence of light precludes primary production by photosynthesis, so groundwater ecosystems are generally reliant on allochthonous organic carbon from surface environments [10,11]. However, particulate organic matter is not always available to groundwater food webs because it is intercepted in the soil and the unsaturated zone [10]. Consequently, the basal food source for stygofauna (obligate groundwater-dweller fauna) consists mainly of bacteria and the organic matter that is assimilated by microorganisms [12–14]. Unlike surface waters, groundwater ecosystems support only a few vertebrate species and the invertebrate assemblages are generally dominated by crustaceans [15,16]. Groundwater invertebrate assemblages also may include mites, oligochaetes and, rarely, insects and mollusks [9,17]. Copepods, amphipods, isopods, and syncarids are crustacean taxa commonly collected in healthy groundwater ecosystems [18,19]. The response of stygobiotic communities to impairments due to anthropic pressures has often been examined through taxonomy-based approaches [20–23]. Recently, some stygofaunal traits have proven to be useful in assessing the response of groundwater communities to groundwater flooding [24], carbon input [25,26], habitat structure [27,28] and chemical contamination [20,29–31]. However, the selection of a set of traits that best characterize groundwater communities is still to come.

Alluvial aquifers are particularly prone to organic pollution as their outcrops are frequently cultivated or occupied by urban settlements [21,32]. As a result, an increase in the concentrations of nitrate, nitrite, and ammonium is often observed in groundwater, with long-lasting exceedances of the legal limits [22]. Organic contamination is known to cause a change in the structure and composition of stygobiotic assemblages. Elevated concentrations of N-compounds, due to contamination through sewage and septic tank leaks, or due to the overuse of natural and synthetic fertilizers, alter the structure of groundwater communities through favoring the proliferation of surface water organisms [22,33,34]. However, the ecological studies conducted on alluvial aquifers contaminated by N-compounds are still rather scarce and reliant on the taxonomy-based approach [21,35].

The aim of this study was to evaluate the performance of a trait-based approach in the ecological characterization of an alluvial aquifer with a ten-year history of N-compound contamination. To this end, we analyzed the biological assemblages of the Vomano alluvial aquifer (central Italy) that, in 2010, was designated as a Nitrate Vulnerable Zone (NVZ) under the Water Framework Directive (EC, 2000) due to a nitrate contamination affecting over 50% of the groundwater body volume [36]. We measured five ecological traits and compared the results with those deriving from the traditional approach based on taxonomic richness and abundance distribution.

2. Materials and Methods

2.1. Study Area

The study was performed on an alluvial aquifer (VO_GWB; Figure 1) located in the hydrographic basin of the River Vomano (Italy). The outcrop of the aquifer is 30 km^2. VO_GWB is mainly fed by the River Vomano waters and rainfalls which reach the base of the aquifer in a few weeks [36,37]. The alluvial deposits of VO_GWB consist mainly of gravel and sand interbedded by silty-clay lenses of limited extension [37]. The basement of the aquifer (at a depth of about 35 m) is made up of silty-clay deposits of very low permeability [37]. The groundwater flow is NW–SE directed, and the aquifer mean discharge is 1.3×10^{-2} m^3/s [36]. VO_GWB is used for drinking and industrial and agricultural purposes. The chemical and quantitative status of VO_GWB is routinely monitored by the local environmental authority according to the requirements of the Water Framework Directive [38] and Groundwater Daughter Directive [39]. In 2010, VO_GWB was designated as a Nitrate Vulnerable Zone (NVZ) under the Water Framework Directive [38] due to a persisting (>10 years) nitrate contamination [36].

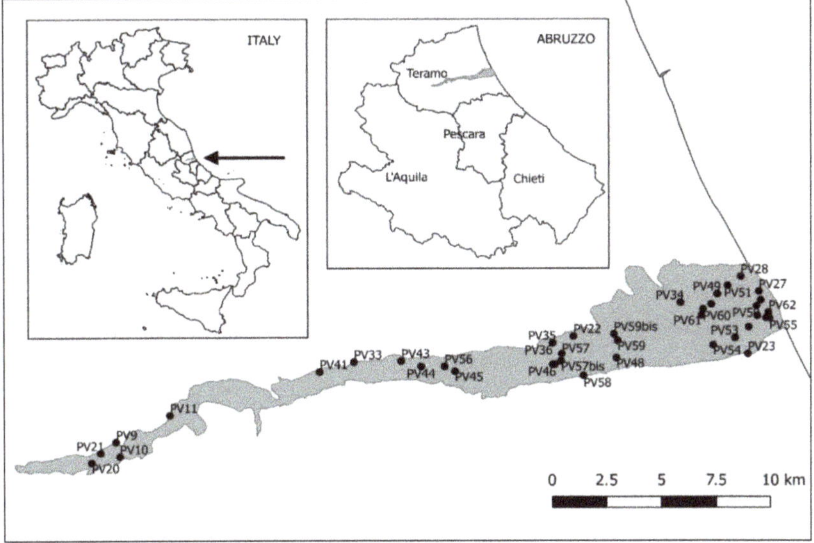

Figure 1. Clockwise: the aquifer VO_GWB in Italy and in the Abruzzo region; sampling sites (black dots) location.

Groundwater sampling occurred in 40 bores used for agricultural purposes (Figure 1). Groundwater organisms and water samples for chemical analyses were collected from shallow bores (up to 30 m deep) in autumn 2014 and spring and winter 2015 for a total of 97 samples. The altitude of the bores varied from 0 to 240 m above sea level and the depth from 2 to 30 m below the soil level. The piezometric level, measured using a freatimeter (PASI, Turin), was in the range 0.5–25.9 m below ground level. The bores' distance from the River Vomano varied from a minimum of 167 m to a maximum of 1733 m.

Groundwater flow of VO_GWB is considered basal, meaning that the groundwater undergoes remixing at all aquifer depths [37]. Preliminary investigations in which the aquifer volume was arbitrarily subdivided into three (0–10 m, 11–20 m, and 21–30 m deep) or two (0–15 m, 16–30 m deep) sectors did not reveal any significant environmental or biological pattern (Supplementary Materials: PRELIMINARY). For this reason, the aquifer was investigated as a unique groundwater unit.

2.2. Biological Survey

Bores were sampled with either an immersion pump (IP), a manual pump (MP), or a modified Cvetkov net [40], depending on the bore type (drilled or hand-dug). Five hundred liters of groundwater were withdrawn using IP and 50 L using MP. The volumes were subsequently filtered by a 60 μm mesh net to collect groundwater fauna. The modified Cvetkov net, equipped with a weight anchored to the tail, was lowered into the hand-dug bores to the bottom and subsequently hauled to filter the entire column of bore water. Ten hauls were performed in each bore [41]. The collected samples were transported in 1 L plastic vials to the laboratory within 6 h and then fixed in 70% ethanol solution. In the laboratory, specimens were sorted under a stereomicroscope at 16× magnification and identified to the lowest taxonomic level possible. Copepods were identified to species level after dissection and subsequent observation with a Leica DM 2500 optical microscope, following dichotomous keys [42,43] and more recent taxonomic papers.

Depending on the degree of adaptation to groundwater, each taxon was classified either as stygobiont (SB: species strictly associated to groundwater habitats, where it completes the whole life cycle; [44]) or non-stygobiont (nSB: epigean species accidentally or occasionally occurring in groundwater). Each specimen of the dominant taxon (Copepoda; see par. 4.) was individually photographed under a Leica M205C stereomicroscope. Length and width were measured with the software LAS (Leica Application Suite, Version 4.7.1 of Leica Microsystems, Wetzlar, Germany). Copepod lengths (in mm) were measured from the tip of the cephalic shield to the end of caudal rami, while width (in mm) was measured at the larger somite bearing legs. Sex was identified in adult individuals. Afterwards, copepod body dimensions (length and width) were converted to species–specific biovolume (in nL) using Equation (1) [45]:

$$BV = a \times b^2 \times CF \tag{1}$$

where BV = biovolume, a = length (mm), b = width (mm), and CF is a correction factor that accounts for the body shape. CF varied from 230 to 750 according to the body shape outlines of each specimen [46]. The biovolumes were then converted into individual fresh weights assuming a specific gravity of 1.1 [24]. The dry mass (mg) was estimated assuming a dry/wet weight ratio of 0.25 and the dry weight carbon content (biomass) was assumed to be 40% of the dry mass [47].

2.3. Environmental Survey

Physicochemical parameters, such as electrical conductivity (μS/cm), pH, dissolved oxygen (mg/L), and temperature (°C), were measured in situ with a multiparametric probe (WTW 3430 SET G). Two liters of groundwater were collected and stored in high-density polyethylene bottles for further chemical analyses aimed at measuring the concentrations of the 110 chemical compounds indicated in the Supplementary Materials (ENV). Particulate organic matter (POM) was obtained as the difference between the sediment weight deprived of the faunal component and dehydrated at 105 °C and that obtained after combustion of the organic fraction at 540 °C.

2.4. Data Analysis: Environmental Variables

As subsequent samples from the same bore must be considered temporal pseudoreplicates producing bias in the statistics [48], the mean values of each abiotic variable were used in the statistical analyses in the place of the raw data. Only 22 variables out of 110 were analyzed, that is, only those showing a standard deviation (SD) $\neq$ 0 (Table 1) so as to exclude irrelevant variables in the statistical models. Potential environmental patterns were investigated through a principal component analysis (PCA) on the basis of the Euclidean distance matrix. Prior to PCA, data were normalized and a Draftsman's plot was used to examine the correlation among the environmental variables [49,50].

Table 1. Mean and standard deviations (SD) of the 22 environmental variables (Var.) retained in the statistical analyses. Alt: altitude (m above sea level); W.t.: water table (m below ground level.); T: temperature (°C); EC: electrical conductivity (µS/cm); O_2: dissolved oxygen (mg/L); POM: particulate organic matter (mg/L); TOC: total organic carbon (mg/L); DOC: dissolved organic carbon (mg/L); NO_2^-, NO_3^-; NH_4^+, SO_4^{2-}; Cl^-, PO_4^{3-}, Ca^{2+}, K^+, Na^+ (mg/L); DIC: cis-1,2-dichloroethylene (µg/L); TCE: 1,1,2,2-tetrachloroethylene (µg/L); CHL: trichloromethane (µg/L); THC: total hydrocarbons (expressed as n-hexane in µg/L). TV: legal threshold value. Mean values exceeding legal limits are indicated in bold.

Var.	Mean	SD	TV
Alt	72.7	72.3	
W.t.	8.3	5.3	
T	16.9	1.7	
EC	1166	231	
pH	7.3	0.2	
O_2	4.7	1.6	
POM	0.73	2.32	
TOC	3.02	2.39	
DOC	2.18	1.70	
NO_2^-	**10.25**	11.19	0.5
NO_3^-	**61.60**	30.07	50
SO_4^{2-}	103.75	51.97	250
Cl^-	55.76	37.49	
PO_4^{3-}	0.27	0.42	
NH_4^+	0.23	0.50	0.5
Ca^{2+}	144.99	30.13	
K^+	5.22	2.312	
Na^+	60.17	33.31	
DIC	0.03	0.043	0.05
TCE	0.07	0.132	1.1
CHL	0.04	0.046	1.5
THC	25.7	4.480	350

2.5. Taxonomy-Based Approach

The exhaustiveness of the biological sampling effort was assessed through taxonomic richness estimations, using nonparametric estimators, namely, Chao1, Chao2, Jackknife1 and Jackknife2, Bootstrap, Michaelis-Menton (MM), and Ugland-Gray-Ellingsen (UGE) [51]. Values were estimated by means of 999 randomizations without replacement.

Abundances were expressed as ind/L dividing the number of individuals of each sample by its volume (liters). Raw data are available in the Supplementary Materials (BIO_VOGWB). Abundances of temporal replicates were summed per each sampling site prior to being used in the statistical analyses. Potential biological patterns were examined using a non-metric multidimensional scaling (nMDS) analysis, incorporating Bray-Curtis similarities after adding a dummy variable equal to 1 [49]. Prior to nMDS, data were $\log(x + 1)$-transformed to approximately conform the data to normality. BIO-ENV analyses [52], based on Spearman's rank correlations, were performed to find the 'best' match between the biological and the environmental data. The statistical significance of the results of the BIO-ENV routine was tested by the global BIO-ENV match permutation test whereby each set of samples was randomly permuted 99 times relative to the others [52]. In addition to multivariate analyses, univariate statistics was performed to assess correlation between individual environmental variables and cumulative abundances per sampling site. The significance was assessed through a permutation test performed by randomly shuffling 999 times the independent variable (i.e., the environmental variables). The permutation test was selected because it is more powerful than tests based on normal theory assumptions when data distribution departs substantially from normality, as it was the case for the data of this study. Regressions were performed with the *ape* library in R 3.0.2 [53].

Finally, the structure of the biological assemblages of N-compound-contaminated alluvial aquifers other than VO_GWB was obtained from literature by using i) ISI Web of Knowledge SM (© 2010 Thomson Reuters, Toronto, ON, CA) with the following strings applied to the topic field: "aquifer" and "stygo*" and "nitrate" or "nitrite" or "ammon*"; ii) Google with the same setting as for i).

2.6. Trait-Based Approach

Potential biological patterns were examined using the following traits and the statistical settings as for the taxonomy-based approach (par. 2.5): Trait#1 (crustaceans vs. non-crustaceans); Trait#2 (SB vs. nSB); Trait#3 (adults vs. juveniles); Trait#4 (males vs. females); Trait#5 (biomass). Since copepod crustaceans accounted for 98% of the aquifer biodiversity, Traits#2–5 were analyzed with respect to copepods only. We analyzed these traits since previous studies have demonstrated that they are reliable indicators of impairment of groundwater communities, with particular references to organic contamination. In detail: i) groundwater assemblages dominated by non-crustacean taxa and a high nSB/SB ratio should be considered impaired [34]; ii) the sex ratio of some populations of copepods may be skewed in favor of female individuals in impaired water bodies [54–56]; iii) the development of juvenile copepods is slowed down in populations exposed to ammonium with respect to control populations [57]; and the survival of copepodids under N-compound exposure is significantly lower than that of adult individuals of the same species [58]. Finally, we analyzed the biomass because its relationship with abundances provides key information about community impairment [49,50]. Distributions of copepod species abundances and biomasses were compared on the same terms with *k*-dominance curves, namely applying the Abundance/Biomass Comparison method (ABC) with the following assumptions [52]: i) in an undisturbed community, the *k*-dominance curve for biomass lies above the curve of abundances for its entire length; ii) as impairment occurs and becomes more severe, the abundance curve lies above the biomass curve, in part, or completely. The W statistic was used to algebraically interpret the position of the curves, the metric values being in the range −1,1 (though neither limit is likely to be attained), with W close to 1 indicating undisturbed condition, W close to 0 indicating intermediate impairment, and W close to −1 indicating severe impairment [52].

Eventual significant patterns of each trait were explored by nMDS plots performed on the Bray–Curtis dissimilarity matrices as in the taxonomy-based approach (par. 2.5). Similarly, BIO-ENV analyses [52] and permutational regressions were used to explore the correlations between the environmental variables and the data of the individual traits.

3. Results

3.1. Environmental Variables

Mean and SD values of the 22 variables with SD ≠ 0 are shown in Table 1.

Mean values of nitrites and nitrates were indicative of contamination according to the legal threshold values (respectively: 0.05 and 50 mg/L) set by Italian regulation [59].

Nitrite contamination was detected in 24 out of 40 bores, 20 of which also showed a nitrate contamination. Overall, bores contaminated by nitrates were 30 out of 40. Fifteen out of 40 bores were contaminated by ionized ammonium. Seven of these bores were also contaminated by nitrates while nine were contaminated by nitrites. Overall, contamination by N-compounds (nitrites, nitrates, and ionized ammonium) involved 36 bores out of 40. The PCA (matrix: 22 variables × 40 sampling sites; plot not shown) explained only 34% of the cumulative variance of the environmental data on the first two axes, failing to highlight a clear pattern of the environmental variables. This result implies that the chemical and physicochemical conditions of VO_GWB were uniform through the aquifer and there were no areas with environmental peculiarities.

3.2. Taxonomy-Based Approach

The stygofauna collected in VO_GWB accounted for 5724 individuals belonging to 38 taxa (35 crustaceans, 2 insect instars, and 1 water mite). Crustaceans were dominant in terms of abundances (99.70%), followed by insects (0.23%) and water mites (0.07%). Among crustaceans, the taxon Copepoda accounted for 97.93% of the total abundances, Amphipoda for 1.52%, Isopoda for 0.02%, Syncarida for 0.16%, and Ostracoda for 0.37%. Two out of seven non-parametric estimators reached the asymptotes, indicating the exhaustiveness of the sampling effort (Supplementary Materials: SRC). The remaining estimators indicated that the sampling effort was such to cover from 67% (Chao1) to 90% (Bootstrap) of the expected biodiversity of the aquifer.

Literature data related to N-compound concentrations and groundwater taxonomic richness in 11 alluvial aquifers worldwide are reported in Table 2.

Table 2. Taxonomic richness (T: number of taxa; SB: number of stygobiotic taxa; nSB: number of non-stygobiotic taxa; C: number of crustacean taxa; Co: number of copepod taxa) and nitrate, nitrite, and ionized ammonium concentrations (mean values in mg/L) relating to 11 alluvial aquifers worldwide obtained from this study and from literature. Ref: references. N-compound exceedances are in bold. (ts): this study.

Country	Alluvial Aquifer	Area	T	SB	nSB	C	Co	NO_3^-	NO_2^-	NH_4^+	REF
Australia	Murray	97,000	11			8	2	2.4		0.04	[60]
Australia	Gwydir	26,600	21		>5	8	2	8.0			[61]
France	Ariège	250	10	10	0	10	0	**90.0**		31	[62]
France	Ariège and Hers	540	36	14	22		4	**>50.0**			[35]
Germany	Rhine	5000	35	16	22	35	23	4.2		0.33	[63]
Italy	Vibrata	48	18	9	9	14	12	**72.5**			[21]
Italy	Adige	45	15	9	6	15	15	7.8	0.01	0.16	[22]
Italy	Lessinia	428		21		20	10	29.7			[64]
Italy	Vomano	30	38	21	17	35	28	**61.6**	**10.25**	0.23	(ts)
Slovenia	Brest	6	38	22	16	23	15	2.5			[65]
Spain	Tajo	201	11	3	8	11	6	11.4	0.02	0.19	[66]

The stygobiotic cyclopoid *Diacyclops cosanus* (Dco) occurred in 15 bores out 40 and was the species with the widest distribution in VO_GWB. The remaining 35 taxa occurred in less than 12 bores out of 40 and 24 taxa (12 of which were stygobiotic) occurred in less than 4 bores. Bore PV28 (13 m deep) showed the highest biodiversity, with 11 taxa, 7 of which were stygobiotic, followed by PV48 (20 m deep) with 10 taxa, 9 of which were stygobiotic. More than 50% of the bores (namely, 23 out of 40) were characterized by a maximum of three taxa and no taxa were collected from three bores, namely, from P26 (18 m deep), PV53 (10 m deep), and PV62 (8 m deep).

The nMDS plot (Figure 2; 2D stress: 0.03; matrix: 38 variables × 40 sampling sites) showed a group of 27 bores of different depths (from 4 to 30 m deep), almost completely overlapping at the center of the plot and surrounded by 11 bores (from 2 to 25 m deep) located at different distances. Thirteen out of 38 taxa were exclusive of the bores overlapping in the center of the plot, namely, the copepods *Paracyclops fimbriatus* (Pfi), *Bryocamptus pygmaeus* (Bpy), *Eucyclops serrulatus* (Ese), *Diacyclops maggii* (Dma), *Graeteriella unisetigera* (Gun), *Nitokra hibernica* (Nhi), *Nitocrella fedelitae* (Nfi), *Nitocrella achaiae* (Nac), *Bryocamptus dentatus* (Bde), *Elaphoidella plutonis* (Epl), *Pseudectinosoma reductum* (Pre), the amphipod *Niphargus* sp. 2 (Nsp2), and a species belonging to the Coleoptera Elmidae (Elm). Three species, namely, the copepods *Diacyclops bicuspidatus* (Dbic) and *Nitocrella kunzi* (Nku), and the amphipod *Niphargus sodalis* (Nso), occurred exclusively in the not-overlapping bores (Figure 2 and Supplementary Materials: BIO_VOGWB).

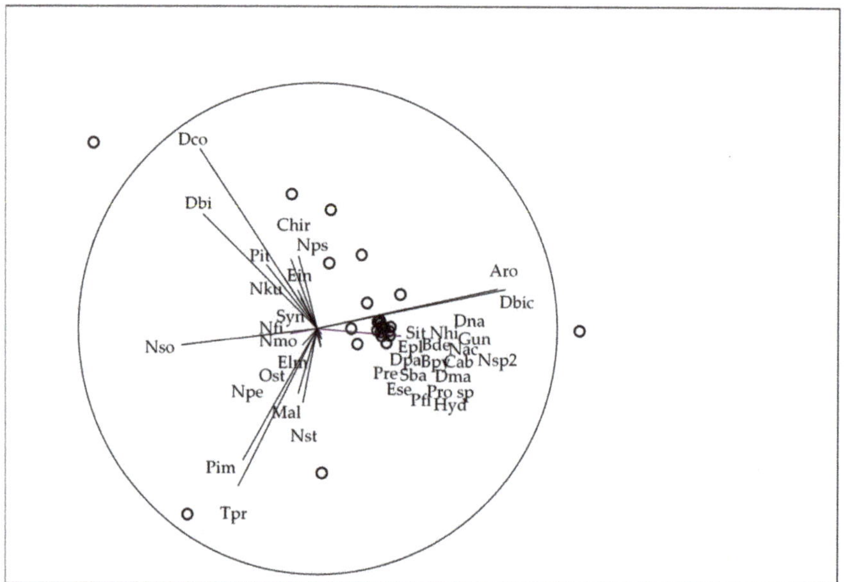

Figure 2. nMDS (2D stress: 0.03) ordination plot following the taxonomy-based approach in VO_GWB. Acronyms are as in the Supplementary Materials: BIO_VOGWB. Vector overlays are reported as exploratory tools to visualize potential linear monotonic relationships, based on Pearson correlation, between the taxa abundances and the nMDS ordination axes [52].

The BIO-ENV analysis showed that there was no significant explanatory relationship between the environmental variables and the matrix of the 38 taxa collected in VO_GWB. In fact, the sole environmental variable which best correlated to the taxa abundances was POM ($\rho = 0.436$) and the best n-dimensional combination consisted of five variables, namely, electrical conductivity, pH, nitrites, nitrates, and ionized ammonium. However, the real value of ρ fell within the 95% confidence limit of the normal distribution of randomly permuted values of ρ (p-value = 14%), indicating that the null hypothesis of no relationship between the best subset of environmental variables and the community structure should not be rejected. Univariate analyses gave significant results. In fact, cumulative densities were negatively correlated, although weakly, to electrical conductivity, dissolved oxygen, nitrates, sulphates, calcium, and sodium (Supplementary Materials: REGR). However, R^2 values were <20% for all parameters except for electrical conductivity ($R^2 = 57\%$), meaning that the regression models had a low predictive potential.

3.3. Trait-Based Approach

Trait#1 (crustacean vs. non-crustacean taxa). VO_GWB was dominated by crustaceans, occurring with 35 out of 38 taxa. The nMDS returned an arched, or horseshoe, pattern. This effect is very common in datasets in which there is a progressive turnover of a dominant variable [67]. In fact, the nMDS bubble plot showed that the arch effect was due to the turnover of crustacean taxa (Supplementary Materials: nMDS_T1). The BIO-ENV analysis indicated no significant relationship between the best subsets of environmental variables (POM: $\rho = 0.435$; p-value = 18%) and the abundances of crustacean and non-crustacean taxa. Univariate analyses did not highlight significant correlation between the cumulative abundances of crustaceans (or non-crustaceans) and the individual environmental variables.

Trait#2 (SB vs. nSB taxa). Seventeen out of 38 taxa were nSB and 21 were SB (Supplementary Materials: BIO_VOGWB). The nMDS (2D Stress: 0.03) showed that SB and nSB taxa occurred almost in all bores (Figure 3a), however, with different abundances.

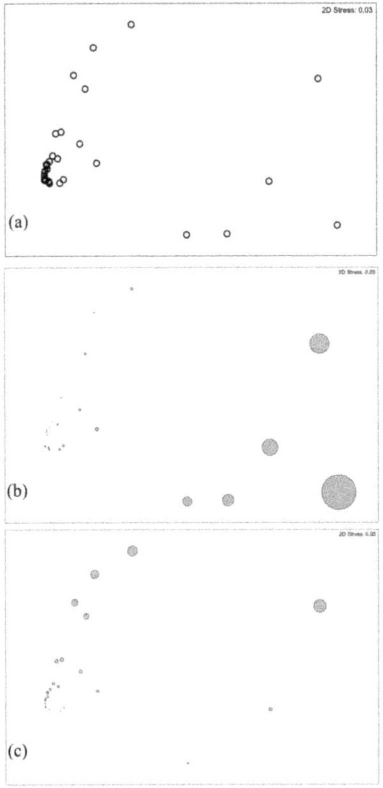

Figure 3. (**a**) nMDS plots (stress: 0.03) performed on copepod abundances data of both nSB and SB species; (**b**) nMDS bubble plot of the nSB species (ind/L); (**c**) nMDS bubble plot of the SB species (ind/L).

SB abundances (Figure 3b) were higher in the eight bores located in the upper sector of the plot while nSB abundances (Figure 3c) were higher in the lower sector of the plot, and in particular, in five bores. The group of eight bores with the highest abundances of SB taxa differed from the group of the five bores in the mean values of some environmental variables. Specifically, the average values of POM (0.44 mg/L), nitrite (9 mg/L), and ionized ammonium (0.031 mg/L) were lower in the bores with high densities of SB than in those with high densities of nSB (POM: 1.33 mg/L, nitrite: 13 mg/L, ionized ammonium: 0.644 mg/L). On the other hand, the mean values of electrical conductivity (1237 µS/cm), nitrates (79 mg/L), sulphates (109 mg/L), chloride (51 mg/L), calcium (156 mg/L), and sodium (53 mg/L) were higher in the sites with high densities of SB than in those with high densities of nSB (electrical conductivity: 835 µS/cm, nitrates: 43 mg/L, sulphate: 45 mg/L, chloride: 30 mg/L, calcium: 105 mg/L, and sodium: 31 mg/L). However, the BIO-ENV analysis indicated no significant relationship between the best subsets of environmental variables (POM: $\rho = 0.443$; p-value = 16%) and the densities of the two ecological categories. The univariate analyses highlighted a significant negative correlation between SB copepod abundances and bore depth (Supplementary Materials: REGR), however, with an R^2 equal to 11%. The abundances of nSB copepods were negatively correlated to electrical conductivity, dissolved oxygen, nitrate, sulphates, calcium, and sodium (Supplementary Materials: REGR). R^2 values were <20% for all parameters except for electrical conductivity ($R^2 = 57\%$).

Trait#3 (adults vs. juveniles). All copepod species were collected with adults and juveniles with a few exceptions, such as the nSB species *Nitokra hibernica* and the SB species *Nitocrella fedelitae*, *Nitocrella morettii*, *Nitocrella kunzi*, *Bryocamptus dentatus*, and *Pseudectinosoma reductum*, which were collected with adults only. The nMDS plot (Supplementary Materials: nMDS_T3) showed no clear separation of the bores, however, the bubble plot highlighted that adult copepods were slightly more numerous than juveniles almost in every bore. The univariate analyses highlighted a significant negative correlation between adult copepod abundances and electrical conductivity, dissolved oxygen, nitrate, sulphates, calcium, and sodium (Supplementary Materials: REGR). A significant negative correlation was assessed between juvenile abundances and electrical conductivity, sulphates, calcium, and sodium as well (Supplementary Materials: REGR). However, R^2 values were all <20%, except for electrical conductivity ($R^2 > 50\%$).

Trait#4 (males vs. females). Twelve copepod species occurred with more females than males, nine species with more males than females, and seven species with the same number of males and females. Overall, the average abundance ± SD of males was 0.26 ± 0.70 ind/L and that of females was 0.67 ± 2.13 ind/L, with a sex ratio equal to 0.38. The nMDS did not show a clear pattern (Supplementary Materials: nMDS_T4). The BIO-ENV analysis indicated no relationship between the best subsets of environmental variables (POM: $\rho = 0.437$; *p*-value = 18%) and the abundances of males and females. The univariate analyses highlighted a significant negative correlation between male abundances and electrical conductivity, dissolved oxygen, nitrate, sulphates, and calcium (Supplementary Materials: REGR). A significant correlation was assessed between female abundances and electrical conductivity, dissolved oxygen, nitrates, sulphates, calcium, and sodium as well (Supplementary Materials: REGR). However, R^2 values were <20% for all parameters and for both males and females, except for electrical conductivity ($R^2 > 43\%$).

Trait#5 (biomass). The total dry weight carbon (mean ± SD) was 0.012 ± 0.061 mg/L C. Multivariate analyses did not reveal significant patterns. The univariate analyses highlighted a significant negative correlation between biomass and electrical conductivity, dissolved oxygen, and calcium (Supplementary Materials: REGR). However, R^2 values were all <20%, except for electrical conductivity ($R^2 = 42\%$). The biomass and abundance curves partially crossed at about the 20th species but since both curves were close to 100% at this point, the crossover was unclear (Figure 4). However, the W metric equal to 0.096 was indicative of a moderate impairment of the copepod assemblages.

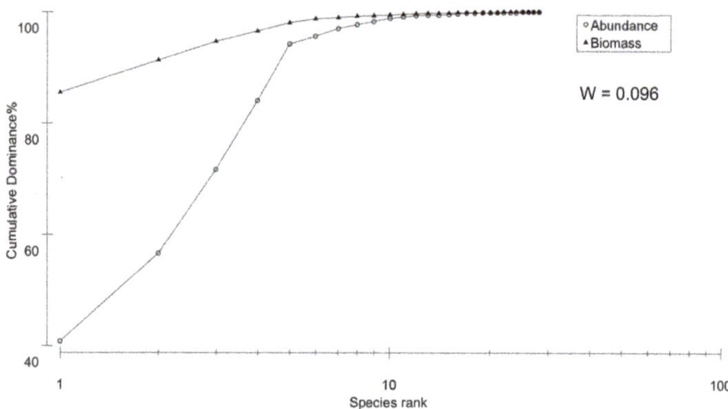

Figure 4. Abundances-biomasses curves of copepod species in GWB_VO aquifer. The two curves' crossover occurring at the 20th species and the metric W = 0.096 were indicative of moderate impairment of the copepod assemblages [52].

4. Discussion

The taxonomic richness of VO_GWB was high and consistent with previous studies in other European and non-European alluvial aquifers [60,64,66,68,69], both at the local and the regional scales [15,70]. However, the number of species collected in VO_GWB is surprisingly high if considering the persisting N-compound contamination which occurred in 90% of the aquifer. This study highlighted the high taxonomic richness and abundances of the assemblages of VO_GWB with respect to those of other alluvial aquifers worldwide contaminated by N-compounds. The taxonomic richness (38 taxa) and the abundances (5725 individuals) of VO_GWB were comparable to those of pristine alluvial aquifers. Such a result is in contrast with previous data showing that freshwater invertebrates are highly sensitive to N-compounds [71]. From this point of view, the results of this study represent a novelty and require further ecotoxicological investigations. In fact, to date, ecotoxicological studies with N-compounds have been focused mainly on non-stygobiotic invertebrates, with the exception of the study carried out by [58], who analyzed the sensitivity of the stygobiotic cyclopoid *Diacyclops belgicus* (not found in VO_GWB) to ionized ammonium. The low number of ecotoxicological data concerning stygobiotic species, and the reasons behind the difficulty in carrying out such studies, have recently been reviewed by [72], who have also claimed as unrealistic the probability of gathering new data in the near future. Thus, the high taxonomic richness and abundances observed in VO_GWB will likely remain unsupported by ecotoxicological considerations for some more time.

The traditional taxonomy-based approach showed that most taxa were collected only in a few bores and showed, therefore, restricted distributions as already observed in other studies (e.g., [64,66]). The most widespread species in VO_GWB, *D. cosanus*, is known to be highly tolerant to elevated values of salinity and electrical conductivity, being collected in other N-contaminated alluvial aquifers, in brackish coastal groundwater in southern Italy, and in sulfidic saturated karst [21]. The occurrence of epigean species in the deeper sectors of VO_GWB is recurrent in alluvial aquifers, especially in those polluted by contaminants of organic origin (e.g., [22,64]). The weak correlations of the environmental and biological variables are a common result in groundwater studies, in particular, in aquifers that are highly homogeneous from a physicochemical point of view [22,66,73]. In this study, the variance explained by the univariate models was low and this indicated that further factors, in addition to those already investigated, must be taken into account in the future to better explain the observed biological patterns. For instance, copepod species interactions could play a role in determining the biological assemblages in VO_GWB as already observed in other aquifers [74]. Similarly, previous studies showed that the grain size composition of sediments [75–77], and the lowering of the groundwater table [21,75], could influence the biological assemblage's patterns in alluvial aquifers.

Trait#1 (crustaceans vs. non-crustaceans) highlighted the predominance of crustaceans in VO_GWB. This result is in contrast with previous studies [21,33,62] which showed that crustacean abundances are lower than those of non-crustaceans in aquifers polluted by N-enriched infiltration waters.

Trait#2 provided additional information to this study, highlighting that stygobiotic species seemed to thrive in the bores where non-stygobiotic species were less abundant. The environmental variables did not entirely explain such a pattern. Hence, additional factors should be investigated such as the isolation of the bores from the surface. In fact, groundwater habitats characterized by intense water exchanges with the surface environment often exhibit a predominance of generalist non-stygobiotic fauna, while isolated groundwater sites show a predominance, or even an exclusive occurrence, of stygobiotic species [19,61,78,79].

Trait#3 showed that juveniles were less abundant than adult copepods and that some species of stygobiotic copepods were collected exclusively with adults. This is a peculiar result because stygobiotic copepod species are known to reproduce in groundwater, albeit showing a low fertility [80]. In unaltered ecosystems, juveniles are usually more abundant than adults, while the opposite was observed in this study, at least for a few species. This result could be related to the N-compound contamination of the aquifer. In fact, juvenile stages are generally more sensitive to pollution than adults [81,82] as it was observed for the stygobiotic cyclopoid *Diacyclops belgicus* when exposed to ammonium nitrate [58].

Trait#4 highlighted that the sex ratio departed from 1 and was equal to 0.38 in favor of females. In diploid populations, sex ratio is usually equal to 1 [83,84], however, surface water populations' sex ratio tends to be skewed as result of seasonality, environmental conditions, and contamination level [54–56].

In general, seasonality does not affect stygobiotic populations [44]. However, N-compounds could have a role in determining a sex ratio <1 in VO_GWB. For example, the sex ratios of the populations of *Cyclops vicinus* and *Thermocyclops crassus* were skewed in favor of female individuals in some altered surface water bodies [54]. Although no correlations were assessed between nitrites, nitrates, or ionized ammonium and abundances of adult male or female copepods in this study, these pollutants may nevertheless have had an effect on the populations in the past, reducing the number of male copepods during the ten-year-long contamination (still persisting in 2015) and inducing the skewed sex ratio <1 observed in 2014–2015.

Trait#5 highlighted that VO_GWB had an average carbon content equal to 0.012 ± 0.061 mg/L C, accounting for 0.049 g C. With respect to the carbon content (0.004 g) measured in an unpolluted chalk aquifer in southern England [24], the amount measured in VO_GWB is high. However, a fair part of this value, namely 0.047 g C, was contributed by the individuals of *Acanthocyclops robustus*, a large non-stygobiotic copepod species collected in bore PV36. Excluding the carbon contributed by this species, the average dry carbon found in VO_GWB was equal to 0.006 g C which is comparable to the amount measured in southern England [24]. However, the partial crossover of the biomasses and abundances curves and the W metric indicated impairment of copepod assemblages although not as severe as expected in a long-term contaminated alluvial aquifer as VO_GWB is.

Based on these results, we can conclude that the trait-based approach provided an added value to this study, allowing a better characterization of the biological assemblages of VO_GWB. The main difficulties that we encountered were i) the high taxonomic expertise required and ii) the long time needed for measuring individuals. In fact, the attribution of the two categories "stygobiotic" and "non-stygobiotic", at least for copepods, necessitated identifications at species level and a solid knowledge of the autecology of each species.

5. Conclusions

In this study, we pursued two objectives, namely, i) to provide new data concerning the biodiversity of an alluvial aquifer affected by a persistent N-compound contamination, and ii) to assess whether an approach based on some selected ecological traits may provide an added value to the traditional taxonomy-based methodology. The alluvial aquifer investigated in this study showed a surprisingly high biodiversity, dominated by crustaceans, comparable to that of alluvial aquifers in good chemical state. This result is in contrast with the observations of previous studies. The results call into question the real sensitivity of stygobiotic species (copepods, in particular) to N-compounds, which turned out to be lower than expected. The trait-based approach provided new details, unveiling signs of impairments of the groundwater community such as low juveniles-to-adults and males-to-females ratios and a crossover of biomasses and abundances curves suggestive of an intermediate alteration of the copepod assemblages. The signs of community alteration were not clearly imputable to the N-compound contamination affecting the aquifer for 10 years so that further investigations are needed, including additional factors such as sediment size and species interactions, in the analyses. The application of a trait-based approach to copepod-dominated groundwater assemblages requires a long time for data acquisition and high taxonomic skills. A trait database could ease the application of such approaches in studies concerning groundwater ecosystems and its setup should be strongly encouraged among researchers.

Supplementary Materials: The Supplementary Materials are available online at http://www.mdpi.com/2073-4441/11/12/2553/s1.

Author Contributions: Conceptualization, T.D.L. and D.M.P.G.; methodology, T.D.L.; validation, T.D.L., B.F. and D.M.P.G.; formal analysis, A.M., A.T.D.C. and M.D.C.; investigation, B.F., A.M. and A.T.D.C.; resources, D.M.P.G.; data curation, B.F., A.M., A.T.D.C. and M.D.C.; writing-original draft preparation, T.D.L. and A.M.; writing-review and editing, T.D.L.; project administration, B.F.; funding acquisition, D.M.P.G.

Funding: This research was funded by the AQUALIFE project LIFE12 BIO/IT/000231 "Development of an innovative and user-friendly indicator system for biodiversity in groundwater dependent ecosystems" of European commission.

Acknowledgments: We thank the students of the MESVA Department of the University of L'Aquila (IT) for the support in sampling activities.

Conflicts of Interest: The authors declare no conflict of interest.

References

1. Culp, J.M.; Armanini, D.G.; Dunbar, M.J.; Orlofske, J.; Poff, L.N.; Pollard, A.I.; Yates, A.G.; Hose, G.C. Incorporating traits in aquatic biomonitoring to enhance causal diagnosis and prediction. *Integr. Environ. Assess. Manag.* **2010**, *7*, 187–197. [CrossRef] [PubMed]

2. Larras, F.; Coulaud, R.; Gautreau, E.; Billoir, E.; Rosebrery, J.; Usseglio-Polatera, P. Assessing anthropogenic pressures on streams: A random forest approach based on benthic diatom communities. *Sci. Total Environ.* **2017**, *15*, 1101–1112. [CrossRef] [PubMed]

3. Van den Brink, P.J.; Alexander, A.; Dessrosiers, M.; Goedkoop, W.; Goethals, P.; Liess, M.; Dyer, S. Traits-based approaches in bioassessment and ecological risk assessment: Strengths, weaknesses, opportunities and threats. *Integr. Environ. Assess. Manag.* **2011**, *7*, 198–208. [CrossRef] [PubMed]

4. McGill, B.J.; Enquist, B.J.; Weiher, E.; Westboy, M. Rebuilding community ecology from functional traits. *Trends Ecol. Evol.* **2006**, *21*, 178–185. [CrossRef]

5. Petchey, O.L.; Gaston, K.J. Functional diversity: Back to basics and looking forward. *Ecol. Lett.* **2006**, *9*, 741–758. [CrossRef]

6. Usseglio-Polatera, P.; Bournaud, M.; Richoux, P.; Tachet, H. Biological and ecological traits of benthic freshwater macroinvertebrates: Relationships and definition of groups with similar traits. *Freshw. Biol.* **2000**, *43*, 175–205. [CrossRef]

7. Carter, J.L.; Resh, V.H. After site selection and before data analysis: Sampling, sorting, and laboratory procedures used in stream benthic macroinvertebrate monitoring programs by USA state agencies. *J. N. Am. Benthol. Soc.* **2001**, *20*, 658–682. [CrossRef]

8. Jones, C.F. Taxonomic sufficiency: The influence of taxonomic resolution on freshwater bioassessments using benthic macroinvertebrates. *Environ. Rev.* **2008**, *16*, 45–69. [CrossRef]

9. Gibert, J.; Culver, M.J.; Dole-Oliver, F.; Malard, F.; Christman, M.; Deharveng, L. Assessing and conserving groundwater biodiversity: Synthesis and perspectives. *Freshw. Biol.* **2009**, *54*, 930–941. [CrossRef]

10. Pabich, W.J.; Valiela, I.; Hemond, H.F. Relationship between DOC concentration and vadose zone thickness and depth below water table in groundwater of Cape Cod, U.S.A. *Biogeochemistry* **2001**, *55*, 247–268. [CrossRef]

11. Shen, Y.; Chapelle, F.H.; Benner, R. Origins and bioavailability of dissolved organic matter in groundwater. *Biogeochemistry* **2015**, *122*, 61–78. [CrossRef]

12. Boulton, A.J. River ecosystem health down under: Assessing ecological condition in riverine groundwater zones in Australia. *Ecosyst. Health* **2001**, *6*, 108–118. [CrossRef]

13. Francois, C.M.; Mamillod Blondin, F.; Malard, F.; Fourel, F.; Lécuyer, C.; Douady, C.J.; Simon, L. Trophic ecology of groundwater species reveals specialization in a low-productivity environment. *Funct. Ecol.* **2016**, *30*, 262–273. [CrossRef]

14. Hofmann, R.; Griebler, C. DOM and bacterial growth efficiency in oligotrophic groundwater: Absence of priming and co-limitation by organic carbon and phosphorus. *Aquat. Microb. Ecol.* **2018**, *81*, 55–71. [CrossRef]

15. Stoch, F.; Galassi, D.M.P. Stygobiotic crustacean species richness: A question of numbers, a matter of scale. *Hydrobiologia* **2010**, *653*, 217–234. [CrossRef]

16. Strona, G.; Fattorini, S.; Fiasca, B.; Di Lorenzo, T.; Di Cicco, M.; Lorenzetti, W.; Galassi, D.M.P. AQUALIFE software: A new tool for a standardized ecological assessment of groundwater dependent ecosystems. *Water* **2019**. (accepted).

17. Humphreys, W. Aquifers: The ultimate groundwater dependent ecosystem. *Aust. J. Bot.* **2006**, *54*, 115–132. [CrossRef]

18. Galassi, D.M.P.; Huys, R.; Reid, J. Diversity ecology and evolution of groundwater copepods. *Freshw. Biol.* **2009**, *54*, 691–708. [CrossRef]

19. Di Lorenzo, T.; Cipriani, D.; Fiasca, B.; Rusi, S.; Galassi, D.M.P. Groundwater drift monitoring as a tool to assess the spatial distribution of groundwater species into karst aquifers. *Hydrobiologia* **2018**, *813*, 137–156. [CrossRef]

20. Malard, F.; Mathieu, J.; Reygrobellet, J.L.; Lafont, M. Biomonitoring groundwater contamination: Application to a karst area in Southern France. *Aquat. Sci.* **1996**, *58*, 158–187. [CrossRef]

21. Di Lorenzo, T.; Galassi, D.M.P. Agricultural impact in Mediterranean alluvial aquifers: Do groundwater communities respond? *Fundam. Appl. Limnol.* **2013**, *182*, 271–282. [CrossRef]

22. Di Lorenzo, T.; Cifoni, M.; Lombardo, P.; Fiasca, B.; Galassi, D.M.P. Ammonium threshold value for groundwater quality in the EU may not protect groundwater fauna: Evidence from an alluvial aquifer in Italy. *Hydrobiologia* **2015**, *743*, 139–150. [CrossRef]

23. Caschetto, M.; Barbieri, M.; Galassi, D.M.P.; Mastrorillo, L.; Rusi, S.; Stoch, F.; Di Cioccio, A.; Petitta, M. Human alteration of groundwater-surface water interactions (Sagittario River, Central Italy): Implication for flow regime, contaminant fate and invertebrate response. *Environ. Earth Sci.* **2014**, *7*, 1791–1807. [CrossRef]

24. Reiss, J.; Perkins, D.M.; Fussmann, K.E.; Krause, S.; Canhoto, C.; Romeijn, P.; Roberton, A.L. Groundwater flooding: Ecosystem structure following an extreme recharge event. *Sci. Total Environ.* **2019**, *625*, 1252–1260. [CrossRef] [PubMed]

25. Wilhelm, F.M.; Taylor, S.J.; Adams, G.L. Comparison of routine metabolic rates of the stygobite, *Gammarus acherondytes* (Amphipoda: Gammaridae) and the stygophile *Gammarus troglophilus*. *Freshw. Biol.* **2006**, *51*, 1162–1174. [CrossRef]

26. Mezek, T.; Simčič, T.; Arts, M.T.; Bracelj, A. Effect of fasting on hypogean (*Niphargus stygius*) and epigean (*Gammarus fossarum*) amphipods: A laboratory study. *Aquat. Ecol.* **2010**, *44*, 397–408. [CrossRef]

27. Descloux, S.; Datry, T.; Usseglio-Polatera, P. Trait-based structure of invertebrates along a gradient of sediment colmation: Benthos versus hyporheos responses. *Sci. Total Environ.* **2014**, *466–467*, 265–276. [CrossRef]

28. Mathers, K.L.; Hill, M.J.; Wood, C.D.; Wood, P.J. The role of fine sediment characteristics and body size on the vertical movement of a freshwater amphipod. *Freshw. Biol.* **2018**, *64*, 152–163. [CrossRef]

29. Sarkka, J.; Levonen, L.; Makela, J. Meiofauna of springs in Finland in relation to environmental factors. *Hydrobiologia* **1997**, *347*, 139–150. [CrossRef]

30. Mauclaire, L.; Gibert, J.; Claret, C. Do bacteria and nutrients control faunal assemblages in alluvial aquifers? *Arch. Hydrobiol.* **2000**, *148*, 85–98. [CrossRef]

31. Pacioglu, O.; Moldovan, O.T. Response of invertebrates from the hyporheic zone of chalk rivers to eutrophication and land use. *Environ. Sci. Pollut. Res.* **2016**, *23*, 4729–4740. [CrossRef] [PubMed]

32. Boy-Roura, M.; Nolan, B.T.; Menció, A.; Mas-Pla, J. Regression model for aquifer vulnerability assessment of nitrate pollution in the Osona region (NE Spain). *J. Hydrol.* **2013**, *505*, 150–162. [CrossRef]

33. Malard, F. Groundwater contamination and ecological monitoring in a Mediterranean karst ecosystem in Southern France. In *Groundwater Ecology: A Tool for Management of Water Resources*; Griebler, C., Danielopol, D., Gibert, J., Nachtnebel, H.P., Notenboom, J., Eds.; Official Publication of the European Communities: Luxembourg, 2001; pp. 183–194.

34. Korbel, K.L.; Hose, G.C. A tiered framework for assessing groundwater ecosystem health. *Hydrobiologia* **2011**, *661*, 329–349. [CrossRef]

35. Marmonier, P.; Maazouzi, C.; Baran, N.; Blanchet, S.; Ritter, A.; Saplairoles, M.; Dole-Olivier, M.J.; Galassi, D.M.; Eme, D.; Dolédec, S.; et al. Ecology-based evaluation of groundwater ecosystems under intensive agriculture: A combination of community analysis and sentinel exposure. *Sci. Total Environ.* **2018**, *613–614*, 1353–1366. [CrossRef] [PubMed]

36. Regione Abruzzo. Piano di Tutela delle Acque. Relazione Generale e Allegati. 2010. Available online: http://www.regione.abruzzo.it/pianoTutelaacque/index.asp?modello=elaboratiPiano&servizio=lista&stileDiv=elaboratiPiano (accessed on 10 October 2019).

37. Desiderio, G.; Nanni, T.; Rusi, S. La pianura del fiume Vomano (Abruzzo): Idrogeologia, antropizzazione e suoi effetti sul depauperamento della falda. *Boll. Soc. Geol. Ital.* **2003**, *122*, 421–434.

38. EC (European Community). Directive 2000/60/EC of the European Parliament and of the Council of 23 October 2000 establishing a framework for Community action in the field of water policy. *Off. J.* **2000**, *L 327*, 1–73.

39. EC (European Community). Directive 2006/118/EC of the European Parliament and of the Council of 12 December 2006 on the protection of groundwater against pollution and deterioration. *Off. J.* **2006**, *L 327/19*, 1–31.

40. Cvetkov, L. Un fi let phréatobiologique. *Bulletin de l'Institut de Zoologie et Musée Sofia* **1968**, *22*, 215–219.

41. Hancock, P.J.; Boulton, A.J. Sampling groundwater fauna: Efficiency of rapid assessment methods tested in bores in eastern Australia. *Freshw. Biol.* **2009**, *54*, 902–917. [CrossRef]

42. Dussart, B.; Defaye, D. *World Directory of Crustacea Copepoda of Inland Waters. II—Cyclopiformes*; Backhuys Publishers: Leiden, The Netherlands, 2009; pp. 1–276.

43. Boxshall, G.A.; Halsey, S.H. *An Introduction to Copepod Diversity*; The Ray Society: London, UK, 2004.

44. Gibert, J.; Stanford, J.A.; Dole-Oliver, M.J.; Ward, J. Basic attributes of groundwater ecosystems and prospects for research. In *Groundwater Ecology*; Gibert, J., Danielopol, D., Stanford, J., Eds.; Academic Press: California, CA, USA, 1994; pp. 7–40.

45. Reiss, J.; Schmid-Araya, J.M. Existing in plenty: Abundance, biomass and diversity of ciliates and meiofauna in small streams. *Freshw. Biol.* **2008**, *53*, 652–668. [CrossRef]

46. Warwick, R.M.; Gee, J.M. Community structure of estuarine meiobenthos. *Mar. Ecol. Prog. Ser.* **1984**, *18*, 97–111. [CrossRef]

47. Feller, R.J.; Warwick, R.M. Introduction to the study of meiofauna. In *Energetics*; Higgins, R.P., Thiel, H., Eds.; Smithsonian Institution Press: Washington, DC, USA, 1988; pp. 181–196.

48. Hurlbert, S.H. Pseudoreplication and the design of ecological field experiments. *Ecol. Monog.* **1984**, *54*, 187–211. [CrossRef]

49. Clarke, K.R.; Gorley, R.N. *PRIMER v6: User Manual/Tutorial*; PRIMER-E: Playmouth, UK, 2006.

50. Clarke, K.R.; Green, R.H. Statistical design and analysis for a 'biological effects' study. *Mar. Ecol.* **1988**, *46*, 213–226. [CrossRef]

51. Magurran, A.E.; McGill, B.J. *Biological Diversity: Frontiers in Measurement and Assessment*; University Press: Oxford, UK, 2011.

52. Clarke, K.R.; Warwick, R.M. *Change in Marine Communities: An Approach to Statistical Analysis and Interpretation*, 2nd ed.; Playmouth Marine Laboratory: Playmouth, UK, 2001.

53. R Core Team. *R: A Language and Environment for Statistical Computing*; R Foundation for Statistical Computing: Vienna, Austria, 2013; Available online: http://www.R-project.org/ (accessed on 2 December 2019).

54. Krupa, E.G. Population densities, sex ratios of adults, and occurrence of malformations in three species of cyclopoid copepods in waterbodies with different degrees of eutrophy and toxic pollution. *J. Mar. Sci. Technol.* **2005**, *13*, 226–237.

55. Peschke, K.; Jonas, G.; Kohler, H.-R.; Karl, W.; Rita, T. Invertebrates as indicators for chemical stress in sewage-influenced stream systems: Toxic and endocrine effects in gammarids and reactions at the community level in two tributaries of Lake Constance, Schussen and Argen. *Ecotoxicol. Environ. Saf.* **2014**, *106*, 115–125. [CrossRef]

56. Ozga, V.A.; da Silva de Castro, V.; da Silva Castiglioni, D. Population structure of two freshwater amphipods (Crustacea: Peracarida: Hyalellidae) from southern Brazil. *Nauplius* **2018**, *26*, e2018025. [CrossRef]

57. Di Marzio, W.D.; Castaldo, D.; Di Lorenzo, T.; Di Cioccio, A.; Sáenz, M.E.; Galassi, D.M.P. Developmental endpoints of chronic exposure to suspected endocrine-disrupting chemicals on benthic and hyporheic freshwater copepods. *Ecotoxicol. Environ. Saf.* **2013**, *96*, 86–92. [CrossRef]

58. Di Lorenzo, T.; Di Marzio, W.D.; Sáenz, M.E.; Baratti, M.; Dedonno, A.A. Iannucci, A.; Cannicci, S.; Messana, G.; Galassi, D.M. Sensitivity of hypogean and epigean freshwater copepods to agricultural pollutants. *Environ. Sci. Pollut. Res.* **2014**, *21*, 4643–4655. [CrossRef]

59. Repubblica Italiana. Decreto Legislativo 16 Marzo 2009, n. 30: Attuazione della direttiva 2006/118/CE, relativa alla protezione delle acque sotterranee dall'inquinamento e dal deterioramento. *Gazzetta Ufficiale* **2009**, *49*, 1–45.

60. Korbel, K.; Chartion, A.; Stephenson, S.; Greenfield, P.; Hose, G.C. Wells provide a distorted view of life in the aquifer: Implications for sampling, monitoring and assessment of groundwater ecosystems. *Sci. Rep.* **2017**, *7*, 40702. [CrossRef] [PubMed]

61. Menció, A.; Korbel, K.L.; Hose, G.C. River–aquifer interactions and their relationship to stygofauna assemblages: A case study of the Gwydir River alluvial aquifer (New South Wales, Australia). *Sci. Total Environ.* **2014**, *479–480*, 292–305. [CrossRef] [PubMed]

62. Dumas, P.; Bou, C.; Gibert, J. Groundwater macrocrustaceans as natural indicators of the Ariège alluvial aquifer. *Int. Rev. Hydrobiol.* **2001**, *86*, 619–633. [CrossRef]

63. Hahn, H.J. The GW-Fauna-Index: A first approach to a quantitative ecological assessment of groundwater habitats. *Limnologica* **2006**, *36*, 119–137. [CrossRef]

64. Galassi, D.M.P.; Stoch, F.; Fiasca, B.; Di Lorenzo, T.; Gattone, E. Groundwater biodiversity patterns in the Lessinian Massif of northern Italy. *Freshw. Biol.* *54*, 830–847. [CrossRef]

65. Brancelj, A.; Zibrat, U.; Jamnik, B. Differences between groundwater fauna in shallow and in deep intergranular aquifers as an indication of different characteristics of habitats and hydraulic connections. *J. Limnol.* **2016**, *75*, 248–261. [CrossRef]

66. Iepure, S.; Rasines-Ladero, R.; Meffe, R.; Carreno, F.; Mostaza, D.; Sundberg, A.; Di Lorenzo, T.; Barroso, J.L. The role of groundwater crustaceans in disentangling aquifer type features—A case study of the Upper Tagus Basin, central Spain. *Ecohydrology* **2017**, *10*, e1876. [CrossRef]

67. Shaw, P.J.A. *Multivariate Statistics for the Environmental Sciences*; Oxford University Press Inc.: Oxford, UK, 2003; pp. 1–231.

68. Deharveng, L.; Stoch, F.; Gibert, J.; Bedos, A.; Galassi, D.M.P.; Zagmajster, M.; Brancelj, A.; Camacho, A.; Fiers, F.; Martin, P.; et al. Groundwater biodiversity in Europe. *Freshw. Biol.* **2009**, *54*, 709–726. [CrossRef]

69. Mahi, A.; Di Lorenzo, T.; Haicha, B.; Belaidi, N.; Taleb, A. Environmental factors determining regional biodiversity patterns of groundwater fauna in semi-arid aquifers of northwest Algeria. *Limnology* **2019**, *20*, 309–320. [CrossRef]

70. Stoch, F.; Artheau, M.; Brancelj, A.; Galassi, D.M.P.; Malard, F. Biodiversity indicators in European ground waters: Towards a predictive model of stygobiotic species richness. *Freshw. Biol.* **2009**, *54*, 745–755. [CrossRef]

71. Camargo, J.A.; Alonso, A. Ecological and toxicological effects of inorganic nitrogen pollution in aquatic ecosystems: A global assessment. *Environ. Int.* **2006**, *32*, 831–849. [CrossRef]

72. Di Lorenzo, T.; Di Marzio, W.D.; Fiasca, B.; Galassi, D.M.P.; Korbel, K.; Iepure, S.; Pereira, J.L.; Reboleira, A.S.; Schmidt, S.I.; Hose, G.C. Recommendations for ecotoxicity testing with stygobiotic species in the framework of groundwater environmental risk assessment. *Sci. Total Environ.* **2019**, *681*, 292–304. [CrossRef] [PubMed]

73. Korbel, K.L.; Hose, G.C. Habitat, water quality, seasonality, or site? Identifying environmental correlates of the distribution of groundwater biota. *Freshw. Sci.* **2015**, *34*, 329–343. [CrossRef]

74. Fattorini, S.; Lombardo, P.; Fiasca, B.; Di Cioccio, A.; Di Lorenzo, T.; Galassi, D.M.P. Earthquake-related changes in species spatial niche overlaps in spring communities. *Sci. Rep.* **2017**, *7*, 443. [CrossRef] [PubMed]

75. Stumpp, C.; Hose, G.C. Groundwater amphipods alter aquifer sediment structure. *Hydrol. Process.* **2017**, *31*, 3452–3454. [CrossRef]

76. Hose, G.C.; Stumpp, C. Architects of the underworld: Bioturbation by groundwater invertebrates influences aquifer hydraulic properties. *Aquat. Sci.* **2019**, *81*, 20. [CrossRef]

77. Korbel, K.L.; Stephenson, S.; Hose, G.C. Sediment size influences habitat selection and use by groundwater macrofauna and meiofauna. *Aquat. Sci.* **2019**, *81*, 39. [CrossRef]

78. Bork, J.; Berkhoff, S.; Bork, S.; Hahn, J.H. Using subsurface metazoan fauna to indicate groundwater–surface water interactions in the Nakdong River floodplain, South Korea. *Hydrogeol. J.* **2009**, *17*, 61–75. [CrossRef]

79. Hahn, H.J.; Fuchs, A. Distribution patterns of groundwater communities across aquifer types in south-western Germany. *Freshw. Biol.* **2008**, *54*, 848–860. [CrossRef]

80. Galassi, D.M.P. Groundwater copepods: Diversity patterns over ecological and evolutionary scales. *Hydrobiologia* **2001**, *454–453*, 227–253. [CrossRef]

81. Migliore, L.; de Nicola Giudici, M. Toxicity of heavy metals to *Asellus aquaticus* (L.) (Crustacea, Isopoda). *Hydrobiologia* **1990**, *203*, 155–164. [CrossRef]

82. Kadiene, U.E.; Bialais, C.; Ouddane, B.; Hwang, J.-S.; Souissi, S. Differences in lethal response between male and female calanoid copepods and life-cycle traits to cadmium toxicity. *Ecotoxicology* **2017**, *26*, 1227–1239. [CrossRef]

83. Fisher, R.A. *The Genetical Theory of Natural Selection*; Clarendon Press: Oxford, UK, 1930.

84. Leigh, E.G. Natural Selection and Mutability. *Am. Nat.* **1970**, *104*, 301–305. [CrossRef]

Article

The Toxicity and Uptake of As, Cr and Zn in a Stygobitic Syncarid (Syncarida: Bathynellidae)

Grant C. Hose [1,*], **Katelyn Symington** [1], **Maria J. Lategan** [2] and **Rainer Siegele** [3]

[1] Department of Biological Sciences, Macquarie University, Sydney, NSW 2109, Australia; katelyn.symington@planning.nsw.gov.au

[2] Department of Molecular Sciences, Macquarie University, Sydney, NSW 2109, Australia; josielategan@bigpond.com

[3] Australian Nuclear Science and Technology Organisation, Locked Bag 2001, Kirrawee DC, NSW 2232, Australia; rns@ansto.gov.au

* Correspondence: grant.hose@mq.edu.au

Received: 22 October 2019; Accepted: 21 November 2019; Published: 28 November 2019

Abstract: Ecotoxicological data for obligate groundwater species are increasingly required to inform environmental protection for groundwater ecosystems. Bathynellid syncarids are one of several crustacean taxa found only in subsurface habitats. The aim of this paper is to assess the sensitivity of an undescribed syncarid (Malacostraca: Syncarida: Bathynellidae) to common groundwater contaminants, arsenic(III), chromium(VI) and zinc, and examine the bioaccumulation of As and Zn in these animals after 14-day exposure. Arsenic was the most toxic to the syncarid (14-day LC50 0.25 mg As/L), followed closely by chromium (14-day LC50 0.51 mg Cr/L) and zinc (14-day LC50 1.77 mg Zn/L). The accumulation of Zn was regulated at exposure concentrations below 1 mg Zn/L above which body concentrations increased, leading to increased mortality. Arsenic was not regulated and was accumulated by the syncarids at all concentrations above the control. These are the first published toxicity data for syncarids and show them to be among the most sensitive of stygobitic crustaceans so far tested, partly due to the low hardness of the groundwater from the aquifer they inhabit and in which they were tested. The ecological significance of the toxicant accumulation and mortality may be significant given the consequent population effects and low capacity for stygobitic populations to recover.

Keywords: groundwater ecology; stygofauna; stygobite; aquifer; syncarida

1. Introduction

Groundwater is vital to society globally. It provides drinking water to over 2 billion people worldwide [1], and, in many areas, is the only reliable water source. Increasing demand for water and risk of contamination from surface activities such as agriculture, urban development and mining threaten the integrity of aquifers, their suitability as a water source, and the biota that inhabit these subterranean ecosystems [2].

Groundwater ecosystems support a unique biota, consisting of microbial assemblages, frequently crustacean-dominated invertebrate assemblages and, somewhat rarely, vertebrates such as fish [3]. Groundwater invertebrates particularly have evolved to dark, stable and low energy conditions of groundwater environments, such that aquifers the world over contain a diversity of unique taxa not found in surface environments [3]. Bathynellid syncarids are archetypal groundwater organisms. Found only in hyporheic or groundwater environments [4], bathynellid syncarids share the common morphological traits of blindness, lack of body pigments, vermiform (wormlike) body shape, and enhanced sensory appendages that have evolved independently across multiple lineages of groundwater fauna [3]. These traits and other adaptations make groundwater fauna morphologically and physiologically different from even related surface water species, and consequently, they may be expected to respond differently to toxicants compared to surface-dwelling species [5].

As in any ecosystem, pollution threatens the health of the groundwater biota, with flow-on effects to other elements of the ecosystem. For groundwater ecosystems, disruption to one element of the ecosystem is likely to have significant effects on the microbial communities and subsequently impact the ability of the ecosystem to self-purify and provide clean groundwater [6]. The risk to groundwater ecosystems is that the fauna has evolved classic k-strategist traits of longevity, low metabolic and reproductive rates [3], meaning that recovery from disturbance is, at best, slow [7].

Contamination of groundwater by metals is a global concern, often occurring as a result of industrial uses, spills and land application of contaminated materials (Castano-Sanchez in press) [8]. Unfortunately, groundwater ecosystems may not be protected by water quality legislation to the same extent as other aquatic ecosystems (such as rivers and estuaries) because water quality guidelines for ecosystem protection are not based on toxicological data for the biota of that ecosystem [9,10]. Instead, existing water quality guidelines for groundwater ecosystems are based on toxicity data for surface waters, but there are a number of reasons why groundwater biota may respond differently to toxicants compared to surface water taxa [5]. In particular, the metabolic rates of groundwater macroinvertebrates tend to be lower than those of related epigean species [11]. As a result, the uptake of toxicants may be lower where uptake of toxicants is an active, metabolic-driven process [5]. Alternatively, the removal of toxicants accumulated by passive processes may be low, leading to greater body burdens in groundwater fauna compared to surface water fauna [5].

The aim of this study is to test the sensitivity of an obligate groundwater crustacean (Malacostraca: Syncarida) to metal and metalloid contaminants (Zn, As and Cr) in groundwater. The first step to mortality and harm from contaminants is the uptake of those contaminants by the organism, so we also examine here the uptake of these metals by the syncarids and relate uptake to exposure and mortality.

2. Materials and Methods

2.1. Test Species

In this study, we test an undescribed species of the family Bathynellidae (Malacostraca: Syncarida). Until recently, all Australian Bathynellidae were assigned to the genus *Bathynella*, but phylogenies using newly collected specimens suggest strong regional separation within the Australian Bathynellidae, with specimens from eastern Australia being a separate clade [12–14], and those tested here likely a novel taxon. Type specimens are held at Macquarie University. Unfortunately, taxonomic keys to allow further identification are currently unavailable. However, multiple specimens from each test were examined to ensure morphological consistency of the test organisms within and between tests. Syncarida do not have morphologically different life stages after hatching [4], making it difficult to determine life stage; however, the syncarids tested were approximately 1000 μm long and of similar size within and between tests.

Test animals were collected from a fractured sandstone aquifer at Somersby, NSW Australia, approximately 80 km north of Sydney (33°22′15.4″ S, 151°18′9″ E, Bore depth = 22 m). The sample collection area is semirural. Groundwater in the region is used heavily by beverage bottling industries and is considered of very high quality. There is no recorded history of groundwater contamination at the site and analysis of the water over time has revealed very low concentrations of As, Cr and Zn, as seen in Table 1. The groundwater is typically low in dissolved solids and has a slightly acidic pH as a consequence of iron in the sandstone aquifer, as seen in Table 1.

Syncarids were collected from groundwater as described in Korbel & Hose [15] as summarised below. Groundwater was pumped using a motorised inertia pump (Waterra, ON, Canada). Approximately 300 L of water was pumped and passed through a 63 μm mesh sieve to collect the invertebrates. The sieve contents were placed in a sealable, 1 L plastic container which was filled with clean groundwater and placed in a portable cooler for transportation to the laboratory.

Table 1. Physicochemical variables and concentrations of common metals in the groundwater at the Somersby (NSW, Australia) collection sites (from [16]).

Water Quality Variable	Units	Concentration
pH		4.2–5.6
Conductivity	µS/cm	131–195
Dissolved Oxygen	% Saturation	59–83
Hardness	mg/L as CaCO$_3$	25–44
Total Organic Carbon	mg/L	3–13
As	mg/L	<0.01
Cr	mg/L	0.03
Zn	mg/L	0.02

In the laboratory, containers containing the invertebrates were placed in a dark environmental cabinet at 18 °C, which approximates the temperature of the groundwater at the time of collection. Syncarids were acclimated to conditions in the environmental cabinet for at least 48 h prior to testing. Tests were conducted within seven days of collection and data from repeat tests were pooled for analysis.

Tests were conducted following the methods reported in Hose et al. [16], which are summarised below. Once acclimated, the syncarids were removed from the collection container using a pipette and placed individually into wells of 24-well plates (Greiner) along with a known volume of clean groundwater from the collection site. Treatment concentrations were randomly allocated to each well and the required volume of clean groundwater added such that the subsequent addition of toxicant solution would give a final test volume of 2.5 mL. The 24-well plates were covered and placed in a dark cabinet at 18 °C for the duration of the test. Mortality was monitored daily and dead animals were removed, rinsed in Milli-Q water and frozen individually for later analysis.

Toxicity tests were conducted over 14 days with the test end point being immobility, defined as failure to move within 15 s of being gently prodded. This test was repeated twice within 15 min to confirm immobility. Immobility was recorded at 96 h and 14 days. Dissolved oxygen, conductivity, pH and temperature were recorded every 48 h from a randomly selected replicate of each treatment using HANNA (Hanna Instruments Inc, Smithfield, RI, USA) handheld meters.

Each test was comprised of a negative control and six or seven test concentrations, each with four or three replicate wells per plate, respectively. The number of animals varied between tests depending on the number in field collections and controls in all tests had ≥13 individuals. The concentration ranges used for each toxicant were 0–10 mg As/L, 0–100 mg Zn/L, and 0–10 mg Cr/L. These concentrations span the ranges reported in groundwater in the literature (e.g., [17–20]). Toxicants were tested individually. Results from a test were not accepted if there was more than 20% mortality of organisms in control treatments [10].

Toxicant stock solutions were made up in Milli-Q water. Chromium(VI) was added as $K_2Cr_2O_7$ (Sigma-Aldrich, 99% purity), arsenic was added as $NaAsO_2$ (Sigma-Aldrich, 98% purity) and zinc was added as $ZnSO_4.7H_2O$ (Ajax Finechem, 99% purity). Test concentrations were measured using the S2 Picofox Total Reflection X-Ray Fluorescence (TRX-RF) Spectrometer (Bruker Nano GMBH, Berlin, Germany), using the method for aqueous samples as outlined in [21]. A 1 mL aliquot of toxicant solution was taken from randomly selected wells of each treatment at the completion of each test. The aliquot of toxicant solution was spiked with approximately 5–10 µL of standard gallium solution, which acts as an internal standard [21,22]. Aliquots of 20 µL of this solution were then pipetted onto clean, hydrophobic carriers and dried before analysis. Analyses of spectra was done using the Spectra software (version 7.5.3.0, Bruker Nano GMBH, Berlin, Germany). Toxicant concentrations were averaged across replicates and these average values were used in subsequent statistical analyses. Merck XVI multielement standard (Merck KGaA, Darmstadt, Germany) was analysed along with the samples, and indicated average (±SD) recoveries of 107 ± 8% for As, 95 ± 5% for Cr, and 80 ± 5% for Zn. Limits of detection were 1 µg/L for As and Zn, and 3 µg/L for Cr.

The accumulation of As and Zn in the syncarids was measured using Particle Induced X-ray Emission (PIXE). Individual syncarids were freeze dried and placed whole, intact on to sample holders (see below). Resource limitations precluded the analysis of Cr.

2.2. PIXE Elemental Analysis

The samples were analysed for trace and bulk element concentrations with μ-PIXE using the Heavy Ion Microprobe (HIMP) at the Australian Nuclear and Science Technology Organisation (ANSTO) [23]. The HIMP uses the Australian National Tandem Research (ANTARES) accelerator to accelerate ions to the MeV energy range. In this analysis, a 3 MeV proton beam with typical spot size of between 5 and 10 μm was used. At this spot size, beam currents between 0.4 and 0.5 nA can be achieved, which is sufficient for μ-PIXE analyses. A high-purity Germanium (Ge) detector was used with a 100 mm^2 active area to measure the characteristic x-rays emerging from the calcification samples. The detector was located 32 mm from the sample for optimum count rate and sensitivity. A 100 μm Mylar foil was used to reduce the low energy x-rays count as well as prevent scattered protons from entering the detector. Samples were fixed to the sample holder by means of double-sided carbon tape. This tape is very pure and does not contain trace elements higher than oxygen ($Z > 8$). The tape does not affect measurements and is effective in holding the samples providing good electric conductivity to allow target current measurement. Each sample was then scanned for 5–30 min for extensive emission data to be obtained. Elemental concentration maps were created using GeoPIXE [24,25] and used to calculate the average elemental concentrations.

2.3. Statistical Analysis

Measured toxicant concentrations were compared with nominal concentrations using least squares linear regression. Concentration response curves were estimated by fitting a two-parameter nonlinear regression function with a binomial error structure using the DRC package [26] in R version 3.5.3 [27]. Weibull, log-logistic and log-normal models were fit to each dataset, with the best fitting model selected based on Akaikes Information Criteria. The model parameters were estimated using maximum likelihood, with starter values determined by the programs self-starter function. Concentrations affecting 10% and 50% of the syncarids (EC10 and EC50 values, respectively) were extrapolated from the fitted curve. Data from both 96 h and 14-day exposures were analysed separately.

The concentrations of As and Zn measured in syncarids after exposure were compared between treatments using one-way analyses of variance. Post-hoc comparisons between treatments were made using SNK tests. Data were log transformed to approximate normality and homogeneity of variance, which were assessed visually using plots of residuals and Q-Q plots. Analyses of variance were done using the NCSS program (Version 10, NCSS LLC, Kaysville, Utah, USA). The significance level (α) for all analyses was 0.05.

3. Results

Measured test concentrations were mostly within 20% of the nominal concentrations, and have been used in all analyses. Dissolved oxygen concentrations remained at 60% saturation or higher throughout all tests. This level of oxygen saturation is in line with the general concentrations observed in the field, as seen in Table 1. Temperature was consistently at or close to the test temperature of 18 °C. The hardness of the diluent water was relatively low, as seen in Table 1, which reflects the low carbonate concentrations in the sandstone aquifer from which the fauna and diluent water was collected. pH values varied little across the tests, and were in the range 5–6 pH units, as expected given the naturally acidic groundwater used as the diluent water.

Control mortality at 96 h was below zero for Cr and As exposures, and 3% for Zn. At 14 days, control mortality was higher across all tests (As: 9%, Cr: 11%, Zn: 11%), but below acceptability criterion of 20%, as seen in Figure S1. A Weibull model was fitted to all data sets, with details provided in Figure S1 and Table S1. Arsenic and Cr were similarly toxic after 96 h, but As was the most toxic after 14-day exposure. Zinc was the least toxic to the syncarids across both exposure periods, as seen in Table 2. As expected, toxicity increased and EC values decreased with exposure period from 96 h to 14 days for all toxicants, as seen in Table 2.

Table 2. Toxicity of As(III), Cr(VI) and Zn to bathynellid syncarids. EC10 and EC50 values indicate the concentration (mg/L) effecting 10% and 50% of the test population, respectively. Values in parentheses indicate 95% fiducial limits, followed by the total number of animals tested (N).

Test Duration	ECx	As(III)		Cr(VI)		Zn(II)	
		Estimate	95% Confidence Interval	Estimate	95% Confidence Interval	Estimate	95% Confidence Interval
		N = 74		N = 71		N = 190	
96 h	10	0.17	(−0.04–0.38)	0.11	(−0.07–0.28)	0.78	(0.30–1.26)
	50	1.99	(0.79–3.19)	1.83	(−0.10–3.77)	3.61	(2.56–4.66)
14 days	10	0.01	(−0.02–0.04)	0.004	(−0.010–0.018)	0.21	(0.02–0.39)
	50	0.25	(0.07–0.44)	0.51	(−0.07–1.09)	1.77	(1.11–2.43)

The mean body concentration of As in the syncarids in controls was 2.8 µg As/g, as seen in Figure 1A. The body concentrations in all exposure treatments were significantly greater than in the controls ($p < 0.05$), and concentrations in syncarids from the 10,000 and 300 mg/L treatments were greater than those in the 100 mg/L treatment ($p < 0.05$), but were not significantly different to each other, as seen in Figure 1A. The PIXE maps showed that As was concentrated around the head, anterior thoracic and abdominal sections, as seen in Figure 2.

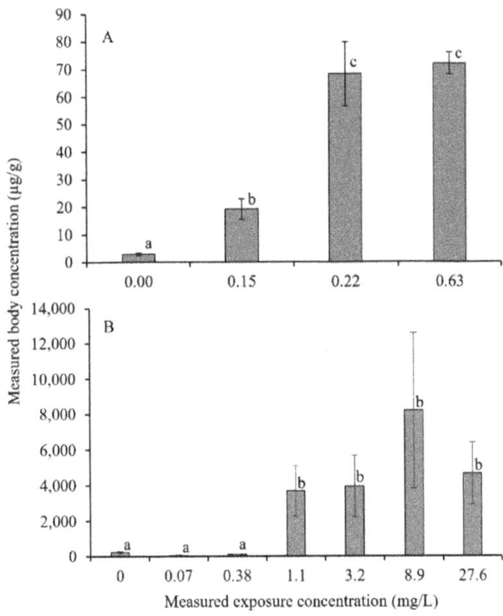

Figure 1. Mean concentration (±SE) between exposure concentration and mean body concentration in bathynellid syncarids exposed to (**A**) As and (**B**) Zn. Common lower case letters indicate no significant difference ($p > 0.05$) between those treatments.

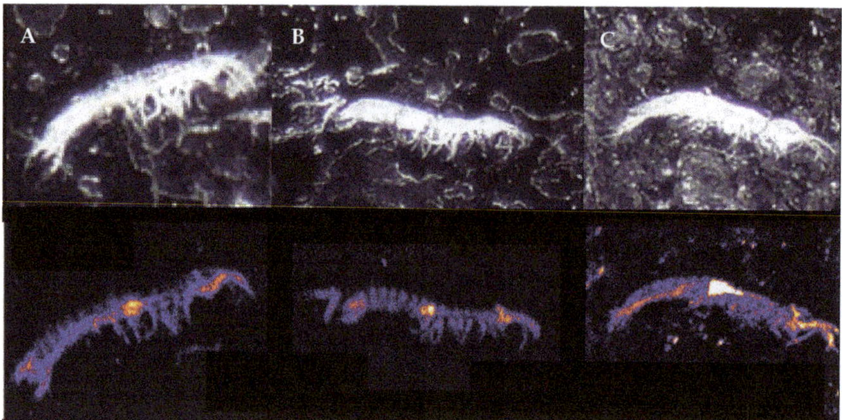

Figure 2. Images of As accumulation in bathynellid syncarids exposed to As in aqueous solution. Mean concentrations of specimens shown were (**A**) 12 ppm, (**B**) 2 ppm, and (**C**) 28 ppm.

Concentrations of Zn measured in the syncarids from the control treatments were on average 228 µg Zn/g, which increased significantly ($p < 0.001$) with exposure to all Zn treatments ≥1 mg Zn/L, as seen in Figure 1B. There was no significant difference ($p > 0.05$) in body concentrations across the Zn treatments, as seen in Figure 1B above 1 mg Zn/L. The accumulation of Zn in the syncarids appears to be concentrated in the anterior thoracic region and abdomen, as seen in Figure 3.

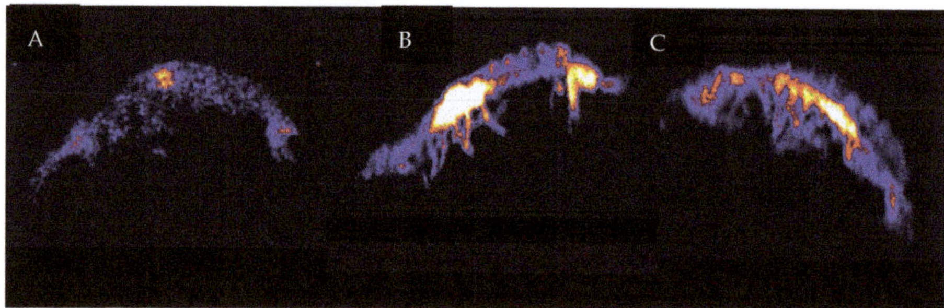

Figure 3. Images of Zn accumulation in bathynellid syncarids exposed to Zn in aqueous solution. Mean concentrations of specimens shown were (**A**) 250 ppm, (**B**) 6400 ppm, and (**C**) 8000 ppm.

4. Discussion

Our tests have provided the first toxicity and accumulation data for the Syncarida, specifically the Bathynellidae, which are an obligate subterranean taxon. We discuss below the sensitivity of the syncarids to the metals and metalloid relative to other stygobiotic taxa and surface water crustaceans. The accumulation of Zn suggests the ability of syncarids to regulate body concentrations at low exposure concentrations, but for As, we found no evidence of regulation at the exposure concentration tested. Tests were conducted in relatively low hardness, low organic content water from a fractured rock aquifer, which likely makes them more susceptible to toxicants than stygobitic taxa from hard, karstic or more organic groundwaters [28]. While this trend is not unexpected, we highlight the need to consider the aquifer types in the management and risk assessment for subterranean ecosystems.

Mortality in controls was ≤11%, below the acceptable threshold of 20%, and water quality conditions were stable during all tests, indicating we had created a suitable testing environment.

Dissolved oxygen (DO) concentrations in the tests were in line with those measured at the collection site in groundwater, as seen in Table 1, but higher than often reported in groundwaters [29]. The relatively high dissolved oxygen concentrations in the static tests is a likely consequence of the small test volume used and large surface area for gas exchange within the 24 well plates. Previous studies suggest that stygobitic crustaceans can survive well in oxygenated laboratory environments (e.g., [11,16]), and do not affect the toxic response of groundwater crustaceans to metals [30].

The toxicity of As to the syncarids after 96 h was similar to that observed for other stygobitic crustaceans. Hose et al. [16] tested the sensitivity of cyclopoid and harpacticoid copepods from the same Somersby location and showed 96-h LC50 values of similar magnitude (5.24, 3.04 mg As/L, respectively) to those estimated here for the syncarids. The syncarids were similar in sensitivity to As as the cladoceran *Daphnia magna* (96-h LC50 1.50–4.34 mg As(III)/L [31]). However, over 14 days, our syncarids (14-day LC50, 0.25 mg As/L) were slightly more sensitive than the cyclopoid and harpacticoid copepods from Somersby (14-day LC50 0.79, 1.46 mg As/L, respectively [16]). The syncarids were also more sensitive to As than the stygobitic amphipod *Niphargus rhenorhodanensis* (240-h LC50, 3.97 mg/L [32]), but similar to other freshwater crustaceans such as *Gammarus fossarum* (LC50, 0.2 mg As/L [32]), which was the most sensitive of a range of taxa in that study. Overall the syncarids were less sensitive to As than Cr, which is consistent with the findings of Canivet et al. [32]. For As and the metals tested, there are few low ECx values in the literature for comparison, nevertheless, LC10 values are reported here so that they may be available to inform the development of environmental quality criteria for groundwaters (e.g., [9]).

Chromium was the most toxic of the elements tested. The acute toxicity of chromium to the syncarids was similar to that recorded for subterranean crustacea reported previously. The ranges of 96-h LC50 values for the styogophilous *Bryocamptus echinatus* (1.26 mg Cr/L [33]), *N. rhenorhodanensis* (1.51–2.07 mg Cr/L [32]), stygobitic copepods (0.7–3.1 mg Cr/L [16]), and the stygobitic isopod *Proassellus* spp. (48-h EC50 0.396–6.35 mg Cr/L [34]) all overlap the 96-h LC50 value for the syncarids (1.8 mg Cr/L). The 96-h LC50 values for *D. magna* exposed to Cr (0.07–0.16 mg Cr/L, [35,36]) were approximately an order of magnitude lower than those for the syncarids tested here, as seen in Table 2. The 14-day LC50 value (0.5 mg Cr/L) is also within the range of values for copepods from the same study site (0.03–1.06 mg Cr/L [16]) and a 10-day LC50 for *N. rhenorhodanensis* (0.23 mg Cr/L [32]) and less than those for a range of epigean taxa [32,36]. The 14-day LC10 for Cr (0.004 mg Cr/L) was below the background concentration at the site (0.03 mg Cr/L). This anomaly is a likely consequence of the exposure-response curve fitting, as seen in Figure S1, which is often less reliable at the extremities of the curve. We do, however, believe that the overall curve fit is reliable and the inherent conservatism in this case is preferable for the application of these data to ecosystem protection.

The syncarids were less sensitive to Zn than the Harpacticoid copepod and more sensitive than the cyclopoid copepod from the same locations over both 96 h and 14 days [16]. Ninety-six hour LC50 values for the syncarids were similar to those determined for the groundwater copepod *Parastenocaris germanica* (96-h LC50 1.7 mg Zn/L) under oxic (10 mg O_2/L) conditions [30], but both *P. germanica* and the syncarid tested here were much more sensitive than several other stygobitic crustaceans such as *Niphargus aquilex* (96-h LC50 180 mg Zn/L [37]), *Asellus cavaticus* (96-h LC50 127 mg Zn/L [37]) and *Niphargus montellianus*, which had no mortality to Zn after 10-day exposure to 25 mg Zn/L [38]. The syncarids were less sensitive to zinc over 96 h than was *D. magna* (96-h LC50 0.15–0.18 mg Zn/L [36,39]).

Overall, the syncarids tested were similar, but slightly more sensitivity than other stygobitic crustaceans. The syncarids were, however, generally less sensitive over 96 h exposure than the standard cladoceran test species, *D. magna*. Current approaches for risk assessment for groundwater include the use of standard test species, such as *D. magna* [40], as a first tier in lieu of toxicity data for groundwater biota. Our results suggest that this may be a suitable interim approach until sufficient data for groundwater taxa become available.

Our syncarids were collected from a fractured rock aquifer which had low hardness, low-organic matter water, contrasting to some previous studies in which biota were collected from karstic aquifers with much greater water hardness. Hardness is a major factor affecting the toxicity of metals, particularly zinc and chromium, to organisms [41–43], and hardness correction has been used to compare organism sensitivity in ecological risk assessments. Interestingly, using the hardness correction of Warne et al. [44], the syncarids tested were the most Zn-sensitive of the stygobitic crustaceans tested to date. No hardness correction algorithm is available for Cr(VI). With likely differences in the hardness of water between aquifer types, it is critical that aquifer types are considered in risk assessments and decision making for subterranean ecosystems and when setting environmental quality standards.

The toxicity of all toxicants increased with exposure period, with decreases in EC values of up to two orders of magnitude between 96 h and 14 days. The significance of such short-term exposures in groundwater ecosystems, with the exception of perhaps karst systems (see [34]), is debatable because of the slow groundwater flow rates in most aquifers. Hose et al. [16] showed that the relative toxicities of As, Cr and Zn were consistent between 96 h, 14-day and 28-day exposures, but with up to a six-fold decrease in ECx values between 14 and 28 days. Di Lorenzo et al. [10] recommend the reporting 96-h LCx values for the purposes of comparison between studies and the existing ecotoxicological databases, but recommend that tests be extended well beyond 96 h. Unfortunately, due to the longevity and low reproductive rates of stygobitic species, continuing tests for durations that are relevant to life cycles of the organisms tested may be difficult. For this reason, incipient LC50 values (sensu [45,46]) would be useful and should be a direction of future research.

The accumulation of both As and Zn appeared greatest in the region at the anterior of the thoracic region. Previous studies using decapods have shown high concentrations of both As [47] and Zn [48,49] in the hepatopancreas, which is consistent with the location of the thoracic hotspot in the syncarids, although we recognise that it is difficult to determine the specific organs with this method, and the uncertainty of the organ location on these small animals. Cresswell et al. [49] suggest that Zn is transported from the gill to the hepatopancreas for detoxication and/or metabolism. Cresswell et al. [49] also report high concentrations of Zn in the antennal gland, which is an important excretion site in crustaceans and which are well developed in the Bathynellacea [50]. However, this was not the case for Zn in this study, although this may explain the high concentration of As in that region. It is unclear which tissues in the posterior of the abdomen are causing the high Zn, but the distribution of As throughout the length of the abdomen may occur in variety of tissues of systems that extend through the syncarid abdomen [51]. Alternatively, it may be a result of water uptake through the abdomen as done by some crustaceans (sensu [52]) but it is unclear whether syncarids do this. Both White and Rainbow [48] and Cresswell et al. [49] report high Zn concentrations in the eye of decapods (associated with zinc metalloenzymes used in vision [53]), which was clearly not the case for our blind (eyeless) stygobitic syncarid.

The body concentration of As in the syncarids was far lower than that recorded for Zn, which reflects the relative abundance of these elements in the environment and most biological systems [54]. The concentration of As in the unexposed syncarids was 2.8 µg As/g and similar to unexposed amphipods and *Daphnia* (<5 µg As/g [55]). The concentrations of As in unexposed *N. rhenorhodanensis* [32] were below detection limit; however, the mean concentration of As in amphipods exposed to 0.1 mg As/L for 10 days (22.6 µg As/g) [32] was of a similar magnitude to the concentrations in syncarids (19 µg As/g) exposed to 0.15 mg As/L for a similar period in this study. The arsenic accumulation data suggest that the syncarids have no ability to regulate As uptake, with the body concentration for syncarids exposed to 0.15 mg As/L being significantly greater than those in the syncarids from controls, as well as mortality occurring at that concentration after 14 days. Arsenic accumulation by the syncarids did not increase significantly as exposure concentration increased beyond 0.22 mg As/L. It is unlikely that this reflects an ability to regulate As uptake at these concentrations since there was no evidence at the lower concentrations, and this has not been reported elsewhere. Rather, the lack of significant increase in body concentrations beyond 0.22 mg As/L is likely a statistical artefact caused by large variability in internal concentrations within the 0.22 mg As/L treatment, as evidenced by the size of the error bars in Figure 1A.

The background concentration of Zn in the syncarids was around 228 µg Zn/g which was similar to that for *N. montellianus* (113 µg Zn/g [38]). The accumulation of Zn by syncarids exposed to 1 mg Zn/L was significantly greater than syncarids in controls and exposed to lower concentrations, as seen in Figure 1B. This trend differs from that observed for *N. montellianus*, for which significant accumulation of Zn did not occur until they were exposed to 10 mg Zn /L [38], but this difference may reflect the greater hardness of the water used in that study (150–165 mg CaCO$_3$/L), and the expected reduced bioavailability and uptake of metals by organisms with increasing hardness [56,57].

Our results suggest an ability of syncarids to regulate Zn at concentrations below 1 mg Zn/L. When considered alongside the mortality data, the increase in accumulation from the 0.38 to the 11 mg Zn/L exposures coincides approximately with a threshold change in mortality after 14 days. Rainbow and Luoma [58] detail several models of metal (particularly Zn) uptake in crustaceans, and that observed here for the syncarids most closely fits that common in the Decapoda (e.g., [59,60]), which might be expected given the phylogenetic proximity of the decapods and syncarids [61]. In this model, Zn concentration in the animals is regulated by excreting Zn until a threshold exposure concentration is reached beyond which the animal is unable to continue regulating their internal concentration [58]. Relatively little metal is stored in detoxified forms [59] and the contribution of Zn adsorbed passively to the exoskeleton to the total body concentration is small [59]. Above the threshold concentration, Zn is accumulated proportionally to the exposure concentration [58] and binds at sites where it will interfere with normal metabolic functioning, leading to toxicity [59]. While for some aquatic organisms the uptake of some metals from solution plateaus at high concentrations, this is has not been reported for invertebrates (e.g., [62]). The lack of apparent increase in mean body concentrations at exposure concentrations above 1 mg Zn/L may be a consequence of the large variability in body concentrations in those animals, which obscures an increasing trend.

Understanding the ecological significance of laboratory response data to population level effects in crustaceans in the field is difficult [63]. Our syncarids are no different, with the ecological significance of their bioaccumulation of Zn and As unclear. The accumulation of Zn by gammarid amphipods at concentrations as low as 0.3 mg Zn/L effected reproduction and energy utilisation in females [64]. At that concentration, the number of aborted broods increased, and the reduced energy absorption and reallocation of that energy away from reproduction resulted in smaller sized offspring [64]. The reproductive strategies of bathynellid syncarids are poorly known, but like other stygobiotic crustaceans they are expected to have low metabolic rates, reproductive rates and small brood sizes (i.e., k-strategy traits), a likely consequence of the low energy groundwater environment (Humphreys 2006). What is the effect of reducing energy absorption and fecundity in a population of organisms that are already at the lower extremes of such rates? Kammenga et al. [65] showed no difference in sensitivity of k-strategist and r-strategist nematodes to cadmium, but whether this holds true for other taxa is unclear.

Maltby and Naylor [64] predict that the smaller offspring resulting from zinc exposure will take longer to reach maturity and may reproduce at a smaller size, which typically means reduced fecundity, which is exacerbated by the increase in aborted broods. The net effect is expected to thus be reduced recruitment and population decline. Importantly for syncarids that feed on detrital matter, bacteria and fungi [66], changes in microbial assemblages, particularly fungal biomass, may also occur at relatively low Zn and As concentrations. Lategan et al. [67] observed 50% declines in biomass of groundwater fungi at 4.4 mg As/L and 0.61 mg Zn/L, thereby exacerbating potential sublethal effects on feeding and energy assimilation. In groundwaters where contamination is typically long-term and difficult to remediate, these impacts will be persistent, and likely lead to a decrease in population viability. The broader consequences of declining invertebrate populations in aquifers is the loss of ecosystem services [6], including changes to aquifer hydraulic properties [68].

Chromium, zinc and arsenic were acutely toxic to the stygobitic bathynellid syncarid. The accumulation of Zn was regulated at exposure concentrations below 1 mg Zn/L, above which body concentrations increased, leading to increased mortality. Arsenic was not regulated and was accumulated by the syncarids at all concentrations above the control. These are the first published toxicity data for syncarids and show them among the most sensitive of stygobitic crustaceans so far tested, likely in part due to the low hardness of the groundwater from the aquifer they inhabit and in which they were tested. The ecological significance of the toxicant accumulation and mortality may be significant given the consequent population effects and low capacity for stygobitic populations to recover.

Supplementary Materials: The following are available online at http://www.mdpi.com/2073-4441/11/12/2508/s1, Figure S1: Dose response curves for bathynellid syncarids exposed to Arsenic (A, B), Chromium (C, D) and Zinc (E, F), for 96 h (A, C, E) and 14 days (B, D, F), Table S1: Fitted dose response curve parameters.

Author Contributions: Conceptualization, G.C.H. and M.J.L.; methodology, G.C.H., M.J.L., and R.S.; formal analysis, K.S., R.S. and G.C.H.; investigation, all; resources, G.C.H. and R.S.; writing—original draft preparation, G.C.H.; writing—review and editing, all; supervision, G.C.H. and M.J.L.; project administration, G.C.H.; funding acquisition, G.C.H. and R.S.

Funding: Bioaccumulation analyses were conducted at the Australian Nuclear Science and Technology Organisation through the support of grants ALNGRA10099 and AINGRA09097P. This project was also supported by CRC CARE project 1-1-08-06/7 and NSW Environmental Trust project 2005/RD/0108.

Acknowledgments: The authors are grateful for the assistance of Ashleigh Keast with the arsenic tests and Sarah Stephenson with field collections. The insightful comments of two anonymous reviewers improved this manuscript.

Conflicts of Interest: The authors declare no conflict of interest.

References

1. Alley, W.M.; Healy, R.W.; LaBaugh, J.W.; Reilly, T.E. Flow and storage in groundwater systems. *Science* **2002**, *296*, 1985. [CrossRef] [PubMed]

2. Morris, B.L.; Lawrence, A.R.L.; Chilton, P.J.C.; Adams, B.; Calow, R.C.; Klinck, B.A. *Groundwater and Its Susceptibility to Degradation: A Global Assessment of the Problem and Options for Management*; United Nations Environment Programme: Nairobi, Kenya, 2003; p. 126.

3. Humphreys, W.F. Aquifers: The ultimate groundwater-dependent ecosystems. *Aust. J. Bot.* **2006**, *54*, 115–132. [CrossRef]

4. Camacho, A.I.; Valdecasas, A.G. Global diversity of syncarids (Syncarida; Crustacea) in freshwater. In *Freshwater Animal Diversity Assessment*; Balian, E.V., Lévêque, C., Segers, H., Martens, K., Eds.; Springer: Dordrecht, The Netherlands, 2008; pp. 257–266.

5. Hose, G.C. Response to Humphreys' (2007) Comments on Hose GC (2005) Assessing the Need for Groundwater Quality Guidelines for Pesticides Using the Species Sensitivity Distribution Approach. *Hum. Ecol. Risk Assess.* **2007**, *13*, 241–246. [CrossRef]

6. Griebler, C.; Avramov, M.; Hose, G. Groundwater ecosystems and their services—Current status and potential risks. In *Atlas of Ecosystem Services—Drivers, Risks, and Societal Responses*; Schröter, M., Bonn, A., Klotz, S., Seppelt, R., Baessler, C., Eds.; Springer: Berlin/Heidelberg, Germany, 2019; pp. 197–203.

7. Hose, G.C.; Asmyhr, M.G.; Cooper, S.J.B.; Humphreys, W.F. Down under Down Under: Austral groundwater life. In *Austral Ark: The State of Wildlife in Australia and New Zealand*; Stow, A., Maclean, N., Holwell, G.I., Eds.; Cambridge University Press: Cambridge, UK, 2015; pp. 512–536.

8. Castaño-Sánchez, A.; Hose, G.; Reboleira, A. Ecotoxicological effects of anthropogenic stressors in subterranean organisms: A review. *Chemosphere* **2019**, in press. [CrossRef]

9. Hose, G.C. Assessing the Need for Groundwater Quality Guidelines for Pesticides Using the Species Sensitivity Distribution Approach. *Hum. Ecol. Risk Assess.* **2005**, *11*, 951–966. [CrossRef]

10. Di Lorenzo, T.; Di Marzio, W.D.; Fiasca, B.; Galassi, D.M.P.; Korbel, K.; Iepure, S.; Pereira, J.L.; Reboleira, A.S.P.S.; Schmidt, S.I.; Hose, G.C. Recommendations for ecotoxicity testing with stygobiotic species in the framework of groundwater environmental risk assessment. *Sci. Total Environ.* **2019**, *681*, 292–304. [CrossRef]

11. Di Lorenzo, T.; Di Marzio, W.D.; Spigoli, D.; Baratti, M.; Messana, G.; Cannicci, S.; Galassi, D.M.P. Metabolic rates of a hypogean and an epigean species of copepod in an alluvial aquifer. *Freshw. Biol.* **2015**, *60*, 426–435. [CrossRef]

12. Little, J.; Schmidt, D.J.; Cook, B.D.; Page, T.J.; Hughes, J.M. Diversity and phylogeny of south-east Queensland Bathynellacea. *J. Aust. J. Zool.* **2016**, *64*, 36–47. [CrossRef]

13. Perina, G.; Camacho, A.I.; Huey, J.; Horwitz, P.; Koenders, A. New Bathynellidae (Crustacea) taxa and their relationships in the Fortescue catchment aquifers of the Pilbara region, Western Australia. *Syst. Biodivers.* **2019**, *17*, 148–164. [CrossRef]

14. Camacho, A.I.; Mas-Peinado, P.; Dorda, B.A.; Casado, A.; Brancelj, A.; Knight, L.R.F.D.; Hutchins, B.; Bou, C.; Perina, G.; Rey, I. Molecular tools unveil an underestimated diversity in a stygofauna family: A preliminary world phylogeny and an updated morphology of Bathynellidae (Crustacea: Bathynellacea). *Zool. J. Linn. Soc.* **2017**, *183*, 70–96. [CrossRef]

15. Korbel, K.L.; Hose, G.C. Habitat, water quality, seasonality, or site? Identifying environmental correlates of the distribution of groundwater biota. *Freshw. Sci.* **2015**, *34*, 329–343. [CrossRef]

16. Hose, G.C.; Symington, K.; Lott, M.J.; Lategan, M.J. The toxicity of arsenic(III), chromium(VI) and zinc to groundwater copepods. *Environ. Sci. Pollut. Res.* **2016**, *23*, 18704–18713. [CrossRef] [PubMed]

17. Fruchter, J. Peer Reviewed: In-Situ Treatment of Chromium-Contaminated Groundwater. *Environ. Sci. Technol.* **2002**, *36*, 464A–472A. [CrossRef] [PubMed]

18. Scheeren, P.J.H.; Koch, R.O.; Buisman, C.J.N.; Barnes, L.J.; Versteegh, J.H. New biological treatment plant for heavy metal contaminated groundwater. In *EMC '91: Non-Ferrous Metallurgy—Present and Future*; Springer: Dordrecht, The Netherlands, 1991; pp. 403–416.

19. Smedley, P.L.; Kinniburgh, D.G. A review of the source, behaviour and distribution of arsenic in natural waters. *Appl. Geochem.* **2002**, *17*, 517–568. [CrossRef]

20. Edmunds, W.M.; Shand, P. Groundwater Baseline Quality. In *Natural Groundwater Quality*; Edmunds, W.M., Shand, P., Eds.; Blackwell Publishing: Oxford, UK, 2009; pp. 1–21.

21. Klockenkämper, R.; von Bohlen, A. Elemental Analysis of Environmental Samples by Total Reflection X-Ray Fluorescence: A Review. *X-ray Spectrom.* **1996**, *25*, 156–162. [CrossRef]

22. Farías, S.S.; Casa, V.A.; Vázquez, C.; Ferpozzi, L.; Pucci, G.N.; Cohen, I.M. Natural contamination with arsenic and other trace elements in ground waters of Argentine Pampean Plain. *Sci. Total Environ.* **2003**, *309*, 187–199. [CrossRef]

23. Siegele, R.; Cohen, D.D.; Dytlewski, N. The ANSTO high energy heavy ion microprobe. *Nucl. Instrum. Methods Phys. Res. Sect. B Beam Interact. Mater. Atoms* **1999**, *158*, 31–38. [CrossRef]

24. Ryan, C.G. Developments in Dynamic Analysis for quantitative PIXE true elemental imaging. *Nucl. Instrum. Methods Phys. Res. Sect. B Beam Interact. Mater. Atoms* **2001**, *181*, 170–179. [CrossRef]

25. Ryan, C.G.; Jamieson, D.N.; Churms, C.L.; Pilcher, J.V. A new method for on-line true-elemental imaging using PIXE and the proton microprobe. *Nucl. Instrum. Methods Phys. Res. Sect. B Beam Interact. Mater. Atoms* **1995**, *104*, 157–165. [CrossRef]

26. Ritz, C.; Streibig, J.C. Bioassay analysis using R. *J. Stat. Softw.* **2005**, *12*, 1–22. [CrossRef]

27. R Core Team. *R: A Language and Environment for Statistical Computing*; R Foundation for Statistical Computing: Vienna, Austria, 2019.

28. Hyne, R.V.; Pablo, F.; Julli, M.; Markich, S.J. Influence of water chemistry on the acute toxicity of copper and zinc to the cladoceran Ceriodaphnia cf dubia. *Environ. Toxicol. Chem.* **2005**, *24*, 1667–1675. [CrossRef] [PubMed]

29. Hahn, H.J. The GW-Fauna-Index: A first approach to a quantitative ecological assessment of groundwater habitats. *Limnol. Ecol. Manag. Inland Waters* **2006**, *36*, 119–137. [CrossRef]

30. Notenboom, J.; Cruys, K.; Hoekstra, J.; Vanbeelen, P. Effect of Ambient Oxygen Concentration Upon the Acute Toxicity of Chlorophenols and Heavy-Metals to the Groundwater Copepod Parastenocaris-Germanica (Crustacea). *Ecotoxicol. Environ. Saf.* **1992**, *24*, 131–143. [CrossRef]

31. Lima, A.; Curtis, C.; Hammermeister, D.; Markee, T.; Northcott, C.E.; Brooke, L.T. Acute and chronic toxicities of arsenic(III) to fathead minnows, flagfish, daphnids, and an amphipod. *Arch. Environ. Contam. Toxicol.* **1984**, *13*, 595–601. [CrossRef]

32. Canivet, V.; Chambon, P.; Gibert, J. Toxicity and bioaccumulation of arsenic and chromium in epigean and hypogean freshwater macroinvertebrates. *Arch. Environ. Contam. Toxicol.* **2001**, *40*, 345–354. [CrossRef]

33. Di Marzio, W.D.; Castaldo, D.; Pantani, C.; Di Cioccio, A.; Di Lorenzo, T.; Saenz, M.E.; Galassi, D.M.P. Relative Sensitivity of Hyporheic Copepods to Chemicals. *Bull. Environ. Contam. Toxicol.* **2009**, *82*, 488–491. [CrossRef]

34. Reboleira, A.S.; Abrantes, N.; Oromí, P.; Gonçalves, F. Acute Toxicity of Copper Sulfate and Potassium Dichromate on Stygobiont Proasellus: General Aspects of Groundwater Ecotoxicology and Future Perspectives. *Water Air Soil Pollut.* **2013**, *224*, 1–9. [CrossRef]

35. Fargasova, A. Toxicity of Metals on Daphnia magna and Tubifex tubifex. *Ecotoxicol. Environ. Saf.* **1994**, *27*, 210–213. [CrossRef]

36. Ewell, W.S.; Gorsuch, J.W.; Kringle, R.O.; Robillard, K.A.; Spiegel, R.C. Simultaneous evaluation of the acute effects of chemicals on seven aquatic species. *Environ. Toxicol. Chem.* **1986**, *5*, 831–840. [CrossRef]

37. Meinel, W.; Krause, R. Zur Korrelation zwischen Zink und verschiedenen pH–Werten in in her toxischen Wirkung auf einige Grundwasser–Organismen. *Z. Für Angew. Zool.* **1988**, *75*, 159–182.

38. Coppellotti Krupa, O.; Toniello, V.; Guidolin, L. Niphargus and Gammarus from karst waters: First data on heavy metal (Cd, Cu, Zn) exposure in a biospeleology laboratory. *Subterr. Biol.* **2004**, *2*, 33–41.

39. Lazorchak, J.M.; Smith, M.E.; Haring, H.J. Development and validation of a Daphnia magna four-day survival and growth test method. *Environ. Toxicol. Chem.* **2009**, *28*, 1028–1034. [CrossRef] [PubMed]

40. Daam, M.A.; Silva, E.; Leitao, S.; Trindade, M.J.; Cerejeira, M.J. Does the actual standard of 0.1 mu g/L overestimate or underestimate the risk of plant protection products to groundwater ecosystems? *Ecotoxicol. Environ. Saf.* **2010**, *73*, 750–756. [CrossRef] [PubMed]

41. Paulauskis, J.D.; Winner, R.W. Effects of water hardness and humic acid on zinc toxicity to *Daphnia magna* Straus. *Aquat. Toxicol.* **1988**, *12*, 273–290. [CrossRef]

42. Pawlisz, A.V.; Kent, R.A.; Schneider, U.A.; Jefferson, C. Canadian water quality guidelines for chromium. *Environ. Toxicol. Water Qual.* **1997**, *12*, 123–183. [CrossRef]

43. Park, E.J.; Jo, H.J.; Jung, J. Combined effects of pH, hardness and dissolved organic carbon on acute metal toxicity to Daphnia magna. *J. Ind. Eng. Chem.* **2009**, *15*, 82–85. [CrossRef]

44. Warne, M.; Batley, G.; van Dam, R.; Chapman, J.; Fox, D.; Hickey, C.; Stauber, J. *Revised Method for Deriving Australian and New Zealand Water Quality Guideline Values for Toxicants—Update of 2015 Version*; Australian and New Zealand Governments and Australian State and Territory Governments: Canberra, Australia, 2018; p. 48.

45. Sprague, J.B. Measurement of pollutant toxicity to fish I. Bioassay methods for acute toxicity. *Water Res.* **1969**, *3*, 793–821. [CrossRef]

46. Avramov, M.; Schmidt, S.I.; Griebler, C. A new bioassay for the ecotoxicological testing of VOCs on groundwater invertebrates and the effects of toluene on Niphargus inopinatus. *Aquat. Toxicol.* **2013**, *130*, 1–8. [CrossRef]

47. Williams, G.; West, J.M.; Snow, E.T. Total arsenic accumulation in yabbies (Cherax destructor clark) exposed to elevated arsenic levels in Victorian gold mining areas, Australia. *Environ. Toxicol. Chem.* **2008**, *27*, 1332–1342. [CrossRef]

48. White, S.L.; Rainbow, P.S. Regulation of zinc concentration by Palaemon elegans (Crustacea: Decapoda): Zinc flux and effects of temperature, zinc concentration and moulting. *Mar. Ecol. Prog. Ser.* **1984**, *16*, 135–147. [CrossRef]

49. Cresswell, T.; Simpson, S.L.; Mazumder, D.; Callaghan, P.D.; Nguyen, A.P. Bioaccumulation Kinetics and Organ Distribution of Cadmium and Zinc in the Freshwater Decapod Crustacean Macrobrachium australiense. *Environ. Sci. Technol.* **2015**, *49*, 1182–1189. [CrossRef] [PubMed]

50. Steenken, B.; Schminke, H.K. Ultrastructure of maxillary gland of Antrobathynella stammeri (Syncarida, Malacostraca). *J. Morphol.* **1996**, *228*, 107–117. [CrossRef]

51. Smith, G. On the Anaspidacea, Living and Fossil. *J. Cell Sci.* **1909**, *s2-53*, 489–578.

52. Fox, H.M. Anal and Oral Intake of Water by Crustacea. *J. Exp. Biol.* **1952**, *29*, 583. [CrossRef]

53. White, S.L.; Rainbow, P.S. Zinc flux in Palaemon elegans (Crustacea: Decapoda): Moulting, individual variation and tissue distribution. *Mar. Ecol. Prog. Ser.* **1984**, *19*, 153–166. [CrossRef]

54. Smith, K.S.; Huyck, H. An overview of the abundance, relative mobility, bioavailabilty, and human toxicity of metals. In *The Environmental Geochemistry of Mineral Deposits: Part, A; Processes, Techniques, and Health Issues*; Plumlee, G., Logsdon, J., Eds.; Society for Economic Geologists: Littleton, CO, USA, 1999; Volume 6A, pp. 29–70.

55. Spehar, R.L.; Fiandt, J.T.; Anderson, R.L.; DeFoe, D. Comparative toxicity of arsenic compounds and their accumulation in invertebrates and fish. *Arch. Environ. Contam. Toxicol.* **1980**, *9*, 53–63. [CrossRef]

56. Winner, R.W.; Gauss, J.D. Relationship between chronic toxicity and bioaccumulation of copper, cadmium and zinc as affected by water hardness and humic acid. *Aquat. Toxicol.* **1986**, *8*, 149–161. [CrossRef]

57. Markich, S.; Brown, P.; Batley, G.; Apte, S.; Stauber, J. Incorporating metal speciation and bioavailability into water quality guidelines for protecting aquatic ecosystems. *Aust. J. Ecotoxicol.* **2001**, *7*, 109–122.

58. Rainbow, P.S.; Luoma, S.N. Metal toxicity, uptake and bioaccumulation in aquatic invertebrates—Modelling zinc in crustaceans. *Aquat. Toxicol.* **2011**, *105*, 455–465. [CrossRef]

59. Rainbow, P.S. Trace metal bioaccumulation: Models, metabolic availability and toxicity. *Environ. Int.* **2007**, *33*, 576–582. [CrossRef]

60. Vijayram, K.; Geraldine, P. Toxicology. Regulation of Essential Heavy Metals (Cu, Cr, and Zn) by the Freshwater Prawn Macrobrachium malcolmsonii (Milne Edwards). *Bull. Environ. Contam. Toxicol.* **1996**, *56*, 335–342. [CrossRef] [PubMed]

61. Koenemann, S.; Jenner, R.A.; Hoenemann, M.; Stemme, T.; von Reumont, B.M. Arthropod phylogeny revisited, with a focus on crustacean relationships. *Arthropod Struct. Dev.* **2010**, *39*, 88–110. [CrossRef] [PubMed]

62. Luoma, S.N.; Rainbow, P.S. *Metal. Contamination in Aquatic Environments: Science and Lateral Management*; Cambridge University Press: Cambridge, UK, 2008. [CrossRef]

63. Kuhn, A.; Munns, W.R., Jr.; Serbst, J.; Edwards, P.; Cantwell, M.G.; Gleason, T.; Pelletier, M.C.; Berry, W. Evaluating the ecological significance of laboratory response data to predict population-level effects for the estruarine amphipod Ampelisca abdita. *Environ. Toxicol. Chem.* **2002**, *21*, 865–874. [CrossRef] [PubMed]

64. Maltby, L.; Naylor, C. Preliminary Observations on the Ecological Relevance of the Gammarus 'Scope for Growth' Assay: Effect of Zinc on Reproduction. *Funct. Ecol.* **1990**, *4*, 393–397. [CrossRef]

65. Kammenga, J.E.; Van Gestel, C.A.M.; Bakker, J. Toxicology. Patterns of sensitivity to cadmium and pentachlorophenol among nematode species from different taxonomic and ecological groups. *Arch. Environ. Contam. Toxicol.* **1994**, *27*, 88–94. [CrossRef] [PubMed]

66. Wellborn, G.A.; Witt, J.D.S.; Cothran, R.D. Chapter 31—Class Malacostraca, Superorders Peracarida and Syncarida. In *Thorp and Covich's Freshwater Invertebrates*, 4th ed.; Thorp, J.H., Rogers, D.C., Eds.; Academic Press: Cambridge, MA, USA, 2015; pp. 781–796. [CrossRef]

67. Lategan, M.J.; Klare, W.; Kidd, S.; Hose, G.C.; Nevalainen, H. The unicellular fungal tool RhoTox for risk assessments in groundwater systems. *Ecotoxicol. Environ. Saf.* **2016**, *132*, 18–25. [CrossRef]

68. Hose, G.C.; Stumpp, C. Architects of the underworld: Bioturbation by groundwater invertebrates influences aquifer hydraulic properties. *Aquat. Sci.* **2019**, *81*, 20. [CrossRef]

Article

Marble Slurry's Impact on Groundwater: The Case Study of the Apuan Alps Karst Aquifers

Leonardo Piccini [1,*], **Tiziana Di Lorenzo** [2], **Pilario Costagliola** [1] **and Diana Maria Paola Galassi** [3]

[1] Department of Earth Science, University of Florence, 50121 Florence, Italy; pilario.costagliola@unifi.it
[2] Research Institute on Terrestrial Ecosystems of the CNR, National Research Council of Italy, 50019 Florence, Italy; tiziana.dilorenzo@cnr.it
[3] Department of Life, Health and Environmental Sciences, University of L'Aquila, 67100 L'Aquila, Italy; dianamariapaola.galassi@univaq.it
[*] Correspondence: leonardo.piccini@unifi.it

Received: 29 October 2019; Accepted: 17 November 2019; Published: 23 November 2019

Abstract: Modern sawing techniques employed in ornamental stones' exploitation produce large amounts of slurry that can be potentially diffused into the environment by runoff water. Slurry produced by limestone and marble quarrying can impact local karst aquifers, negatively affecting the groundwater quality and generating a remarkable environmental and economic damage. A very representative case-study is that of the Apuan Alps (north-western Tuscany, Italy) because of the intensive marble quarrying activity. The Apuan Alps region extends over about 650 km^2; it hosts several quarries, known all over the world for the quality of the marble extracted, and a karst aquifer producing about 70,000 m^3/day of high-quality water used directly for domestic purposes almost without treatments. In addition, Apuan Alps are an extraordinary area of natural and cultural heritage hosting many caves (about 1200), karst springs and geosites of international and national interest. During intense rain events, carbonate slurry systematically reaches the karst springs, making them temporarily unsuitable for domestic uses. In addition, the deterioration of the water quality threatens all the hypogean fauna living in the caves. This paper provides preliminary insights of the hydrological and biological indicators that can offer information about the impact of the marble quarrying activities on groundwater resources, karst habitats and their biodiversity.

Keywords: groundwater; karst aquifer; pollution; quarry; Apuan Alps

---

1. Introduction

Karst areas constitute about 10% of the land surface worldwide and there is a widespread concern about the effects that human activities have upon karst environments [1,2]. Karst aquifers often host water resources that are very important, in terms of both quantity and quality, usually not requiring any treatment before being distributed through civil aqueducts. For this reason, karst groundwater sources are attaining a growing economic and social value. According to UNESCO's International Geoscience Programme (IGCP) 513, concerning "karst aquifers and water resources" and their preservation, about 25% of the world's population relies on drinking water supplies from karst aquifers. On the other hand, karst areas are extremely vulnerable and could be easily (and potentially irreversibly) damaged by human activities such as deforestation, agricultural practices, urbanization, tourism, military activities, water exploitation, mining and quarrying (e.g., [3], and reference therein). Climate change poses an additional risk to karst groundwater, exacerbating the major challenges in the next future.

An unexpectedly high diversity of living forms (micro, meio and macro-organisms) have been discovered in karst aquifers in the second half of the 20th century. Karst stygobionts are often geographically rare and phylogenetically isolated taxa, with ancient origins which date back to the Early Tertiary period [4]. The stygofauna show morphological, physiological and behavioral characteristics suitable for life in absence of light and scarcity of trophic resources (e.g., [5–8]). The stygofauna of karst habitats are represented by vertebrates as well as by small invertebrates, among which crustaceans are dominant [9,10]. Beyond the importance that such highly specialized fauna have in terms of global biodiversity, stygofauna provides a wide range of ecosystem services, including those related to the degradation of pollutants [11,12].

Known since the Roman age for the extraction of the precious white marble, the Apuan Alps, in north-western Tuscany, are a very peculiar mountain range and an extraordinary region of natural and cultural heritage in the Mediterranean basin, with many geological and biological features of national and international interest. This mountain range, whose maximum elevation is 1942 m a.s.l. at Pisanino Mount, consists of both metamorphosed and non-metamorphosed carbonate rocks belonging to three different tectonic units (Figure 1). Apuan metamorphic unit is mainly represented by meta-dolostone, marble and dolomitic marble (about 115 km^2 as outcrops) and by cherty limestones (about 20 km^2) ([13] and references therein).

In an area of about 650 km^2, the carbonate rock assemblages display very different structural and lithological traits. For this reason, they offer a wide variety of hydrogeological conditions displayed in many springs with very different hydrodynamics and hydro-chemical features [13]. The carbonate rocks contain important groundwater resources due to the high rainfall, and the high infiltration coefficient, that can reach 75% of total precipitation. Most of the major springs drain metamorphic carbonate aquifers, and supply drinking water mainly for inhabitants of the coastal plain (where Carrara and Massa towns are located).

The marble exploitation in the Apuan Alps is presently carried out in about 150 quarries that are frequently located at high altitudes (Figure 2a). Line drilling and sawing are the most modern techniques to produce large rectangular blocks suitable for being cut into smaller, regularly shaped products and slabs (Figure 2b). Sawing can be accomplished by a variety of tools, including wire, belt and chain saws. Pollutants resulting from quarrying, including the marble dust (locally known as "*marmettola*") that is a product of rock sawing, easily infiltrate by means of meteoric waters into the underlying karst aquifers through solution cavities. This condition is responsible for the high vulnerability of the groundwater bodies, which is particularly critical for metamorphic carbonate rocks (i.e., marbles and meta-dolostone), due to the absence of retention of pollutants in the unsaturated zone (epikarst and vadose zone) and to the high hydraulic conductivity in the saturated one [14]. As a matter of fact, this fine-grained material is easily mobilized by rainwater and dispersed into and through the subterranean karst systems [15].

In the last decade, the annual marble production in the Apuan Alps quarries has been about 1.2 Mt/year [16]. The extraction and squaring of marble blocks produce about 4% dust per unit [15], in respect to marble production, which so amounts to about 0.05 Mt/year, or, as volume, about 25,000 m^3/year, assuming a porosity of 35%. Most of this powder is disposed on quarry yards and stored in controlled repository areas but a substantial part of it is dispersed into the surrounding areas by wind and runoff before collecting. Furthermore, it is quite frequent that quarries intercept open fissures and solution pits through which marble powder enters the subterranean drainage usually driven by high-intensity rainfall events [17,18].

While the marble quarries in the Apuan Alps have been long and widely studied in many environmental respects, including marble formation volumes [19], rock mass stability [20] and restoration perspectives [21], the real effects of marble slurry pollution, and in particular its impact on groundwater and the processes governing its transport through the karst aquifers, have not yet been systematically examined, despite the fact that this aspect is crucial for the development of management measures aimed at preserving the water quality and the biodiversity of karst systems.

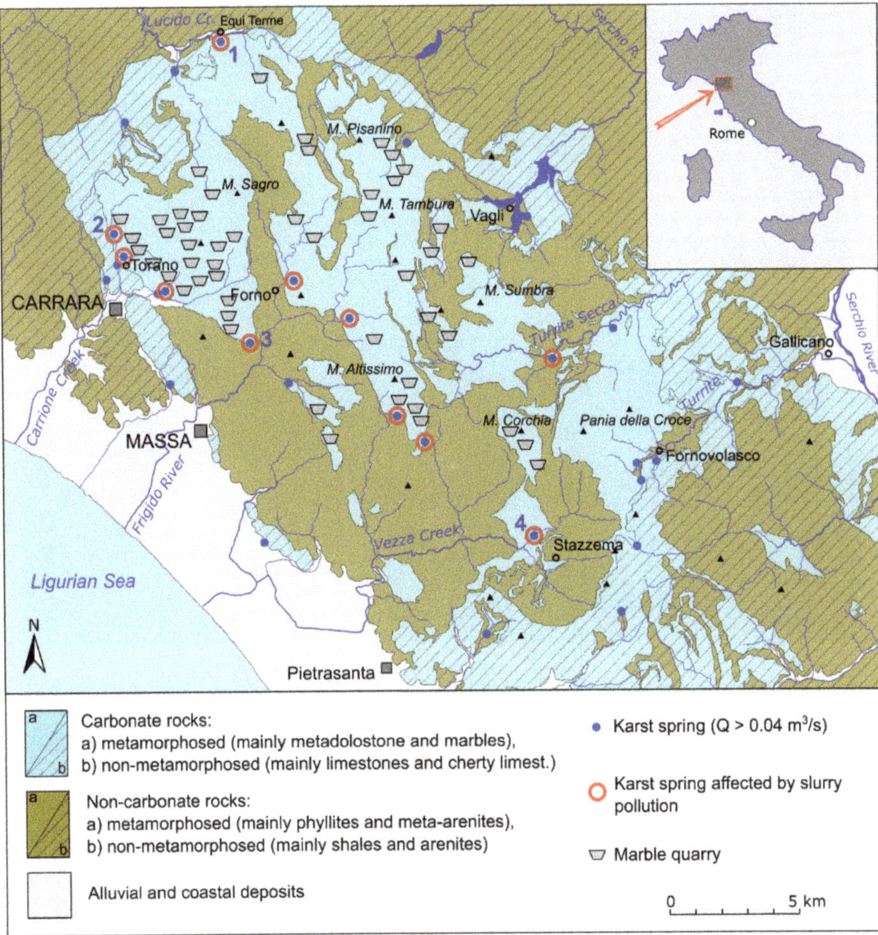

Figure 1. Hydrogeological sketch map of the Apuan Alps with the location of the major karst springs (average discharge >0.04 m^3/s) fed by carbonate aquifers and marble quarries. Numbers refer to monitored karst systems: (1) Equi springs; (2) Carbonera spring; (3) Cartaro spring; (4) Monte Corchia cave stream.

For these reasons, the Apuan Alps can be considered a relevant case-study area, where the impact of slurries produced by marble quarries on karst aquifers can be investigated in order to obtain an indication of general interest and worldwide application. The main purpose of this article is to provide the information available up to now regarding *marmettola* pollution, including its physical and chemical features, and the potential impacts that this particular waste product can cause on the Apuan Alps karst ecosystem, and more generally, in similar karst areas heavily affected by marble quarrying. The article also provides new data on both the hydro-physical properties of groundwater bodies and their biodiversity.

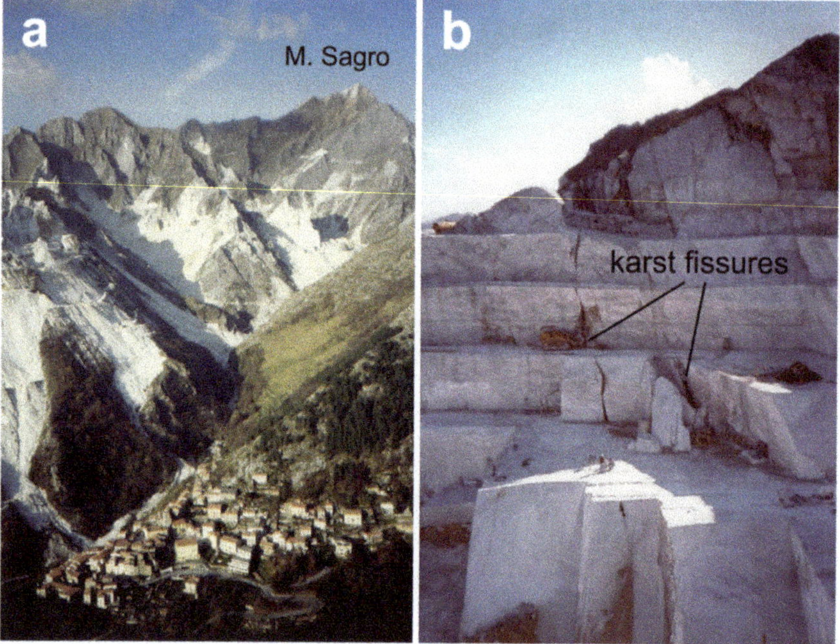

Figure 2. (**a**) A view of Colonnata valley, in Northern Apuan Alps, showing the intense quarrying that concerns high-altitude areas too. (**b**) Quarrying is performed through sawing large blocks with chain or diamond-wire saws; karst fissures or solution pit interception is very frequent (Focolaccia quarry, M. Tambura, central Apuan Alps).

2. Chemical and Physical Features of *Marmettola* and Karst Spring Sediments

Marmettola is a fine, white powder produced during the extraction and squaring of marble blocks by way of chain and diamond-wire saws. Minor sources are represented by the dust produced by the intense traffic of heavy trucks, bulldozers and tractors transporting and moving heavy marble blocks along quarry roads and service areas. The chemical composition of the powder reflects the mineralogical constituents of the different marble varieties. The grains are essentially composed by calcite (usually 95–99%), mainly in sand-to-silt sizes, with a small percentage (1–3%) of dolomite [15,22,23]. Silicates, primarily quartz, muscovite and/or chlorite and clay minerals, are often present in minor amounts [15,21].

In order to investigate textural features, we examined some slurry samples collected in 2017, coming from a quarry at Mount Sagro, by scanning electron microscope (SEM) analysis. Grains produced by chain saws were very angular in shape and with planar surfaces formed along calcite cleavage planes (Figure 3a). SEM image processing revealed that marble slurry is mainly composed by weight of a fine to very-fine sand (Figure 3b), whereas most of the particle sizes (expressed as equivalent circle diameter = ECD) range between 8 and 32 μm, if we consider the number of particles (Figure 3c). A consistent amount of the finer silt-sized fraction is possibly derived from consumption of larger grains during sawing [15]. The powder produced by diamond-wire cutting is generally finer (most grains <62.5 μm in size) but with grains showing an angular shape like those coming from chain saws [15].

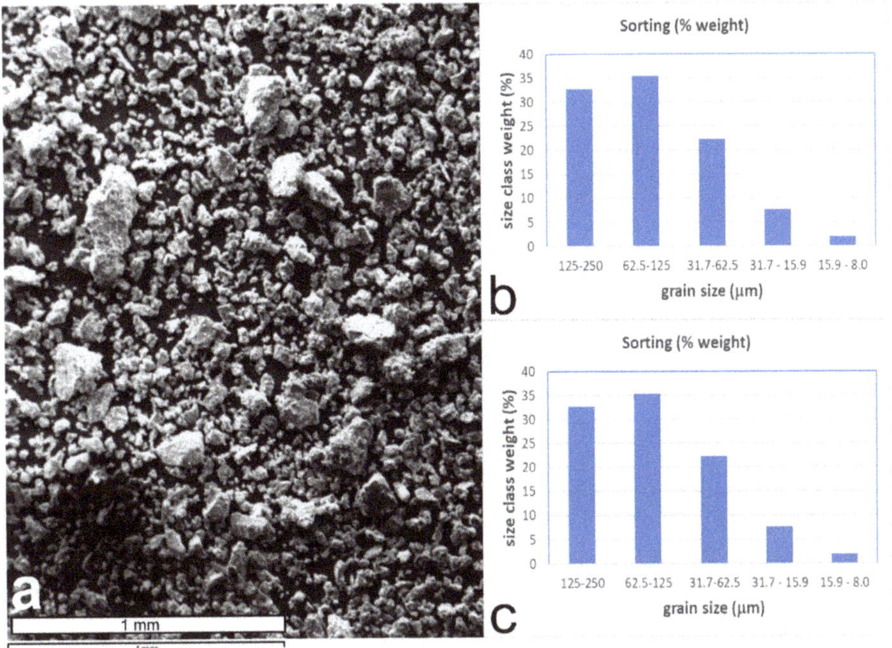

Figure 3. (**a**) SEM image of *marmettola* produced by chain sawing collected at a marble quarry in the Mount Sagro area (note the very angular shape of particles). (**b**) Weight sorting of the sample collected (weight percent for each grain size class). (**c**) Frequency sorting (number of grains) of the same sample.

By means of direct investigation of karst systems contaminated by marble slurry, such that of Monte Corchia cave, we realized that suspended sediments found in springs and cave-streams consisted of fine sand and mud having a light grey or ochre color. From a chemical and mineralogical point of view, these deposits had a higher amount of Mg and Si respect to *marmettola*, due to the occurrence of dolomite and silicates of natural origin. In a sample collected in 2012 at Tana dei Tufi, a captured spring located near Torano village (Carrara marble district) often affected by turbidity, SEM analysis showed that about 75% of the grains were carbonate, the remaining being silicate particles. Grain morphometric analysis performed on SEM images revealed that grains' ECDs mainly ranged from 15.9 to 62.5 µm, with the silicate particles a little finer than carbonate ones (Figure 4a,b). Other morphometric parameters, such as the aspect ratio (length/breadth ratio of each) and the roundness (calculated as perimeter^2/4π * grain area), were similar in both carbonate and silicate grains (Figure 4c–f). SEM images confirmed also that most of this sediment consisted of calcite angular grains like those forming *marmettola*. Grains finer than 16 µm were poorly represented, probably because they were washed away as suspended load during floods. These preliminary data suggest that the transport through the karst network of the Apuan Alps produces an effective sorting and a mixing with natural sediments.

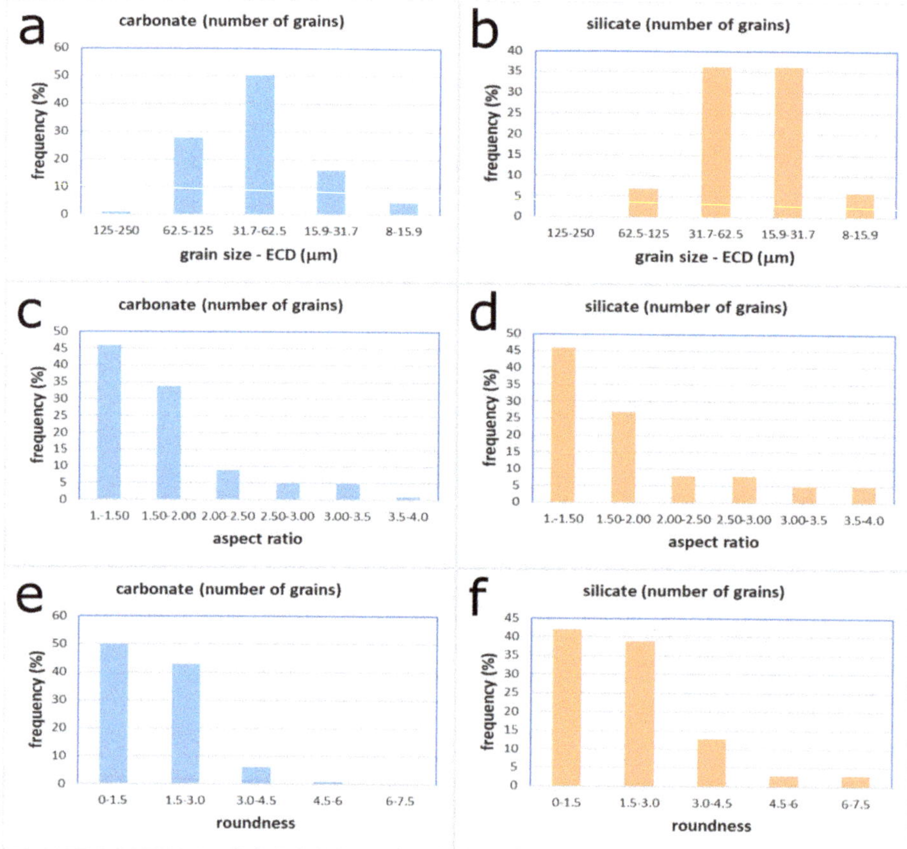

Figure 4. ECD (equivalent circular diameter; in μm), aspect ratio and roundness of carbonate and silicate fractions of a sediment collected at the Tana dei Tufi karst spring (Torano, Northern Apuan Alps) (see text for explanation).

3. Impact of Marble Slurry on Karst Groundwater

Once marble powder mixes up with water, it produces a fluid mud that rapidly disperses through the karst groundwater network [15,24]. This slurry negatively affects the groundwater quality, and its persistence in the karst system may produce remarkable environmental and economic damage [25,26]. In the Apuan Alps, the *marmettola* slurry has been observed in cave streams and karst springs (Figure 5).

Although several studies have described the environmental impacts of human activities on karst systems [1,27], there are relatively a few papers that specifically refer to the effects from quarrying on karst drainage systems. The relationship between quarrying activities and the environmental damage caused on carbonate rock has been well documented for over sixty years [28]. On the contrary, there are only a few reports that include major discussions of the environmental impacts on the karst aquifer of slurry produced by in-situ stone cutting and almost none referring to the impact on the karst stygofauna [29–31].

Figure 5. (**a**) *Marmettola* deposited in the occasionally flooded Buca di Equi cave. (**b**) Whitish, turbid water due to the marble slurry flowing out from the Equi karst spring observed several days after a strong flood in October 2013.

One of the most effective ways to investigate the dynamics of marble slurry transfer through karst subterranean networks is by monitoring of main spring hydro-physical parameters (flow rate, temperature, electrical conductivity and turbidity) coupled with sampling of suspended load. In the Apuan Alps, the turbidity observed at springs during floods can be often related to the suspended transport of marble slurry produced by quarries [15]. In fact, we have frequently observed that springs whose catchment areas do not embrace quarry districts have usually clear or slightly muddy water even during floods, apart from some springs that are fed by surface sinking water. On the contrary, turbidity due to carbonate slurry is systematically observed in water springs located in quarrying districts (Figure 5b).

In order to study the effects of the marble slurry on the karst aquifers of the Apuan Alps, a monitoring network has been recently (starting from April 2018) installed by the authority in charge for the environmental monitoring in Tuscany (Agenzia Regionale per la Protezione Ambientale Toscana-ARPAT). The monitoring settlement concerns three karst springs (Carbonera, Cartaro and Equi), and one cave stream (Monte Corchia Cave; see Figure 1 for location). Stream and spring monitoring is performed by automatic probes acquiring water level (resolution 1 mm), water temperature (resolution 0.1 °C), electrical conductivity (accuracy 1%) and turbidity (resolution 1—NTU (Nephelometric Turbidity Unit)); data are freely available at http://sira.arpat.toscana.it. Samples of *marmettola* were collected in quarries, whereas sediments were collected in a cave stream (Monte Corchia Cave) and karst springs, in order to compare the physical features of both marble sawing powder and springs suspended load.

One of the clearest examples of the marble slurry's impact on the Apuan Alps karst systems is provided by the springs located close to Equi Terme village (Massa) (see Figure 1 for location). Equi springs are fed by the Northern sector of the Apuan Alps karst aquifer, consisting mainly of marble and meta-dolostones. Marble quarries are the main human activity occurring in the catchment area. This multi-spring is characterized by high and very rapid fluctuations of water flow. During the dry season, the discharge can decrease down to 0.03–0.04 m³/s, whereas major floods can increase it to a total flow rate of 15–20 m³/s [32]. During floods and some days after, the water is highly turbid, but a slight whitish opalescence can persist for several days afterwards. Figure 6 shows the flow rate and turbidity diagram from April 2018 to April 2019. Floods are characterized by tight peaks followed by a rapid decrease of flow rate, indicating a quick transmission of hydraulic pressure through the karst network due to massive infiltration [33]. Turbidity peaks, which can reach 400 or more NTU, are not perfectly overlaid to the discharge peaks being shifted some days afterward.

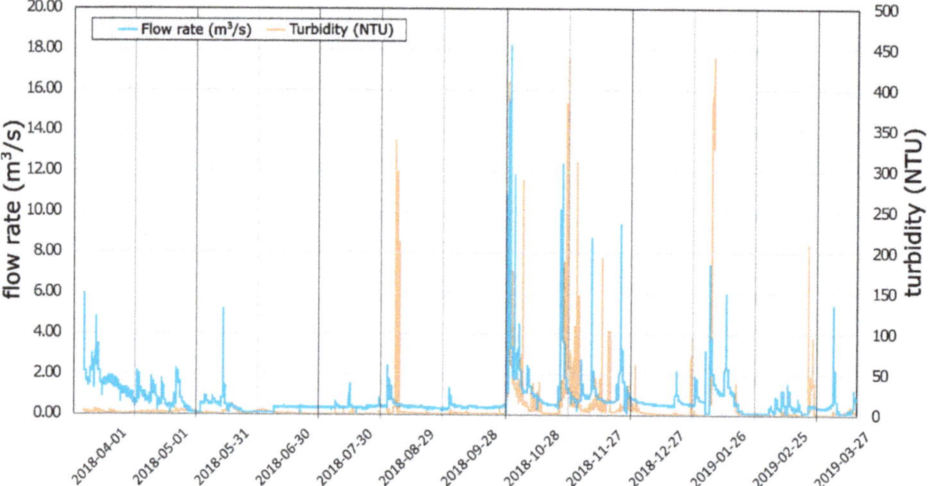

Figure 6. Flow rate and turbidity of Equi springs from April 2018 to April 2019 (dates are expressed as yyyy-mm-dd).

A detail graph (Figure 7) related to one minor flood (occurred on 1 September 2018 at Equi springs) following a long period of low flow rate, shows some features that clarify the origin of turbidity. The turbidity peak occurred around 48–50 h after the first flow peak, when the discharge decreased almost to the pre-flood level. This means that during minor floods the water remains clear until the *marmettola* slurry, washed away from the quarries and infiltrated in the catchment area, reaches the spring through a high-conductivity conduits network. In this case, the turbidity is not due to the mobilization of fine sediments contained in the phreatic sector of the aquifer but to the infiltration of surface runoff contaminated by marble slurry.

During major floods, the Equi springs displayed a different behavior of the turbidity parameters with respect to flow rate fluctuation (Figure 8). Turbidity showed a slight but significant increase just a few hours after the discharge rising, whereas the first substantial turbidity peak occurred with a delay of about 48 h after flood. Anyway, we observed that most of the solid load, measured as total suspended sediment (TSS), according to the methods proposed by Drysdale et al. [15], occurred during the maximum flow rate phase. Despite the high values of NTU, the turbidity peaks following with about two days of delay the first flood peak, contributed only for about the 40% to the amount of TSS, since discharge was low. Actually, the total TSS mobilized during this flood was exceptionally high, approaching 37 tons (37,000 kg).

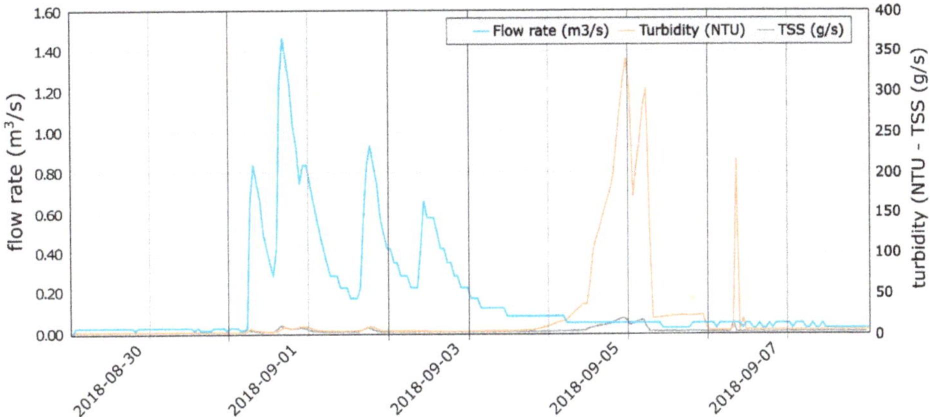

Figure 7. Flow rate and solid transport diagram of Equi springs during a minor flood on 1 September 2018 (dates are expressed as yyyy-mm-dd). The turbidity peak occurred more than four days after the flow peak.

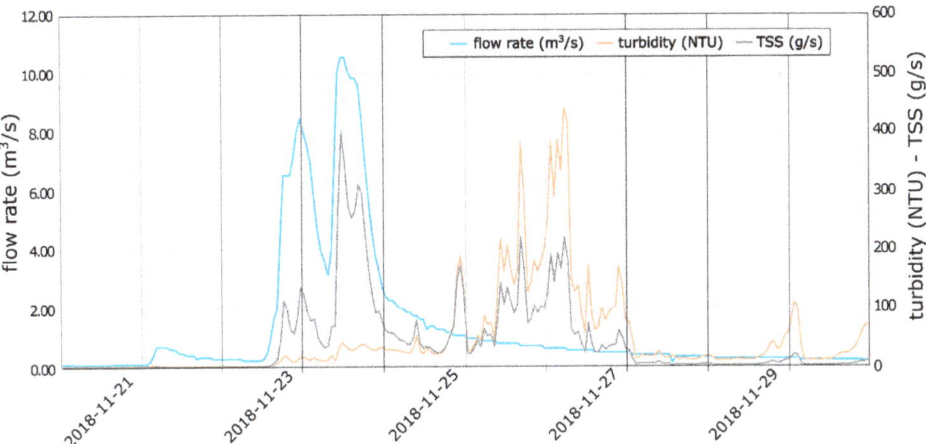

Figure 8. Flow rate and solid transport diagram of Equi springs during a major flood on 23 November 2018 (dates are expressed as yyyy-mm-dd). The major turbidity peak occurred about three days after the flow rate increase, but the TSS reached its greatest values during the second flood peak.

Comparing the different turbidity responses of these spring group during minor floods (Figure 7) with respect to major floods (Figure 8), we can infer that the TSS mobilized during the phase of major flow rate probably consists of sediments already stored in the phreatic system, which may be dragged away by water only in the case of a high flow velocity.

In conclusion, if we consider the effective TSS load instead of turbidity, we observe two major transport phases, the first during the flood peaks and the second one during the flood decremental phase (Figure 8). Considering the pattern of water and sediment fluxes, it may be argued that the second solid transport wave corresponded to the finest portion of *marmettola* washed away directly on quarry areas.

4. Impact of Marble Slurry on Karst Stygofauna

It is not an easy task assessing which are the main impacts of marble slurry on the organisms that live in karst aquifers. The main difficulty is the limited accessibility of these environments and the consequent impossibility of directly inspecting the habitats where stygobiotic species dwell [34]. Very appropriately, karst systems have been compared to black boxes in which one can assess exactly the input and the output but cannot directly know what is happening inside [35]. This similitude comes from hydrogeology but actually fits ecology very well. Karst stygofauna may be sampled in the epikarst (mostly caves) or at the karst outlets, through water filtration at the springs. In both cases, however, there is no direct inspection of the habitats except, perhaps, for gours, concretion pools, ponds and rivers, when present.

Despite the sampling difficulties, groundwater ecologists have collected enough information to outline the distributional patterns of many stygobiotic taxa. We know, for instance, that the distribution of stygofauna is not uniform in the aquifer, but that most of the species, which do not have adaptations to withstand fast flows, tend to colonize the habitats where the groundwater flow is low ([36–38] and references therein). Most of the species are collected from the small fractures and dissolution fissures (apertures from about one millimeter to a few centimeters) of the karst network, rather than in the fast drains [37,39–41]. Groundwater flow is therefore the main organizing factor for karst communities [42] and in extreme cases, such as in the event of an earthquake, the massive earthquake-induced flow discharge can trigger a huge washout of groundwater fauna with a destructive effect on the biological assemblages [36].

In normal conditions, however, groundwater flow acts together with other factors, such as behavioral adaptations. In fact, the occurrence of juveniles of some stygobiotic species at spring-fed karst outlets during a flood event [43] is suggestive of an adaptive strategy: some species release the eggs during the low groundwater level period and use the flood to disperse the juveniles so as to favor the colonization of new habitats [43]. Competition for trophic resources and predation are additional factors that can shape the distribution of communities within karst aquifers, as already observed in previous field and laboratory studies [44]. Finally, the chemical contamination of karst aquifers also influences the distribution of species within the karst network. An organic pollution (indicated by high nitrates and ammonium concentrations and particulate organic matter) may favor, for instance, the permanence of epigean species which, having high metabolic rates, exert a strong competitive pressure on the stygobiotic species, often replacing them completely within the community [45].

Unfortunately, there are no available data on the effect of marble quarry wastes on karst communities. Based on the observations made at Equi springs, we can realistically expect that a substantial quantity of *marmettola* settles in the smaller fractures of the karst systems, where the flow velocities are low and where, as mentioned above, most species are present. A complete or partial filling of these habitats likely induces chemical variations influencing the availability of oxygen, nutrients and habitat space reduction. This scenario recalls the phenomenon of clogging observed in hyporheic zones (transitional ecotones between surface water and groundwater bodies). In fact, the deposition of fine sediments in the hyporheic zone of the Alpine stream Noce Bianco in Northern Italy due to hydropeaking caused a reduction of the interstitial habitats available to stygobiotic taxa [46]. Similarly, Mori et al. [47] observed that the stygobiotic community of the hyporheic zones of the Sava River (Slovenia) was strongly affected after severe flooding with clogging of interstitial spaces by fine sediments. Microhabitat clogging may select species with certain physiological or trophic characteristics. For instance, in the reduced pore spaces of the hyporheic zones of the River Rhone catchment, only small-sized and slow-moving organisms, mainly copepods, ostracods and cladocerans, occurred [48]. This phenomenon has been confirmed recently by laboratory experiments. Korbel et al. [49] observed that stygobiotic crustaceans, such as the small-sized copepods and the large-sized amphipods and syncarids, are able to move easily in sand substrates ($\varphi = 0.3$–0.7 mm) and gravel ($\varphi = 2$–4 mm). However, only copepods were able to move in silts ($\varphi < 60$ μm), whereas the larger amphipods and syncarids remained on the sediment surface without being able to dig in.

The inability to cope with fine sediments for syncarids and amphipods increases when the groundwater level decreases. In microcosm experiments, none of these animals managed to dig in the mud layers when the overlying water was removed [49]. This phenomenon has serious consequences in terms of loss of ecosystem services. In fact, laboratory experiments have shown that syncarids and amphipods are crucial for maintaining the effective porosity of fine sediment-filled habitats. In the absence of these animals, porosity decreases significantly in two months, causing a further reduction of pore spaces [50]. The diameter of pore spaces in silts is predicted to be <25 µm [51] and the same is expected in quarry waste deposits because most of the particles we measured were from 8 to 32 µm in diameter. The intergranular spaces of *marmettola* deposits are, therefore, smaller than the smallest dimension of most of stygobiotic crustaceans. Copepods manage to dig in these deposits, and it has to be expected that fine sediments adhere tenaciously to the body of the animal (Figure 9). We do not know how this sort of "coating" will negatively affect the physiology of these organisms, but fissured aquifers filled with marble slurry are likely inhospitable habitats for most of karst stygofauna.

Figure 9. Silt-coating of an individual of *Elaphoidella phreatica* (Crustacea Copepoda) from a clay-sandy aquitard in Tuscany (Italy). The photos were obtained by scanning electron microscopy. The grains of silt adhere tenaciously to the body of the animal also covering the setae and appendages. In the detail of the photo, the coating of the fifth thoracic leg involved in mating behavior of these animals is visible.

In order to test this hypothesis, we carried out biological surveys in the Monte Corchia cave, located in one of the most intensely exploited areas with marble quarries in the Apuan Alps. The Monte Corchia karst system is located in the south-western part of the Apuan Alps and is developed inside the carbonate core of a syncline, almost completely enclosed by a low-permeable basement (Figure 1). Monte Corchia cave is about 65 km long and 1185 m deep and it is presently one the largest and deepest caves in Italy [52].

Specimens were collected from the vadose zone of the cave, located between 900 and 800 m a.s.l., which consists of several passages that have been part of the touristic path since 2001. We sampled six percolation water drips with permanent water flow and six concretional pools. The water in the concretional pools was collected by syringes at approximately quarterly intervals from June 2016 to October 2019. Percolation water drips were directed into a 500 mL plastic filtering bottles with partly removed walls covered by a net of 60 μm mesh a few cm above the bottom. The animals coming with the drips were retained in the small amount of water in the filtering bottles and collected on a quarterly basis. Altogether we collected about 700 specimens, 85% of which came from the drips. All the animals were copepods, as also observed in other studies (e.g., [35,53,54]). However, all the animals collected in our surveys belonged to a single taxon, a new undescribed species of the canthocamptid harpacticoid genus *Moraria*.

This result contrasts with the data of Meleg et al. [55], who reported 11 stygobiotic copepod species from drips and drip pools of five Romanian caves; of Galassi et al. [37], who reported up to eight copepod species from two drips of Frasassi cave; and especially, with the data (up to 33 species per cave) recorded in the Slovenian caves and reviewed by Pipan and Culver [53]. That of Monte Corchia Cave seems, therefore, to be an unusual situation, being similar only to two non-Dinaric caves in Slovenia; namely, Huda luknja, occurring in an isolated karst area, and Snežna jama, an ice cave in the Kamnik–Savinja Alps (two stygobiotic copepod species per cave) [53]. At present, data do not allow us to know whether the persistently low biodiversity observed in Monte Corchia Cave is due to the presence of *marmettola* and its clogging effects, or to other factors, such as the isolation of the cave that prevents colonization of groundwater similarly to the conditions of Huda luknja in Slovenia [54].

5. Discussion and Conclusions

This study did not aim at proposing strategies to contain the production of marble slurry or to prevent its dispersion in subterranean waterways but had the objective to denounce the fact that too many aspects of this environmental issue have not yet been adequately addressed. For example, the priority issue for the Italian authorities is the correct identification of *marmettola* as waste or as a by-product. This is not a matter of little importance, since the management regime of by-products is much less restrictive from a legislative point of view. Although the marble powder produced by the quarrying activities in the Apuan Alps is managed as waste, the cross-checks carried out by the competent authorities have highlighted that a relevant amount of it is dispersed in the environment. It would therefore be appropriate to apply an economic sanction that compensates for the environmental damage, including the loss of ecosystem services, consequently to *marmettola* dispersion. However, the quantification of this penalty, although expressly demanded by the Water Framework Directive in Europe, is difficult to determine now. The main impediment is due to the scarce knowledge of the impacts of marble slurry on the fauna dwelling in karst aquifers. With this review we aimed at disseminating information on the possible biological impacts, and secondly, on the poor biodiversity detected in some parts of the Apuan Alps karst system, such as in Monte Corchia cave. Stygofauna are "out of sight" and it is, therefore, easy to maintain these species as "out of mind" [56].

However, this short-sighted way to perceive karst ecosystems is an oversight before which we cannot remain silent. Fortunately, the turbidity due to *marmettola* detected at the springs is visible to the naked eye and has an immediate impact on the public, which is shocked by the idea of drinking "dirty" water. The quantification of the environmental impact in this case is easier, since it is possible to estimate the damage caused to the aqueduct filters by the high turbidity or the temporary closure of the distribution network. However, not only the springs presently captured for civil aqueducts but those that can be potentially captured in the future, must be monitored and possibly subjected to safeguarding measures.

Based on a first set of data, this study documented the worsening of the water quality during flood events due to the *marmettola* transport in the karst system of the Apuan Alps and a supposed negative effect of suspended solid deposition on the local groundwater community. The study highlighted that, although marble slurry does not constitute an immediate danger to public health, it is nevertheless responsible for a serious impact on the Apuan surface and subterranean waterways. We suspect that an unknown amount of *marmettola* has been already accumulated in the karst aquifers and it is discharged only in conjunction with high-intensity rain events. As already stated, the effects of such a physical pollution, apart the obvious deterioration of the quality of the water intended for human consumption, are still not known in detail. It can be expected that the marble slurry has potentially reduced the water exchanges among karst conduits and fractures and impacted the rock mass porosity, changing the hydrodynamics of the karst networks by also reducing their storage capacity. If so, it could be expected that the local environment will remain damaged (at some extent) for an unknown time period, even if the *marmettola* dispersion were to be completely stopped. Resilience will be hindered by the depletion of the groundwater biological communities that, otherwise, would be able to provide ecological services of great importance, most of them related to the aquifer self-purification.

In conclusion, the first results of this study have highlighted that the present-day footprint of quarrying activities in the Apuan Alps is greater than that indicated on maps or aerial photographs. It is necessary, therefore, to boost participatory processes involving economic stakeholders and environmental experts, and speleologists who are often the most informed about the conditions within karst systems too. Due to their long history of exploitation, the Apuan Alps represent a red flag that we hope will trigger biological and hydrogeological studies in other areas.

Author Contributions: Conceptualization, L.P., P.C. and T.D.L.; methodology, L.P., P.C. and T.D.L.; validation, L.P. and D.M.P.G.; formal analysis, L.P.; investigation, L.P., P.C. and T.D.L.; species identification: T.D.L. and D.M.P.G.: resources, D.M.P.G.; data curation, L.P., P.C., T.D.L. and D.M.P.G.; writing—original draft preparation, L.P.; writing—review and editing, L.P., P.C., T.D.L. and D.M.P.G.; visualization, L.P.; supervision, L.P. and D.M.P.G.; project administration, L.P. and D.M.P.G.; funding acquisition, D.M.P.G.

Funding: This research was partially funded by the AQUALIFE project LIFE12 BIO/IT/000231 "Development of an innovative and user-friendly indicator system for biodiversity in groundwater dependent ecosystems" of the European commission.

Acknowledgments: We wish to thank all the institutions that in different ways have encouraged and supported this research, with special mentions to: Agenzia Regionale per la Protezione dell'Ambiente, Toscana—ARPAT; Parco Regionale delle Alpi Apuane, Federazione Speleologica Toscana.

Conflicts of Interest: The authors declare no conflict of interest.

References

1. Williams, P.W. Environmental change and human impact on karst terrains: An introduction. *Catena* **1993**, *25*, 1–19.
2. Van Beymen, P.; Townsend, K. A disturbance Index for karst environments. *Environ. Manag.* **2005**, *36*, 101–116. [CrossRef] [PubMed]
3. Hamdan, I.; Margane, A.; Ptak, T.; Wiegand, B.; Sauter, M. Groundwater vulnerability assessment for the karstic aquifers of Tanour and Rasoun springs catchment area (NW-Jordan) using COP and EPIK intrinsic methods. *Environ. Earth Sci.* **2016**, *75*, 1474. [CrossRef]
4. Gibert, J.; Culver, D.C. Assessing and conserving groundwater biodiversity: An introduction. *Freshw. Biol.* **2009**, *54*, 639–648. [CrossRef]
5. Gibert, J.; Danielopol, D.L.; Stanford, J.A. *Groundwater Ecology*; Academic Press: New York, NY, USA, 1994; p. 571.
6. Coineau, N. Adaptations to interstitial groundwater life. In *Ecosystems of the World, 30: Subterranean Ecosystems'*; Wilkens, H., Culver, D.C., Humphreys, W.F., Eds.; Elsevier: Amsterdam, The Netherlands, 2000; pp. 189–210.
7. Di Lorenzo, T.; Di Marzio, W.D.; Spigoli, D.; Baratti, M.; Messana, G.; Cannicci, S.; Galassi, D.M.P. Metabolic rates of a hypogean and an epigean species of copepod in an alluvial aquifer. *Freshw. Biol.* **2015**, *60*, 426–435. [CrossRef]

8.   Iepure, S.; Rasines-Ladero, R.; Meffe, R.; Carreno, F.; Mostaza, D.; Sundberg, A.; Di Lorenzo, T.; Barroso, J.L. The role of groundwater crustaceans in disentangling aquifer type features—A case study of the Upper Tagus Basin, central Spain. *Ecohydrology* **2017**, *10*, e1876. [CrossRef]

9.   Galassi, D.M.P.; Stoch, F.; Fiasca, B.; Di Lorenzo, T.; Gattone, E. Groundwater biodiversity patterns in the Lessinian Massif of northern Italy. *Freshw. Biol.* **2009**, *54*, 830–847.

10.   Di Lorenzo, T.; Cipriani, D.; Fiasca, B.; Rusi, S.; Galassi, D.M.P. Groundwater drift monitoring as a tool to assess the spatial distribution of groundwater species into karst aquifers. *Hydrobiologia* **2018**, *813*, 137–156. [CrossRef]

11.   Boulton, A.J.; Fenwick, G.; Hancock, P.J.; Harvey, M.S. Biodiversity, functional roles and ecosystem services of groundwater invertebrates. *Invertebr. Syst.* **2008**, *22*, 103–116. [CrossRef]

12.   Griebler, C.; Avramov, M. Groundwater ecosystem services: A review. *Freshwat. Sci.* **2014**, *34*, 355–367. [CrossRef]

13.   Doveri, M.; Piccini, L.; Menichini, M. Hydrodynamic and geochemical features of metamorphic carbonate aquifers and implications for water management: The Apuan Alps (NW Tuscany, Italy) case study. In *Karst Water Environment*; The Handbook of Environmental Chemistry; Younos, T., Schreiber, M., Kosic Ficco, C., Eds.; Springer: Cham, Swizterland, 2019; Volume 68, pp. 209–249. [CrossRef]

14.   Civita, M.; Forti, P.; Marini, P.; Meccheri, M.; Micheli, L.; Piccini, L.; Pranzini, G. *Pollution Vulnerability Map for the Aquifers of the Apuane Alps—A Brief Guide*; Gr. Naz. per la Difesa dalle Catastrofi Idrogeologiche; NRC: Firenze, Italy, 1991; p. 56.

15.   Drysdale, R.N.; Pierotti, L.; Piccini, L.; Baldacci, F. Suspended sediments in karst spring waters near Massa (Tuscany), Italy. *Environ. Geol.* **2001**, *40*, 1037–1050.

16.   Regione Toscana. Praer: Piano Regionale Delle Attività Estrattive. Available online: http://www.regione.toscana.it/-/praer-piano-regionale-delle-attivita-estrattive (accessed on 1 September 2019).

17.   Gunn, J.; Hardman, D.; Lindesay, W. Problems of limestone quarrying in an adjacent to the Peak District National Park. *Annales de la Société géologique de Belgique* **1985**, *108*, 59–63.

18.   Ekmekci, M. *Impact of Quarries on Karst Groundwater Systems. Hydrogeological Processes in Karst Terranes*; IAHS Publication: Wallingford, Oxfordshire, UK, 1993; Volume 207, pp. 3–6.

19.   Mastrorocco, G.; Salvini, R.; Vanneschi, C. Fracture mapping in challenging environment: A 3D virtual reality approach combining terrestrial LiDAR and high definition images. *Bull. Eng. Geol. Environ.* **2018**, *77*, 691–707. [CrossRef]

20.   Gulli, D.; Pellegri, M.; Marchetti, D. Mechanical behaviour of Carrara marble rock mass related to geo-structural conditions and in-situ stress. In Proceedings of the 15th Pan-American Conference on Soil Mechanics and Geotechnical Engineering (PCSMGE)/8th South American Congress on Rock Mechanics (SCRM), Buenos Aires, Argentina, 15–18 November 2015; pp. 253–260.

21.   Gentili, R.; Sgorbati, S.; Baroni, C. Plant species patterns and restoration perspectives in the highly disturbed environment of the Carrara marble quarries (Apuan Alps, Italy). *Restor. Ecol.* **2011**, *19*, 32–42. [CrossRef]

22.   Cantisani, E.; Fratini, F.; Malesani, P.; Molli, G. Mineralogical and petrophysical characterisation of white Apuan Marble. *Periodico di Mineralogia* **2005**, *74*, 117–140.

23.   Cantisani, E.; Pecchioni, E.; Fratini, F.; Garzonio, C.A.; Malesani, P.; Molli, G. Thermal stress in the Apuan marbles: Relationship between microstructure and petrophysical characteristics. *Int. J. Rock Mech. Min. Sci.* **2009**, *46*, 128–137. [CrossRef]

24.   Mantelli, F.; Montigiani, A.; Lotti, L.; Piccini, L.; Bianucci, P.; Malcapi, V. Le acque sotterranee del sistema carsico del M. Corchia: Valorizzazione, salvaguardia e rischi di inquinamento. In Proceedings of the 3 Convegno Nazionale "Protezione e gestione delle acque sotterranee per il III millennio", Parma, Italy, 13–15 October 1999; pp. 115–125.

25.   Rizzo, G.; D'Agostino, F.; Ercoli, L. Problems of soil and groundwater pollution in the disposable of "marble" slurries in NW Sicily. *Environ. Geol.* **2008**, *55*, 929–935. [CrossRef]

26.   Celik, M.Y.; Sabah, E. Geological and technical characterisation of Iscehisar (Afyon-Turkey) marble deposits and the impact of marble waste on environmental pollution. *J. Environ. Manag.* **2008**, *87*, 106–116. [CrossRef]

27.   Drew, D.H.H.; Hotzl, H. *Karst Hydrogeology and Human Activities: Impact, Consequences, and Implications*; IAH International contributions to hydrogeology, 20; A.A. Balkema: Rotterdam, The Netherlands, 1999.

28.   Foose, R.M. Ground-water behavior in the Hershey Valley, Pennsylvania. *Geol. Soc. Am. Bull.* **1953**, *64*, 623–645. [CrossRef]

29. Whitten, T. Applying ecology for cave management in China and neighbouring countries. *J. Appl. Ecol.* **2009**, *46*, 520–523. [CrossRef]

30. Souza-Silva, M.; Martins, R.P.; Ferreira, R.L. Cave conservation priority index to adopt a rapid protection strategy: A case study in Brazilian Atlantic rain forest. *Environ. Manag.* **2015**, *55*, 279–295. [CrossRef] [PubMed]

31. Sugai, L.S.M.; Ochoa-Quintero, J.M.; Costa-Pereira, R.; Roque, F.O. Beyond above ground. *Biodivers. Conserv.* **2015**, *24*, 2109–2112. [CrossRef]

32. Worthington, S.R.H. How preferential flow delivers pre-event groundwater rapidly to streams. *Hydrol. Proc.* **2019**, *33*, 2373–2380. [CrossRef]

33. Poggetti, E.; Lazzaroni, M.; Verole, M.; Piccini, L. Risultati e interpretazione idrodinamica del monitoraggio in continuo delle sorgenti carsiche di Equi Terme (Alpi Apuane). In Proceedings of the 22 Congresso Nazionale di Speleologia, Pertosa, Auletta (Salerno), Italy, 30 May–2 June 2015; pp. 369–374.

34. Korbel, K.; Chariton, A.; Stephenson, S.; Greenfield, P.; Hose, G.C. Wells provide a distorted view of life in the aquifer: Implications for sampling, monitoring and assessment of groundwater ecosystems. *Sci. Reports* **2017**, *7*, 40702. [CrossRef]

35. Mangin, A. L'approche systémique du karst, conséquences conceptuelles et méthodologiques. *Bulletin de la Société Méridionale de Spéléologie et de Préhistoire* **1984**, *24*, 131–145.

36. Galassi, D.M.P.; Lombardo, P.; Fiasca, B.; Di Cioccio, A.; Di Lorenzo, T.; Petitta, M.; Di Carlo, P. Earthquakes trigger the loss of groundwater biodiversity. *Sci. Rep.* **2014**, *4*, 6273. [CrossRef]

37. Galassi, D.M.P.; Fiasca, B.; Di Lorenzo, T.; Montanari, A.; Porfirio, S.; Fattorini, S. Groundwater biodiversity in a chemoautotrophic cave ecosystem: How geochemistry regulates microcrustacean community structure. *Aquat. Ecol.* **2017**, *51*, 75–90. [CrossRef]

38. Stoch, F.; Fiasca, B.; Di Lorenzo, T.; Porfirio, S.; Petitta, M.; Galassi, D.M.P. Exploring copepod distribution patterns at three nested spatial scales in a spring system: Habitat partitioning and potential for hydrological bioindication. *J. Limnol.* **2016**, *75*, 1–13. [CrossRef]

39. Brancelj, A. Microdistribution and high diversity of Copepoda (Crustacea) in a small cave in central Slovenia. *Hydrobiologia* **2002**, *477*, 59–72. [CrossRef]

40. Brancelj, A. Fauna of an unsaturated karstic zone in Central Slovenia: Two new species of Harpacticoida (Crustacea: Copepoda), *Elaphoidella millennii* n. sp. and E. tarmani n. sp., their ecology and morphological adaptations. *Hydrobiologia* **2009**, *621*, 85–104. [CrossRef]

41. Culver, D.C.; Christman, M.C.; Sket, B.; Trontelj, P. Sampling adequacy in an extreme environment: Species richness patterns in Slovenian caves. *Biodivers. Conserv.* **2004**, *13*, 1209–1229. [CrossRef]

42. Gibert, J.; Vervier, P.; Malard, F.; Laurent, R.; Reygrobellet, J.L. Dynamics of communities and ecology of karst ecosystems: Example of three karsts in Eastern and Southern France. In *Groundwater Ecology*; Gibert, J., Danielopol, D.L., Stanford, J.A., Eds.; Academic Press: San Diego, CA, USA, 1994; pp. 425–450.

43. Turquin, M.J. The tactics of dispersal of two species of Niphargus (perennial, Troglobitic amphipoda). In Proceedings of the 8th International Congress of Speleology, Bowling-Green, KY, USA, 18–24 July 1981; Volume 1, pp. 353–355.

44. Fattorini, S.; Lombardo, P.; Fiasca, B.; Di Cioccio, A.; Di Lorenzo, T.; Galassi, D.M.P. Earthquake-related changes in species spatial niche overlaps in spring communities. *Sci. Rep.* **2017**, *7*, 443. [CrossRef] [PubMed]

45. Malard, F.; Reygrobellet, J.L.; Mathieu, J.; Lafont, M. The use of invertebrate communities to describe groundwater flow and contaminat transport in a fractured rock aquifer. *Arch. Hydrobiol.* **1994**, *131*, 93–110.

46. Bruno, M.C.; Maiolini, B.; Carolli, M.; Silveri, L. Impact of hydropeaking on hyporheic invertebrates in an Alpine stream (Trentino, Italy). *Ann. Limnol.* **2009**, *45*, 157–170. [CrossRef]

47. Mori, N.; Simčič, T.; Žibrat, U.; Brancelj, A. The role of river flow dynamics and food availability for the hyporheic microcrustaceans: A reach scale study. *Fund. Appl. Limnol. Arch. Hydrobiol.* **2012**, *180*, 335–349. [CrossRef]

48. Descloux, S.; Datry, T.; Usseglio-Polatera, P. Trait-based structure of invertebrates along a gradient of sediment colmation: Benthos versus hyporheos responses. *Sci. Total Environ.* **2014**, *466*, 265–276. [CrossRef]

49. Korbel, K.L.; Stephenson, S.; Hose, G.C. Sediment size influences habitat selection and use by groundwater macrofauna and meiofauna. *Aquat. Sci.* **2019**, *81*, 39. [CrossRef]

50. Stumpp, C.; Hose, G.C. The impact of water table drawdown and drying on subterranean aquatic fauna in in vitro experiments. *PLoS ONE* **2013**, *8*, e78502. [CrossRef]

51. Stakman, W.P. Determination of pore size by the air bubbling pressure method. In Proceedings of the Wageningen Symposium "Water in the unsaturated zone"; IASH/AIHS: Unesco, Belgium, 1966; pp. 366–372. Available online: http://hydrologie.org/redbooks/a082/RB082.pdf (accessed on 1 November 2019).

52. Piccini, L.; Zanchetta, G.; Drysdale, R.N.; Hellstrom, J.; Isola, I.; Fallick, A.E.; Leone, G.; Doveri, M.; Mussi, M.; Mantelli, F.; et al. The environmental features of the Monte Corchia cave system (Apuan Alps, central Italy) and their effects on speleothem growth. *Int. J. Spel.* **2008**, *37*, 153–172. [CrossRef]

53. Pipan, T.; Culver, D.C. Organic carbon in shallow subterranean habitats. *Acta Carsologica* **2013**, *42*, 291–300. [CrossRef]

54. Pipan, T.; Navodnik, V.; Janzekovic, F.; Novak, T. Studies of the fauna of percolation water of Huda Luknja, a cave in isolated karst in Northeast Slovenie. *Acta Carsologica* **2008**, *37*, 141–151. [CrossRef]

55. Meleg, I.N.; Moldovan, O.T.; Iepure, S.; Fiers, F.; Brad, T. Diversity patterns of fauna in dripping water of caves from Transylvania. *Ann. Limnol. Int. J. Lim.* **2011**, *47*, 185–197. [CrossRef]

56. Wood, P.J.; Gunn, J.; Perkins, J. The impact of pollution on aquatic invertebrates within a subterranean ecosystem—Out of sight out of mind. *Archiv Fur Hydrobiol.* **2002**, *155*, 223–237. [CrossRef]

Article

Food Shortage Amplifies Negative Sublethal Impacts of Low-Level Exposure to the Neonicotinoid Insecticide Imidacloprid on Stream Mayfly Nymphs

Julia G. Hunn *, Samuel J. Macaulay and Christoph D. Matthaei

Department of Zoology, University of Otago, Dunedin 9016, New Zealand; sam.macaulay@otago.ac.nz (S.J.M.); christoph.matthaei@otago.ac.nz (C.D.M.)
* Correspondence: juliagracehunn@gmail.com; Tel.: +64-27-727-7825

Received: 17 September 2019; Accepted: 10 October 2019; Published: 15 October 2019

Abstract: Interactions of pesticides with biotic or anthropogenic stressors affecting stream invertebrates are still poorly understood. In a three-factor laboratory experiment, we investigated effects of the neonicotinoid imidacloprid, food availability, and population density on the New Zealand mayfly *Deleatidium* spp. (Leptophlebiidae). Larval mayflies (10 or 20 individuals) were exposed to environmentally realistic concentrations of imidacloprid (controls, 0.97 and 2.67 µg L^{-1}) for nine days following five days during which individuals were either starved or fed with stream algae. Imidacloprid exposure had severe lethal and sublethal effects on *Deleatidium*, with effects of the lower concentration occurring later in the experiment. The starvation period had delayed interactive effects, with prior starvation amplifying imidacloprid-induced increases in mayfly impairment (inability to swim or right themselves) and immobility (no signs of movement besides twitching appendages). Few studies have investigated interactions with other stressors that may worsen neonicotinoid impacts on non-target freshwater organisms, and experiments manipulating food availability or density-dependent processes are especially rare. Therefore, we encourage longer-term multiple-stressor experiments that build on our study, including mesocosm experiments involving realistic stream food webs.

Keywords: multiple stressors; pesticides; freshwater ecology; ecotoxicology; synergism; resource limitation; population density

---

1. Introduction

Neonicotinoids have become the most commonly used insecticides worldwide [1,2]. Because they persist in soil for many months and are highly water-soluble [3], they enter freshwaters predominantly by leaching into groundwater and surface runoff after rain, with the latter causing high-concentration pulses in surface waters [4–6]. Consequently, they have been frequently found in surface waters draining agricultural areas worldwide [4,7–11]. In North America and Europe, concentrations of imidacloprid—the most commonly detected neonicotinoid—often exceed safety benchmarks for freshwaters such as the United States acute and chronic invertebrate Aquatic Life Benchmarks (0.385 and 0.01 µg L^{-1}, respectively) [12].

Neonicotinoids are neuro-active compounds that are highly specific to invertebrates, particularly insects [13]. This means they have been perceived as relatively safe for vertebrate wildlife, human consumers of treated crops and operators applying these pesticides to crops or seeds. However, this specificity also means neonicotinoids pose a risk to non-target invertebrates. The pesticides bind to the postsynaptic nicotinic acetylcholine receptors (nAChRs), resulting in continuous nervous system stimulation, which is lethal at sufficient concentrations [14,15]. Larvae of aquatic insects are particularly vulnerable to pesticides, with uptake occurring via gills and trachea during respiration, through feeding, and through the epidermis [16]. A number of recent studies have investigated toxic effects of neonicotinoids on freshwater invertebrates [5,16–21]. Among the organisms studied, mayflies (order *Ephemeroptera*) were the most sensitive to neonicotinoids [22,23]. This is a concerning finding, considering mayfly larvae are critical for supporting many aquatic and terrestrial food webs [5,6] and are important biological indicators of stream health [24,25].

Existing pesticide risk assessment programmes rely heavily on toxicity data produced by highly standardised laboratory experiments that can lack ecological realism [26–29]. To account for uncertainties related to the projection of toxicity assessments from benign laboratory conditions towards harsher field conditions and to predict regulatory acceptable concentrations (RACs), an assessment factor of 100 below the acute LC50 (the concentration lethal to 50% of the test organisms) has been established (see e.g., [30]). However, measured insecticide concentrations in surface waters often exceed their respective RACs, as observed, for example, in a meta-analysis of peer-reviewed literature on agricultural insecticide concentrations in EU surface waters [31]. Furthermore, detrimental impacts of pesticides on structure and/or biodiversity of agricultural streams have been observed frequently [32–34]. Indeed, an increasing number of recent studies have concluded that current pesticide regulations do not sufficiently protect the aquatic environment and that insecticides threaten aquatic biodiversity [5,6,31,35,36].

A basic requirement of valid toxicity tests is high survivorship of test organisms in control treatments, which means suboptimal conditions due to 'natural stressors' are usually avoided [18,37]. By contrast, in the real world these environmental variables fluctuate, often drastically, and optimal conditions for animals are rare [27]. Therefore, interactive effects of contaminants and natural stressors on non-target organisms are a possibility that should be more regularly considered in ecological risk assessment [38,39]. Abiotic factors such as water chemistry and temperature are increasingly investigated in order to improve our understanding of their effects on the toxicity of contaminants to aquatic organisms [40,41]. However, biotic factors such as food availability, competition, or predation are rarely included in such studies [18], and interactions between these natural stressors and contaminants such as pesticides remain largely unknown.

Laboratory studies typically provide test organisms with food ad libitum, whereas in realistic conditions nutritional resources are often restricted [42]. During development, animals can face transient periods of food shortage [43], causing them to lower their metabolic rate to increase their chance of surviving until more favourable conditions are reached [44]. Some studies suggest that the energy deficits resulting from resource limitations can increase an organism's sensitivity to a chemical [45,46]. Alternatively, the presence of a toxicant can impair foraging ability, possibly resulting in decreased survivorship [46]. A further environmental challenge for animals in natural systems is population density, which is often linked to food availability. Competitive processes have been shown to alter the outcome of pesticide exposure to invertebrates; for example, it has been suggested that competitive relationships may prolong the recovery of invertebrate populations following pulse exposure to pesticides [47], and that some populations may not recover at all in the presence of a less-sensitive competitor species [48]. Currently, there are no studies investigating the combined effects of food limitation, population density, and neonicotinoid exposure on freshwater invertebrates.

To reduce the abovementioned knowledge gaps, we employed a full factorial design to investigate the individual and combined effects of low-level exposure to the neonicotinoid imidacloprid, food availability, and density of individuals on the larvae of a common endemic New Zealand mayfly (*Deleatidium* spp.). *Deleatidium* are periphyton grazers that feed on the surfaces of rocks, including upper surfaces, and constitute an important food source for fish including introduced brown trout (*Salmo trutta*) [49]. We tested the following hypotheses:

1. Exposure to low concentrations of imidacloprid results in decreased survivorship, increased occurrence of impairment and immobility [6], and reduced moulting frequency in *Deleatidium*. Moulting is controlled by the endocrine and nervous systems; therefore, neurological disruption

due to imidacloprid may interfere with moulting [50], which has been observed in a previous 96-h imidacloprid exposure involving *Deleatidium* nymphs [41].

2. Due to the potentially suppressed metabolic rate of individuals [44], which may increase mayfly sensitivity to pesticides [42,45], a period of starvation prior to pesticide exposure causes amplified negative responses of survivorship, impairment, immobility, and moulting frequency.

3. A higher density of mayfly individuals intensifies the effect of the starvation period, with the resulting increase in competition for resources also amplifying the predicted adverse effects of imidacloprid on all four mayfly responses, similarly to above. Density-dependent increases in competition intensity are common in ecological systems [51].

2. Materials and Methods

2.1. Mayfly Food Supply

Our experiment was run in Austral winter in a temperature-controlled room at the University of Otago in Dunedin, New Zealand. Mayfly larvae were fed stream periphyton grown on unglazed terracotta tiles ($10 \times 10 \times 1.4$ cm). Roughly 300 tiles were placed in Lindsay Creek, a second-order stream, at a site in suburban Dunedin (45.8420 °S, 170.5408 °E). The site was concealed from public view, had *Deleatidium* mayflies present in its benthic invertebrate community, and was located in a small stream with a forested catchment that was unlikely to flood (causing bed disturbance) due to high rainfall. Additionally, because there is little urban influence and no agriculture upstream of the site (the stream drains a nature reserve clad in native forest culminating in Mt. Cargill at 680 m a.s.l.), it was unlikely that imidacloprid or any other pesticide was present in the stream. Tiles were exposed on 1 July 2016, two weeks prior to the start of the experiment, and collected on the morning of each day they were required.

2.2. Mayfly Collection

Roughly 750 *Deleatidium* larvae were collected on 15 July 2016 by electric fishing from a reach of Silver Stream (45.8096 °S, 170.4211 °E), a fourth-order stream near Dunedin. Electric fishing strongly increases invertebrate drift rates and is a fast, efficient method for collecting stream invertebrates. This technique has been used successfully in New Zealand and North America for ecological experiments where large numbers of live, unharmed invertebrate specimens were needed [49,52].

The stream reach is located in a native forest catchment without agriculture, thus the test organisms had no prior exposure to imidacloprid. A section of this reach measuring 75 m was repeatedly sampled using a pulsed DC backpack Electric Fishing Machine (Kainga EFM300; NIWA, Christchurch, New Zealand). *Deleatidium* larvae included a range of instars (sizes ranging 5–15 mm in length excluding tail filaments); late instar larvae with visible black wing pads were excluded due to an increased chance of emerging during the experiment.

Mayflies were returned to a climate-controlled room with a 16:8-h light:dark photoperiod and left overnight in aerated 4-L containers filled with stream water, to allow water temperature to adjust to the ambient temperature. Over the next 48 h, room temperature was gradually raised from 9 to 15 °C and then kept at 15 ± 1 °C for the entire experimental period (the same temperature as in a previous study on stream mayflies exposed to imidacloprid [53]). All mayflies were provided with periphyton tiles for food during this period to ensure they were not subjected to any food limitation prior to the experimental starvation period (see below). After the first 24 h, mayflies were transferred to aerated 4-L containers containing ASTM (American Society for Testing and Materials) artificial soft water (ASW) at climate-room temperature and then left to acclimate for a further 24 h. ASW consists of reconstituted deionized water containing known quantities of dissolved solutes (0.57 mM $NaHCO_3$, 0.17 mM $CaSO_4 \bullet 2H_2O$, 0.25 mM $MgSO_4$, and 0.03 mM KCl).

2.3. Experimental Design

Our experiment was conducted over 14 days in July 2016 (following the 2-day acclimation period), comprising a 5-day starvation period and 9 days of imidacloprid exposure. At the end of the acclimation period, 36 tiles with periphyton (see above) were collected from Lindsay Creek. Eighteen of these were scrubbed to remove all accrued algae so that half the experimental units were subjected to a starvation period while being provided with the same substratum as the non-starved group. Thirty-six 1.16-L glass containers measuring $19.9 \times 14.4 \times 6.3$ cm (Keep'N'Box 116 cl; Luminarc, Arques, France) were filled with 500 mL ASW. One tile was added per container (tiles with and without algae were randomly allocated), and 20 mayflies were placed in each. Containers were placed randomly in an aeration tube setup using four air pumps (Aqua One Precision 7500; Aqua One, Ingleburn, NSW, Australia) to aerate all containers, and left for five days. This duration was chosen for the starvation period because in a previous related experiment, *Deleatidium* showed high survivorship (90%–100%) over 96 h plus an acclimation period, all with no feeding [41]. During previous pilot studies, we had also observed that the mayfly larvae showed clear signs of hunger after 48 h of no feeding: almost all individuals started feeding as soon as a tile with periphyton was placed into the container for the first time [54], a behaviour which was not observed when tiles were subsequently replaced. For these reasons, we were confident that most mayflies would survive the starvation period of five days while still experiencing some stress due to hunger.

In order to determine how the other two stressors—starvation and mayfly density—might interact with different levels of imidacloprid toxicity, our experiment required two distinctly different imidacloprid concentrations. Accordingly, we used one control concentration ($0\ \mu g\ L^{-1}$), one 'lower' concentration expected to cause some sublethal adverse effects (in this case impairment; see below for a definition of this sublethal endpoint), and one 'higher' concentration expected to cause some mortality. These concentrations, 0.9 and $2.1\ \mu g\ L^{-1}$ respectively, were determined from the results of two separate 28-day concentration-response experiments examining imidacloprid toxicity on *Deleatidium* larvae [55,56].

Following the starvation period, the three imidacloprid treatments were randomly allocated to the 36 experimental units, with 500 mL of solution in each container. Solutions were made using ASW and analytical standard grade imidacloprid (99.9% purity Imidacloprid PESTANAL®; Sigma-Aldrich, Castle Hill, NSW, Australia). A working imidacloprid stock solution ($10\ \mu g\ L^{-1}$) was used to produce the treatment solutions. Control containers received 500 mL each of pesticide-free ASW. All tiles were replaced with a fresh tile (with periphyton food) from the stream, and all units were provided with periphyton food for the remainder of the experiment (the 9-day imidacloprid exposure period, see Section 2.4). To test the potential effects of mayfly density, containers were then manipulated so that half the units held 10 mayfly individuals and half held 20, by removing excess individuals from the '10' treatments and replacing any that had died in the '20' treatments while ensuring they were from the correct starvation treatment.

2.4. Determination of Mayfly Responses

Every three days during the imidacloprid exposure period (Days 3, 6, and 9), mayflies were observed and the numbers of dead, alive, impaired, and immobile individuals (out of those alive) and the mean number of moults per individual in each container were determined. In this experiment, we defined 'impairment' and 'immobility' as purely sublethal responses, to avoid confounding sublethal and lethal responses by combining the former with mayfly survivorship, and therefore only considered individuals that were still alive for these two sublethal responses. Impairment was defined as the inability of an individual to swim or to right itself. All individuals were gently lifted into the water column to test swimming ability—the natural response for a healthy *Deleatidium* larva is to swim toward a solid surface to cling to. If unable to swim, the individual was tested for righting ability by gently placing it ventral side up. An individual that was impaired could also be immobile, a more severe sublethal response. In order to test for immobility, an impaired individual was placed onto a piece of stainless-steel mesh to test its ability to walk. If no sign of movement ability was evident besides some twitching of appendages, the individual was recorded as immobile.

Dead individuals and shed exuviae were removed from the containers daily. Periphyton tiles and solutions were replaced once during the pesticide exposure period, on Day 6, with tiles collected from the stream on the same morning. Also on Day 6, prior to replacing the tiles and solution, one 1000-µL water sample was taken from each treatment combination by sampling one of the three replicate containers at random (n = 12 in total) and freezing at −21 °C for later determination of the actual imidacloprid concentrations using the ELISA method (see Section 2.5). Three of the four water samples per imidacloprid addition treatment (0.9 and 2.1 µg L^{-1}; six in total) were randomly selected for ELISA testing.

2.5. Validation of Imidacloprid Concentrations

Concentrations of imidacloprid in the experimental units were verified using an enzyme-linked immunosorbent assay (ELISA) kit. This kit consists of a microplate with wells containing an antibody with a known binding efficiency for imidacloprid. Concentrations are determined by analysing the amount of imidacloprid that binds to these antibodies. A set of 'standards', with known imidacloprid concentrations, are analysed in the same way as the samples being tested. The kit (Imidacloprid ELISA Kit, 96 Test Microtiter Plate (Product No. 500800); ABRAXIS LLC, Warminster, PA, USA) had a detection range of 0.06 µg L^{-1} to 1.2 µg L^{-1}. According to the kit's instructions, concentrations around 0.4 µg L^{-1} (the middle of the calibration range) yield the most accurate results; therefore, samples were diluted to 0.4 µg L^{-1} prior to testing. Each sample was tested in either duplicates or triplicates. Immediately following the ELISA procedure, the microplate was analysed using a microplate photometer (FLUOstar Omega Microplate reader; BMG LABTECH, Allmendgruen, Ortenberg, Germany).

2.6. Data Analysis

Effects of the experimental manipulations on survivorship, impairment, immobility, and moulting in *Deleatidium* larvae were assessed using separate three-way ANOVAs in SPSS version 24 (IBM SPSS Statistics; IBM Corp., Armonk, NY, USA). The ANOVA model included three categorical fixed factors, 'imidacloprid concentration' (0, 0.9, and 2.1 µg L^{-1}), 'starved' (yes or no), and 'density' (10 or 20 individuals), plus all factor interactions. The resulting model was intercept (degrees of freedom 1) + imidacloprid concentration (d.f. 2) + starved (1) + density (1) + starved × concentration (2) + density × concentration (2) + starved × density (1) + starved × density × concentration (2) + error (24) (n = 36). For the survivorship response, total sample size (n) was 36 on all three sampling dates. For the sublethal responses impairment, immobility, and moulting, n differed depending on the experimental units remaining on a given date (n decreased with time as mayflies died; see Section 3.2). Because of this decrease, each sampling date was analysed individually for all mayfly response variables. ANOVAs were performed for all sampling dates where the study design was considered balanced enough. The significance level α for all tests was $p = 0.05$, and all response patterns summarised in the text were significant. For all significant main effects of imidacloprid concentration, pairwise comparisons were conducted with Tukey's HSD post-hoc tests. Standardised effect sizes (partial η^2 values, range 0–1; [57]) are presented for all p-values < 0.1 to allow our readers to evaluate the likely biological relevance of the results [58]. Effect sizes can be classified as: <0.10 very small, ≥0.10 small, ≥0.30 medium, and ≥0.50 large [59]. Because factor main effects can be difficult to interpret when significant interactions are present, for any results where an interaction had a larger effect size than the corresponding significant main effect(s), the interaction alone was interpreted [60].

3. Results

3.1. Achieved Imidacloprid Concentrations

Actual concentrations were (mean ± SE) 108 ± 5% (0.97 ± 0.05 µg L^{-1}) and 127 ± 10% (2.67 ± 0.2 µg L^{-1}) of the target concentrations of 0.9 and 2.1 µg L^{-1}, respectively. All figures, tables, and discussion are based on these achieved concentrations.

3.2. Sample Size

For mayfly survivorship, the full sample size of 36 experimental units remained across all three sampling dates, while sample sizes for the sublethal responses (impairment, immobility, and moulting frequency) decreased with time in some treatment combinations (Table 1). Because of this decrease, the experimental design became increasingly unbalanced from Day 6 onwards. Therefore, statistical results for the latter three responses are presented only for sampling dates where outputs were deemed robust enough to interpret. For illustrative purposes, response patterns for all three dates are shown for all four variables (Figures 1 and 2). However, for immobility and moulting frequency, only the statistically robust results of the ANOVAs for Days 3 and 6 are presented in tabular form (the data on Day 9 for these two responses contained many zero values and some missing values—see Figure 2).

Table 1. Numbers of replicate experimental units (with at least one alive animal) remaining for each treatment combination across sampling dates. Note that the response 'moulting frequency' was calculated using the sample size of the previous sampling date (three days earlier); therefore, sample sizes for this response variable are equal to that of the previous date.

Conc. (µg L^{-1})	Starved Y/N	Density	Day 3	Day 6	Day 9
0.0	N	10	3	3	3
0.0	N	20	3	3	3
0.0	Y	10	3	3	3
0.0	Y	20	3	3	3
0.97	N	10	3	3	2
0.97	N	20	3	3	3
0.97	Y	10	3	3	2
0.97	Y	20	3	3	3
2.67	N	10	3	2	0
2.67	N	20	3	3	1
2.67	Y	10	3	3	1
2.67	Y	20	3	3	2
Total	-	-	36	35	26

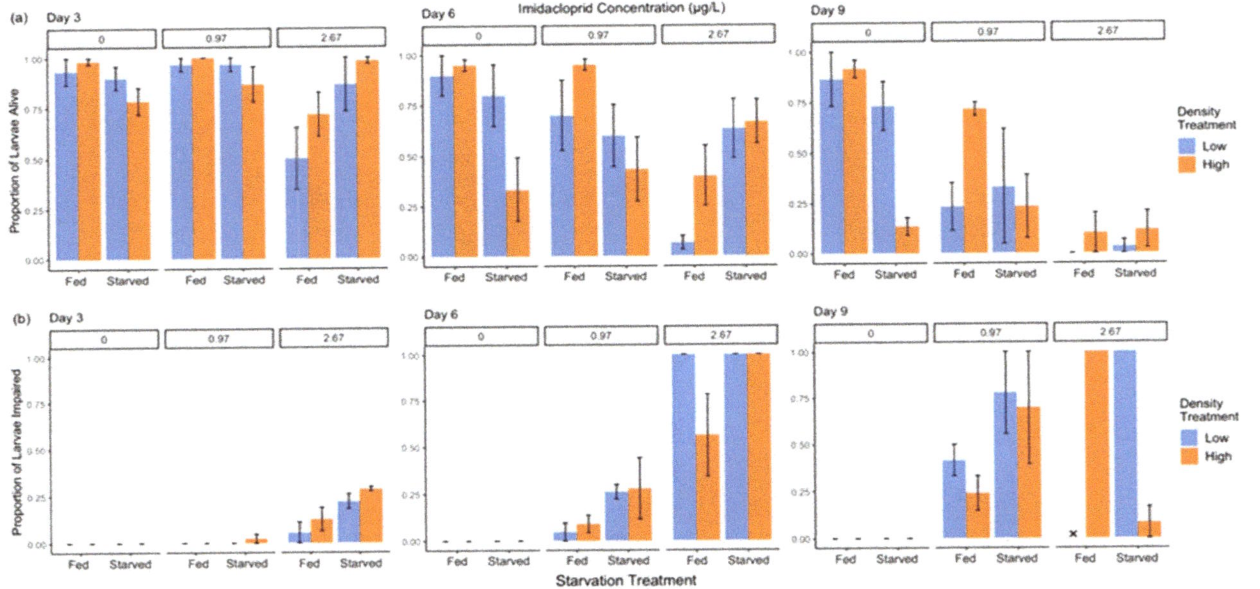

Figure 1. Responses of (**a**) survivorship and (**b**) impairment to the three imidacloprid treatments at the two mayfly nymph densities (10 and 20) and in the two starvation treatments (starved = starved for five days immediately prior to the imidacloprid exposure period). Bars represent means ± SEs. In cases where bars are missing, 'X' represents a treatment group in which all mayflies had died in all replicates—i.e., no sublethal responses were measured—whereas '_' represents a zero value. See Tables 2 and 3 for statistical results.

3.3. Mayfly Responses

3.3.1. Survivorship

Deleatidium survivorship generally decreased in one or both imidacloprid addition treatments on all days of the imidacloprid exposure period (Figure 1a, Table 2). On Days 3 and 6, the main effect of imidacloprid on survivorship was overridden by a stronger starvation × imidacloprid interaction (see below). Effects of imidacloprid exposure increased with time. The largest effect of all treatment combinations on mayfly survival was the imidacloprid main effect on Day 9 (partial $\eta^2 = 0.67$), when survivorship decreased in a concentration-dependent manner.

Survivorship was also influenced by a starvation × imidacloprid interaction on all three dates (Table 2). The general pattern across all dates was that in treatments without imidacloprid addition, survivorship was lower in starvation treatments (Figure 1a). In the lower imidacloprid treatment, this pattern was also evident, whereas in the higher treatment the pattern was reversed (except for Day 9 when survivorship was generally low in the higher treatment).

Mayfly density alone had no effect on survivorship; however, on Day 9 there was a weak interaction between density and imidacloprid (Table 2). In treatments without imidacloprid addition, mean survivorship (across both starvation treatments) was lower in mayfly groups with 20 individuals than in those with 10 individuals, whereas this pattern was reversed in the two imidacloprid addition treatments (Figure 1a). Finally, on Days 6 and 9, there was an overall (in the control and lower imidacloprid treatments) negative effect of starvation on mayfly groups with 20 individuals, whereas starvation had little to no effect on groups with 10 individuals (starvation × density interaction).

Table 2. Results of three-way ANOVAs comparing the 'survivorship' response between experimental treatments. p-values < 0.05 are in bold text, and effect sizes (ES; partial η^2 values; range 0–1) are given in brackets where $p < 0.10$. Two interactions that overrode main effects of imidacloprid (due to a larger effect size) are underlined. Effect size categories are: 'trivial' < 0.1, 'small' > 0.1, 'medium' > 0.3, 'large' > 0.5 [59]. Treatment level rankings for significant main effects of imidacloprid are based on significant differences determined in pairwise comparisons using post-hoc tests. Imidacloprid concentrations in these rankings are in μg L^{-1}.

Sampling Day	Imidacloprid Concentration (IMI) p (ES) Ranking	Starvation p (ES)	Density p	Starved × IMI p (ES)	Density × IMI p (ES)	Starved × Density p (ES)	Starved × Density × IMI p
Day 3	**0.01** (0.32) 0.97 > 2.67	0.34	0.48	<u>**0.001** (0.43)</u>	0.14	0.16	0.96
Day 6	**0.007** (0.34) (0 = 0.97) > 2.67	0.27	0.94	<u>**<0.001** (0.49)</u>	0.11	**0.01** (0.25)	0.84
Day 9	**<0.001** (0.67) 0 > 0.097 > 2.67	**0.01** (0.27)	0.97	**0.03** (0.25)	**0.03** (0.25)	**0.01** (0.27)	0.15

3.3.2. Impairment

The proportion of impaired mayfly nymphs (out of those alive) in treatments without added imidacloprid was zero on all three sampling dates, and imidacloprid addition strongly increased impairment on all dates (Figure 1b, Table 3). On Day 3, imidacloprid only increased *Deleatidium* impairment at the highest concentration (2.67 μg L^{-1}), whereas on Day 6, impairment increased in a concentration-dependent manner. By Day 9 (when few surviving mayflies remained, see Figure 1a), impairment was similar at 0.97 μg L^{-1} and at 2.67 μg L^{-1}.

Starvation weakly increased impairment on Day 6, and on Day 3 there was a moderate main effect of starvation but a stronger starvation × concentration interaction overrode it (Table 3). On Day 3, starved mayflies showed greater impairment than non-starved mayflies, but only in the higher imidacloprid treatment (Figure 1b). A strong starvation × concentration interaction also occurred on Day 9. This effect was similar to that observed on Day 3, except that on Day 9 the increase in impairment with starvation occurred only in the lower imidacloprid treatment (note that the number of experimental units in the higher imidacloprid treatment had decreased considerably by Day 9 due to many mayfly deaths, see Table 1).

Also on Day 9, mayfly density had a moderate main effect on impairment, but a slightly stronger interaction between density and concentration overrode this main effect (Table 3). Impairment was somewhat higher in containers with 10 individuals than in those with 20, but only in the lower imidacloprid treatment (Figure 1b). In the higher imidacloprid treatment, the much smaller, unequal sample sizes (with one treatment combination missing completely) did not allow reliable interpretation of this interaction.

Table 3. Results of three-way ANOVAs comparing the 'impairment' response between experimental treatments. p-values < 0.05 are in bold text, and effect sizes (ES; partial η^2 values; range 0–1) are given in brackets where $p < 0.10$. Two interactions that overrode main effects of starvation or density (due to a larger effect size) are underlined. See Table 2 for further details.

Sampling Day	Imidacloprid Concentration p (ES) Ranking	Starvation p (ES)	Density p (ES)	Starved × IMI p (ES)	Density × IMI p (ES)	Starved × Density p	Starved × Density × IMI p
Day 3	**<0.001** (0.80) (0 = 0.97) < 2.67	**0.001** (0.36)	0.113	**<0.001** (0.47)	0.21	0.84	0.93
Day 6	**<0.001** (0.91) 0 < 0.97 < 2.67	**0.01** (0.26)	0.22	0.16	0.12	0.18	0.13
Day 9	**<0.001** (0.76) 0 < (0.97 = 2.67)	0.40	**0.01** (0.35)	**0.002** (0.56)	**0.03** (0.38)	0.80	0.80

3.3.3. Immobility and Moulting

For these two sublethal responses, only the results of the ANOVAs for Days 3 and 6 were statistically robust, due to the reduced, strongly unequal sample sizes on Day 9 (see Section 3.2). As for mayfly impairment, the proportion of immobile nymphs (out of those alive) in treatments without added imidacloprid was zero on all sampling dates (Figure 2a). On Day 3, immobility was either zero or very low and unaffected by the experimental manipulations (Table 4). On Day 6, immobility increased strongly at the higher imidacloprid concentration (Table 4, Figure 2a).

A moderate effect of starvation, also on Day 6, was overridden by a slightly larger starvation × concentration interaction (Table 4). In this interaction, the increase in immobility in treatments with added imidacloprid compared to controls was amplified by starvation, especially at the higher imidacloprid concentration.

In contrast to the other two sublethal responses, moulting frequency of mayfly nymphs (always calculated using the sample size of the collection date three days earlier, see Table 1) was unaffected by all manipulated factors and their interactions (Table 4, Figure 2b).

Table 4. Results of three-way ANOVAs comparing responses between experimental treatments for 'immobility' (grey) and 'moulting frequency' (white). p-values < 0.05 are in bold text, and effect sizes (ES; partial η^2 values; range 0–1) are given in brackets where $p < 0.10$. One interaction that overrode a main effect of starvation (due to a larger effect size) is underlined. See Table 2 for further details.

Sampling Day	Imidacloprid Concentration p (ES) Ranking	Starvation p (ES)	Density p	Starved × IMI p (ES)	Density × IMI p	Starved × Density p	Starved × Density × IMI p
Day 3	0.16	0.87	0.17	0.97	0.16	0.87	0.97
	0.60	0.61	0.50	0.76	0.47	0.87	0.38
Day 6	**<0.001** (0.54) (0 = 0.97) < 2.67	**0.004** (0.31)	0.81	**0.01** (0.34)	0.32	0.16	0.58
	0.56	0.54	0.78	0.93	0.78	0.31	0.29

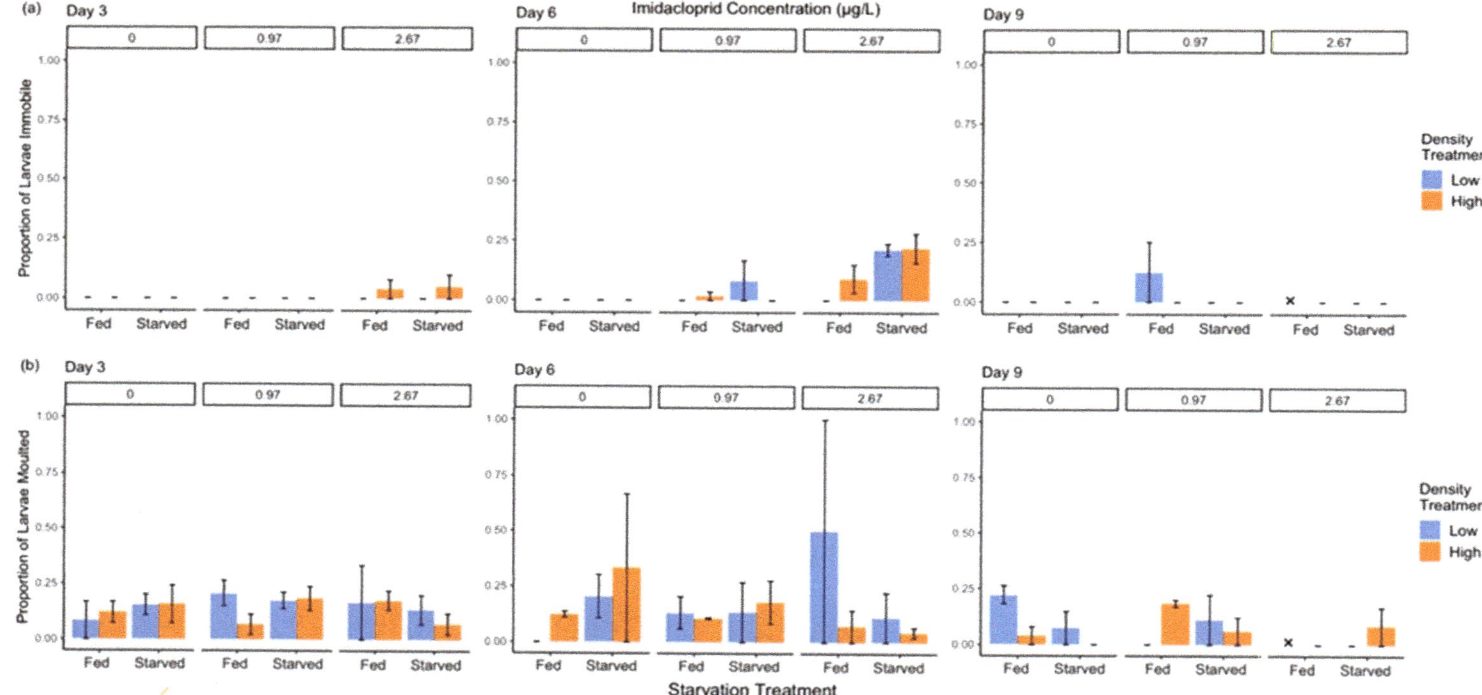

Figure 2. Responses of (**a**) immobility and (**b**) moulting frequency to the three imidacloprid treatments at the two mayfly nymph densities (10 and 20) and in the two starvation treatments (Starved = starved for five days immediately prior to the imidacloprid exposure period). Bars represent means ± SEs. In cases where bars are missing, '×' represents a treatment group in which all mayflies had died in all replicates—i.e., no sublethal responses were measured—whereas '_' represents a zero value. See Table 4 for statistical results.

4. Discussion

4.1. Imidacloprid Effects on Mayflies

Our experiment investigated the individual and interactive effects of exposure to field-realistic concentrations of imidacloprid, a starvation period, and population density on nymphs of the mayfly *Deleatidium*. Overall, imidacloprid exposure strongly reduced mayfly survivorship and strongly increased impairment, supporting our first hypothesis. Imidacloprid also resulted in strongly increased levels of mayfly immobility, again as predicted, although only on one sampling date (Day 6), most likely because all individuals immobilized on Day 6 had died by Day 9. Our higher imidacloprid concentration (2.67 µg L^{-1}) had relatively immediate effects on both survivorship and impairment, whereas the lower concentration (0.97 µg L^{-1}) had little effect immediately but 'caught up' with the higher concentration by the ninth day of exposure. The delayed effects on mayfly responses at our lower concentration are not unexpected because the toxicity of neonicotinoids is time-dependent [6,61] and the blockage of nAChRs in the invertebrate central nervous system may be largely irreversible [62,63] (see Section 4.4 for further discussion of this point).

4.2. Starvation Effects and Interactions with Imidacloprid

The importance of strengthening the integration of ecological principles into the field of ecotoxicology is increasingly being recognized [27,64–66]. Our study was designed to make a step towards this goal by determining the effects of exposure to imidacloprid in combination with two biotic 'natural stressors' commonly affecting mayfly larvae in streams. Robust main effects of the prior 5-day starvation period on Day 6 (increased impairment) and Day 9 (reduced survivorship) of the pesticide exposure period indicated that starvation had delayed effects on both mayfly responses. In another multiple-stressor experiment involving insecticide exposure with a prior starvation period, Dinh, Janssens, and Stoks [67] found that starvation had both immediate and delayed negative sublethal effects on the growth rate and physiology of larvae of the European damselfly *Coenagrion puella*. A six-day starvation period (prior to seven days of exposure to the neuro-active insecticide chlorpyrifos) reduced growth rate and various proxies of insect immune system activity, indicating some suppression of metabolic rate and immune function, both during the starvation period and continuing into the pesticide exposure period. This finding parallels our own results. Nevertheless, starvation did not interact with the pesticide to reduce *C. puella* survivorship, which was only significantly reduced when individuals were exposed to chlorpyrifos and a combination of prior starvation and a prior heat wave (three-way interaction) [67]. In our study, interactions between starvation and imidacloprid concentration occurred on several sampling dates for survivorship, impairment and immobility. However, in the case of survivorship, this was not usually because the prior starvation period amplified the adverse effects of the subsequent imidacloprid exposure, as we had predicted in our second hypothesis. Instead, such an amplified negative effect of starvation occurred only at the lower imidacloprid concentration, whereas starvation appeared to have a positive effect on survivorship at the higher imidacloprid concentration. By contrast, the sublethal mayfly responses largely supported our hypothesis. Both mayfly impairment (on Day 3) and immobility (on Day 6) were amplified in starvation treatments in the earlier stages of the imidacloprid exposure period, and these effects were ultimately drowned out by the more severe lethal impacts of imidacloprid on Day 9.

Experiments investigating the potential interactions between nutritional conditions of non-target aquatic organisms and pesticides are scarce [39]. However, a 2013 study on the damselfly *Enallagma cyathigerum* by Janssens and Stoks [68] revealed that nutritional conditions can influence the outcome of exposure to chlorpyrifos. The damselfly larvae responded to the pesticide by accelerating their development—possibly an adaptive response to shorten larval lifespan—but could do so only under optimal nutritional conditions. Immune function in invertebrates is known to be variable depending not only on food availability but also a range of other fluctuating abiotic factors such as salinity and temperature [69,70], and it is likely that there are physiological trade-offs involved which may alter the outcome of exposure to contaminants such as neonicotinoids. Consequently, while support for our hypothesis regarding adverse synergistic interactions between starvation and imidacloprid exposure was evident only for sublethal responses, our findings along with evidence from previous studies suggest that further research on this topic is warranted.

4.3. Mayfly Density Effects and Interactions with Starvation or Imidacloprid

Mayfly population density (10 or 20 individuals per container) had no robust factor main effects on any of the studied mayfly responses, but significant interactions of density with starvation occurred for survivorship on Days 6 and 9. On these days, survivorship in the lower imidacloprid treatment and especially in controls was overall lower in starved treatments. However, this effect was more pronounced in starved containers with 20 individuals, particularly in controls on Day 9 (where survivorship was only about 10%, compared to about 70% in starved containers with 10 individuals). This response pattern suggests that mayfly larvae found it harder to recover from the starvation period at the higher population density, perhaps due to increased competition for the periphyton resources provided on the pre-colonised tiles. Interestingly, this potential consequence of intra-specific competition did not become evident until six days after the end of the five-day starvation period. Furthermore, there were weak to moderate interactions between mayfly density and imidacloprid concentration on Day 9 for survivorship and impairment. However, the generally low survivorship at the higher imidacloprid concentration and the small, unequal sample sizes for impairment on this day prevented biologically meaningful interpretations of these two interactions.

Currently, no other studies have investigated the possible combined effects of population density and neonicotinoids on non-target freshwater invertebrates. The few existing studies investigating density-dependent or competitive processes and their interactions with pollutants [47,48,71,72] involved interspecific interactions or looked at very different responses such as individuals' longevity. Consequently, there is little to compare these findings of our study to. In fact, population density of test organisms was not mentioned in two recent reviews of pollutant interactions with natural variables or stressors [38,39].

In our experiment, the higher mayfly population density led to increased mayfly mortality when combined with a preceding starvation period (as discussed above), whereas population density had no effects on the studied sublethal mayfly responses. It is worth noting that density of *Deleatidium* larvae in the bed substrata of New Zealand streams can exceed 6000 individuals per m^2 [73]; this is more than 8.5 times the density in our 20-nymph treatment (which is equivalent to 698 individuals per m^2). Thus, intra-specific competition for periphyton food resources among *Deleatidium* individuals during periods of food shortage could be considerable in real streams.

4.4. Which Are the Most Informative Sublethal Mayfly Responses?

While quite a few studies have now assessed the toxicity of imidacloprid to mayfly larvae [22,23, 41,53,74–79], the most reliable indicators of sublethal toxic effects on these particular organisms have not been well established. In our study, 'impairment' and 'immobility' were treated as distinct, purely sublethal responses, with neither response including mortality and impairment occurring at lower concentrations than immobility. A recent study on the North American mayfly *Isonychia bicolor* also documented several sublethal responses to imidacloprid that occurred prior to total immobilization, including muscle spasms and lethargy [74]. While the blockage of nAChRs in the invertebrate central nervous system by neonicotinoids is widely regarded as irreversible [63,64], there is some recent evidence of the potential for reversible binding, with recovery of two freshwater invertebrates (including the North American mayfly *Neocloeon triangulifer*) from an immobilized state observed following one short-term pulse of imidacloprid [76]. For mayflies at least, it therefore seems that 'impairment' as defined in our study (without including mortality) is a crucial sublethal endpoint to consider, as such impairments most likely alter an individual's ability to feed, evade predators, and therefore survive in real stream communities. Thus, studies that do not address sublethal endpoints, such as impairment, may underestimate the true toxicity of imidacloprid [74].

None of our three experimental factors or their interactions affected *Deleatidium* moulting frequency. This was generally very low, and statistical analysis was possible only on Days 3 and 6 of the pesticide exposure period. These results contrast with those from the 96-h experiment by Macaulay et al. [41], the only other existing study involving *Deleatidium* nymphs and imidacloprid. In that study, *Deleatidium* moulting frequency was generally much higher than in the present study, and imidacloprid exposure decreased moulting frequency. The lack of consistency across these studies, in spite of both being run during Austral winter, suggests that moulting frequency may not be a reliable sublethal response for *Deleatidium* nymphs, although it could be for other mayfly taxa [80,81].

4.5. Limitations of Our Study

We acknowledge that, compared to the relatively few existing mesocosm experiments in which entire stream communities were exposed to pesticides combined with other stressors, e.g., [66,82,83], our multiple-stressor laboratory study has low environmental realism, which limits the extrapolation of our findings to real streams. Nevertheless, similar to the laboratory experiment on damselfly larvae by Dinh et al. [67] in which prior exposure to a simulated heat wave under food limitation made the insecticide chlorpyrifos more lethal, our experiment provides a 'proof-of-principle' that synergistic interactions between a neonicotinoid and food limitation can worsen sublethal impacts on stream insects. Our findings warrant further tests of this 'proof-of-principle' in more field-realistic experiments.

Second, our selected two imidacloprid concentrations were successful in producing different lethal and sublethal mayfly responses, but the higher achieved concentration (2.67 µg L^{-1}) caused rather severe effects on some of the responses from Day 6 of the pesticide exposure period onwards. This potentially made detection of interactions between the three experimental factors more difficult because the generally strong effect of imidacloprid exposure may have hidden weaker effects of the other two factors.

Finally, the concentration ranges of neonicotinoids in New Zealand's freshwaters are still poorly known. Neonicotinoids are widely applied in agriculture and horticulture [84] and have recently been detected in agricultural streams [85] and in groundwater [86], and several other pesticides have been found in sheep/beef farming streams [87,88]. However, because the first stream survey in New Zealand for neonicotinoids was conducted during a long drought and the limited number of sites sampled may not have included any sites with high pesticide use in their catchments, the concentrations measured are probably serious underestimates of the concentrations that can occur [85]. Consequently, our experimental imidacloprid concentrations can be mainly compared to the commonly detected concentrations outside New Zealand. A 2016 meta-analysis of neonicotinoid residues in freshwaters across 11 countries by Sánchez-Bayo et al. [6] found an average detected imidacloprid concentration of 0.73 µg L^{-1}, with maximum concentrations reaching up to 320 µg L^{-1}. Of the 27 surveys testing for imidacloprid in this meta-analysis, the average concentrations exceeded our low concentration of 0.97 µg L^{-1} in six studies, and the highest concentrations detected exceeded both our treatment concentrations in 11 studies. In another survey of streams around Sydney, Australia, imidacloprid concentrations were mostly between 0.05–1.0 µg L^{-1} but reached up to 4.6 µg L^{-1} [4]. Based on these data, the imidacloprid concentrations used in our study (0.97 and 2.67 µg L^{-1}) seem likely to occur in New Zealand streams draining agricultural or urban landscapes to which neonicotinoids are applied. However, future surveys of New Zealand streams during periods of variable flows, ideally focusing on catchments where pesticide applications on land are known to be high, are required to confirm this.

5. Conclusions

Improving the ecological realism of ecotoxicological studies is crucial for providing the scientific knowledge needed to regulate pesticide use more safely in the future. Our study starts to address this challenge by combining low, field-realistic concentrations of imidacloprid with two natural stressors.

Our findings suggest that periods of food shortage may worsen the impact of exposure to imidacloprid on stream mayfly populations and, ultimately, the stream and terrestrial food webs they are vital to [6]. Thus, we demonstrate the importance of integrating ecological concepts (i.e., the effects of natural stressors such as food shortages) into studies of contaminant effects on freshwater ecosystems.

Worldwide, there are an increasing number of studies assessing the toxicity of neonicotinoids to non-target organisms, but very few that explore potential interactions with natural or anthropogenic stressors that may worsen the impacts of neonicotinoids on these organisms in real ecosystems. Many more such multiple-stressor experiments are required, including experiments involving realistic stream food webs in open-system mesocosms that permit natural immigration and emigration of stream biota, in order to assess the broader effects of neonicotinoids on community structure and ecosystem processes.

It is also necessary to perform more ecological multiple-stressor field studies aimed at evaluating to which extent the current pesticide regulation may in fact be protective, for example by surveying benthic invertebrate community compositions in multiple streams spanning wide gradients of pesticide concentrations combined with other agricultural stressors. Such studies would progress real integration between the fields of freshwater ecology and ecotoxicology, a long-term goal we believe to be of utmost importance if the future biological integrity of our freshwater ecosystems is to be maintained or improved.

Author Contributions: Conceptualization, J.G.H. and C.D.M.; Methodology, J.G.H., S.J.M., and C.D.M; Formal analysis, J.G.H. and C.D.M.; Investigation, J.G.H. and S.J.M.; Writing—original draft preparation, J.G.H.; Writing—review and editing, J.G.H., S.J.M., and C.D.M.; Visualization, J.G.H. and S.J.M.

Funding: This research received no external funding.

Acknowledgments: We thank Kim Garrett, Matt Jarvis, Matt Ward, and Yvonne Kahlert for their assistance with field trips and laboratory work, and Nicky McHugh for helping with the logistics of the laboratory experiments.

Conflicts of Interest: The authors declare no conflict of interest.

References

1. Jeschke, P.; Nauen, R.; Schindler, M.; Elbert, A. Overview of the status and global strategy for neonicotinoids. *J. Agric. Food Chem.* **2010**, *59*, 2897–2908. [CrossRef] [PubMed]
2. Goulson, D. An overview of the environmental risks posed by neonicotinoid insecticides. *J. Appl. Ecol.* **2013**, *50*, 977–987. [CrossRef]
3. Lewis, K.A.; Tzilivakis, J.; Warner, D.; Green, A. An international database for pesticide risk assessments and management. *Hum. Ecol. Risk Assess.* **2016**, *22*, 1050–1064. [CrossRef]
4. Sánchez-Bayo, F.; Hyne, R.V. Detection and analysis of neonicotinoids in river waters—Development of a passive sampler for three commonly used insecticides. *Chemosphere* **2014**, 143–151. [CrossRef]
5. Morrissey, C.A.; Mineau, P.; Devries, J.H.; Sánchez-Bayo, F.; Liess, M.; Cavallaro, M.C.; Liber, K. Neonicotinoid contamination of global surface waters and associated risk to aquatic invertebrates: A review. *Environ. Int.* **2015**, *74*, 291–303. [CrossRef]
6. Sánchez-Bayo, F.; Goka, K.; Hayasaka, D. Contamination of the aquatic environment with neonicotinoids and its implication for ecosystems. *Front. Environ. Sci.* **2016**, *4*, 1–14. [CrossRef]
7. Van Dijk, T.C.; Van Staalduinen, M.A.; Van der Sluijs, J.P. Macro-invertebrate decline in surface water polluted with imidacloprid. *PLoS ONE* **2013**, *8*, e62374. [CrossRef]
8. Kreuger, J.; Graaf, S.; Patring, J.; Adielsson, S. Pesticides in Surface Water in Areas with Open Ground and Greenhouse Horticultural Crops In Sweden 2008. Technical Report. Swedish University of Agricultural Sciences. Available online: https://pub.epsilon.slu.se/5413/ (accessed on 1 May 2019).
9. Starner, K.; Goh, K.S. Detections of the neonicotinoid insecticide imidacloprid in surface waters of three agricultural regions of California, USA, 2010–2011. *Bull. Environ. Contam. Toxicol.* **2012**, *88*, 316–321. [CrossRef]
10. Hladik, M.L.; Kolpin, D.W. First national-scale reconnaissance of neonicotinoid insecticides in streams across the USA. *Environ. Chem.* **2015**, *13*, 12–20. [CrossRef]

11. Stehle, S.; Bub, S.; Schulz, R. Compilation and analysis of global surface water concentrations for individual insecticide compounds. *Sci. Total Environ.* **2018**, *639*, 516–525. [CrossRef]

12. Aquatic Life Benchmarks and Ecological Risk Assessments for Registered Pesticides. United States Environmental Protection Agency. Available online: https://www.epa.gov/pesticide-science-and-assessing-pesticide-risks/aquatic-life-benchmarks-and-ecological-risk (accessed on 1 September 2019).

13. Simon-Delso, N.; Amaral-Rogers, V.; Belzunces, L.P.; Bonmatin, J.M.; Chagnon, M.; Downs, C.; Furlan, L.; Gibbons, D.W.; Giorio, C.; Girolami, V.; et al. Systemic insecticides (neonicotinoids and fipronil): Trends, uses, mode of action and metabolites. *Environ. Sci. Pollut. Res.* **2015**, *22*, 5–34. [CrossRef] [PubMed]

14. Tomizawa, M.; Casida, J.E. Neonicotinoid insecticide toxicology: Mechanisms of selective action. *Annu. Rev. Pharmacol. Toxicol.* **2005**, *45*, 247–265. [CrossRef] [PubMed]

15. Bonmatin, J.M.; Giorio, C.; Girolami, V.; Goulson, D.; Kreutzweiser, D.P.; Krupke, C.; Liess, M.; Long, M.; Marzaro., M.; Mitchell, E.A.D.; et al. Environmental fate and exposure; neonicotinoids and fipronil. *Environ. Sci. Pollut. Res.* **2015**, *22*, 35–67. [CrossRef] [PubMed]

16. Pisa, L.W.; Amaral-Rogers, V.; Belzunces, L.P.; Bonmatin, J.M.; Downs, C.A.; Goulson, D.; Kreutzweiser, D.P.; Krupke, C.; Liess, M.; McField, M.; et al. Effects of neonicotinoids and fipronil on non-target invertebrates. *Environ. Sci. Pollut. Res.* **2015**, *22*, 68–102. [CrossRef]

17. Stoughton, S.J.; Liber, K.; Culp, J.; Cessna, A. Acute and chronic toxicity of imidacloprid to the aquatic invertebrates *Chironomus tentans* and *Hyalella azteca* under constant- and pulse-exposure conditions. *Arch. Environ. Contam. Toxicol.* **2008**, *54*, 662–673. [CrossRef]

18. Pestana, J.L.; Loureiro, S.; Baird, D.J.; Soares, A.M. Fear and loathing in the benthos: Responses of aquatic insect larvae to the pesticide imidacloprid in the presence of chemical signals of predation risk. *Aquat. Toxicol.* **2009**, *93*, 138–149. [CrossRef]

19. LeBlanc, H.M.K.; Culp, J.M.; Baird, D.J.; Alexander, A.C.; Cessna, A.J. Single versus combined lethal effects of three agricultural insecticides on larvae of the freshwater insect *Chironomus dilutus*. *Arch. Environ. Contam. Toxicol.* **2012**, *63*, 378–390. [CrossRef]

20. Malev, O.; Klobucar, R.S.; Fabbretti, E.; Trebse, P. Comparative toxicity of imidacloprid and its transformation product 6-chloronicotinic acid to non-target aquatic organisms: Microalgae *Desmodesmus subspicatus* and amphipod *Gammarus fossarum*. *Pestic. Biochem. Physiol.* **2012**, *104*, 178–186. [CrossRef]

21. Nyman, A.; Hintermeister, A.; Schirmer, K.; Ashauer, R. The insecticide imidacloprid causes mortality of the freshwater amphipod *Gammarus pulex* by interfering with feeding behavior. *PLoS ONE* **2013**, *8*, e62472. [CrossRef]

22. Alexander, A.C.; Culp, J.M.; Liber, K.; Cessna, A.J. Effects of insecticide exposure on feeding inhibition in mayflies and oligochaetes. *Environ. Toxicol. Chem.* **2007**, *26*, 1726–1732. [CrossRef]

23. Roessink, I.; Merga, L.B.; Zweers, H.J.; Van den Brink, P.J. The neonicotinoid imidacloprid shows high chronic toxicity to mayfly nymphs. *Environ. Toxicol. Chem.* **2013**, *32*, 1096–1100. [CrossRef] [PubMed]

24. Maxted, J.R.; Barbour, M.T.; Gerritsen, J.; Proetti, V.; Primrose, N.; Silvia, A.; Penrose, D.; Renfrow, R. Assessment framework for mid-Atlantic coastal plain streams using benthic macroinvertebrates. *J. Am. Benthol. Soc.* **2000**, *19*, 128–144. [CrossRef]

25. Klemm, D.J.; Blocksom, K.A.; Fulk, F.A.; Herlihy, A.T.; Hughes, R.M.; Kaufmann, P.R.; Peck, D.V.; Stoddard, J.L.; Thoeny, W.T.; Griffith, M.B.; et al. Development and evaluation of a macroinvertebrate biotic integrity index (MBII) for regionally assessing mid–Atlantic highlands streams. *Environ. Manag.* **2003**, *31*, 656–669. [CrossRef] [PubMed]

26. Vignati, D.A.; Ferrari, B.J.; Dominik, J. Laboratory–to–field extrapolation in aquatic sciences. *Environ. Sci. Technol.* **2007**, *41*, 1067–1073. [CrossRef] [PubMed]

27. Bednarska, A.J.; Jevtić, D.M.; Laskowski, R. More ecological ERA: Incorporating natural environmental factors and animal behavior. *Integr. Environ. Assess. Manag.* **2013**, *9*, e39–e46. [CrossRef]

28. Sánchez-Bayo, F.; Tennekes, H.A. Environmental risk assessment of agrochemicals—A critical appraisal of current approaches. In *Toxicity and Hazard of Agrochemicals*; Larramendy, M., Soloneski, S., Eds.; InTechOpen: London, UK, 2015. Available online: https://www.intechopen.com/books/toxicity-and-hazard-of-agrochemicals/environmental-risk-assessment-of-agrochemicals-a-critical-appraisal-of-current-approaches (accessed on 1 October 2019). [CrossRef]

29. Buchwalter, D.B.; Clements, W.H.; Luoma, S.N. Modernizing water quality criteria in the United States: A need to expand the definition of acceptable data. *Environ. Toxicol. Chem.* **2017**, *36*, 285–291. [CrossRef]

30. EFSA Panel on Plant Protection Products and their Residues (PPR). Guidance on tiered risk assessment for plant protection products for aquatic organisms in edge-of-field surface waters. *EFSA J.* **2013**, *11*, 3290. [CrossRef]

31. Stehle, S.; Schulz, R. Pesticide authorization in the EU—Environment unprotected? *Environ. Sci. Pollut. Res.* **2015**, *22*, 19632–19647. [CrossRef]

32. Liess, M.; von der Ohe, P.C. Analyzing effects of pesticides on invertebrate communities in streams. *Environ. Toxicol. Chem.* **2005**, *24*, 954–965. [CrossRef]

33. Schäfer, R.B.; von der Ohe, P.C.; Rasmussen, J.; Kefford, B.J.; Beketov, M.A.; Schulz, R.; Liess, M. Thresholds for the effects of pesticides on invertebrate communities and leaf breakdown in stream ecosystems. *Environ. Sci. Technol.* **2012**, *46*, 5134–5142. [CrossRef]

34. Beketov, M.A.; Kefford, B.J.; Schäfer, R.B.; Liess, M. Pesticides reduce regional biodiversity of stream invertebrates. *Proc. Natl. Acad. Sci. USA* **2013**, *110*, 11039–11043. [CrossRef] [PubMed]

35. Szöcs, E.; Brinke, M.; Karaoglan, B.; Schäfer, R.B. Large scale risks from agricultural pesticides in small streams. *Environ. Sci. Technol.* **2017**, *51*, 7378–7385. [CrossRef] [PubMed]

36. Russo, R.; Becker, J.M.; Liess, M. Sequential exposure to low levels of pesticides and temperature stress increase toxicological sensitivity of crustaceans. *Sci. Total Environ.* **2018**, *610*, 563–569. [CrossRef] [PubMed]

37. Liess, M. Population response to toxicants is altered by intraspecific interaction. *Environ. Toxicol. Chem.* **2002**, *21*, 138–142. [CrossRef]

38. Holmstrup, M.; Bindesbøl, A.M.; Oostingh, G.J.; Duschl, A.; Scheil, V.; Köhler, H.R.; Loureiro, S.; Soares, A.M.V.M.; Ferreira, A.L.G.; Kienle, C.; et al. Interactions between effects of environmental chemicals and natural stressors: A review. *Sci. Total Environ.* **2010**, *408*, 3746–3762. [CrossRef]

39. Laskowski, R.; Bednarska, A.J.; Kramarz, P.E.; Loureiro, S.; Scheil, V.; Kudłek, J.; Holmstrup, M. Interactions between toxic chemicals and natural environmental factors—A meta–analysis and case studies. *Sci. Total Environ.* **2010**, *408*, 3763–3774. [CrossRef]

40. Heugens, E.H.; Hendriks, A.J.; Dekker, T.; Straalen, N.M.V.; Admiraal, W. A review of the effects of multiple stressors on aquatic organisms and analysis of uncertainty factors for use in risk assessment. *Crit. Rev. Toxicol.* **2001**, *31*, 247–284. [CrossRef]

41. Macaulay, S.J.; Buchwalter, D.B.; Matthaei, C.D. Water temperature interacts with the insecticide imidacloprid to increase lethal and sublethal toxicity to mayfly larvae. *N. Z. J. Mar. Freshw. Res.* **2019**. [CrossRef]

42. Hopkins, W.A.; Snodgrass, J.W.; Roe, J.H.; Staub, B.P.; Jackson, B.P.; Congdon, J.D. Effects of food ration on survival and sublethal responses of lake chubsuckers (*Erimyzon sucetta*) exposed to coal combustion wastes. *Aquat. Toxicol.* **2002**, *57*, 191–202. [CrossRef]

43. Metcalfe, N.B.; Monaghan, P. Compensation for a bad start: Grow now, pay later? *Trends Ecol. Evol.* **2001**, *16*, 254–260. [CrossRef]

44. Storey, K.B. Regulation of hypometabolism: Insights into epigenetic controls. *J. Exp. Biol.* **2015**, *218*, 150–159. [CrossRef] [PubMed]

45. Bridges, T.S.; Farrar, J.D.; Duke, B.M. The influence of food ration on sediment toxicity in *Neanthes arenaceodentata* (Annelida: Polychaeta). *Environ. Toxicol. Chem.* **1997**, *16*, 1659–1665. [CrossRef]

46. Mommaerts, V.; Reynders, S.; Boulet, J.; Besard, L.; Sterk, G.; Smagghe, G. Risk assessment for side–effects of neonicotinoids against bumblebees with and without impairing foraging behavior. *Ecotoxicology* **2010**, *19*, 207–215. [CrossRef] [PubMed]

47. Foit, K.; Kaske, O.; Liess, M. Competition increases toxicant sensitivity and delays the recovery of two interacting populations. *Aquat. Toxicol.* **2012**, *106*, 25–31. [CrossRef] [PubMed]

48. Liess, M.; Foit, K.; Becker, A.; Hassold, E.; Dolciotti, I.; Kattwinkel, M.; Duquesne, S. Culmination of low–dose pesticide effects. *Environ. Sci. Technol.* **2013**, *47*, 8862–8868. [CrossRef]

49. McIntosh, A.R.; Townsend, C.R. Impacts of an introduced predatory fish on mayfly grazing in New Zealand streams. *Limnol. Oceanogr.* **1995**, *40*, 1508–1512. [CrossRef]

50. Song, M.Y.; Stark, J.D.; Brown, J.J. Comparative toxicity of four insecticides, including imidacloprid and tebufenozide, to four aquatic arthropods. *Environ. Toxicol. Chem.* **1997**, *16*, 2494–2500. [CrossRef]

51. Gurevitch, J.; Morrow, L.L.; Wallace, A.; Walsh, J.S. A metaanalysis of competition in field experiments. *Am. Nat.* **1992**, *140*, 539–572. [CrossRef]

52. Peckarsky, B.L.; McIntosh, A.R. Fitness and community consequences of avoiding multiple predators. *Oecologia* **1998**, *113*, 565–576. [CrossRef]

53. Beketov, M.; Liess, M. Potential of 11 pesticides to initiate downstream drift of stream macroinvertebrates. *Arch. Environ. Contam. Toxicol.* **2008**, *55*, 247–253. [CrossRef]

54. Hunn, J.G.; (University of Otago, Dunedin, NZ). Personal Communication, 2016.

55. Hunn, J.G. Imidacloprid Effects on New Zealand Mayfly Nymphs: Acute and Chronic Exposure and Interactions with Natural Stressors. Bachelor Thesis, University of Otago, Dunedin, New Zealand, October 2016.

56. Macaulay, S.J. Insecticide Contamination of Streams in a Warming Climate. Ph.D. Thesis, University of Otago, Dunedin, New Zealand, June 2019.

57. Garson, D. Multivariate GLM, MANOVA, and MANCOVA 2015 Edition. Available online: www.statisticalassociates.com (accessed on 5 October 2019).

58. Nakagawa, S. A farewell to Bonferroni: The problems of low statistical power and publication bias. *Behav. Ecol.* **2004**, *15*, 1044–1045. [CrossRef]

59. Nakagawa, S.; Cuthill, I.C. Effect size, confidence interval and statistical significance: A practical guide for biologists. *Biol. Rev.* **2007**, *82*, 591–605. [CrossRef] [PubMed]

60. Quinn, G.P.; Keough, M.J. *Experimental Design and Data Analysis for Biologists*; Cambridge University Press: Cambridge, UK, 2002.

61. Tennekes, H.A.; Sánchez-Bayo, F. The molecular basis of simple relationships between exposure concentration and toxic effects with time. *Toxicology* **2013**, *309*, 39–51. [CrossRef] [PubMed]

62. Buckingham, S.D.; Lapied, B.; le Corronc, H.L.; Sattelle, F. Imidacloprid actions on insect neuronal acetylcholine receptors. *J. Exp. Biol.* **1997**, *200*, 2685–2692. [PubMed]

63. Zhang, A.; Kaiser, H.; Maienfisch, P.; Casida, J.E. Insect nicotinic acetylcholine receptor: Conserved neonicotinoid specificity of [^{3}H] imidacloprid binding site. *J. Neurochem.* **2000**, *75*, 1294–1303. [CrossRef] [PubMed]

64. Chapman, P.M. Integrating toxicology and ecology: Putting the "eco" into ecotoxicology. *Mar. Pollut. Bull.* **2002**, *44*, 7–15. [CrossRef]

65. Gessner, M.O.; Tlili, A. Fostering integration of freshwater ecology with ecotoxicology. *Freshw. Biol.* **2016**, *61*, 1991–2001. [CrossRef]

66. Maloney, E.M. How do we take the pulse of an aquatic ecosystem? Current and historical approaches to measuring ecosystem integrity. *Environ. Toxicol. Chem.* **2019**, *38*, 289–301. [CrossRef]

67. Dinh, K.V.; Janssens, L.; Stoks, R. Exposure to a heat wave under food limitation makes an agricultural insecticide lethal: A mechanistic laboratory experiment. *Glob. Chang. Biol.* **2016**, *22*, 3361–3372. [CrossRef]

68. Janssens, L.; Stoks, R. Fitness effects of chlorpyrifos in the damselfly *Enallagma cyathigerum* strongly depend upon temperature, food level and can bridge metamorphosis. *PLoS ONE* **2013**, *8*, e68107. [CrossRef]

69. Mydlarz, L.D.; Jones, L.E.; Harvell, C.D. Innate immunity, environmental drivers, and disease ecology of marine and freshwater invertebrates. *Annu. Rev. Ecol. Evol. Syst.* **2006**, *37*, 251–288. [CrossRef]

70. DeBlock, M.; Stoks, R. Short-term larval food stress and associated compensatory growth reduce adult immune function in a damselfly. *Ecol. Entomol.* **2008**, *33*, 796–801. [CrossRef]

71. Muturi, E.J.; Costanzo, K.; Kesavaraju, B.; Lampman, R.; Alto, B.W. Interaction of a pesticide and larval competition on life history traits of *Culex pipiens. Acta Trop.* **2010**, *116*, 141–146. [CrossRef] [PubMed]

72. Alexander, A.C.; Culp, J.M.; Baird, D.J.; Cessna, A.J. Nutrient–insecticide interactions decouple density–dependent predation pressure in aquatic insects. *Freshw. Biol.* **2016**, *61*, 2090–2101. [CrossRef]

73. Effenberger, M.; Sailer, G.; Townsend, C.R.; Matthaei, C.D. Local disturbance history and habitat parameters influence the microdistribution of stream invertebrates. *Freshw. Biol.* **2006**, *51*, 312–332. [CrossRef]

74. Camp, A.A.; Buchwalter, D.B. Can't take the heat: Temperature-enhanced toxicity in the mayfly *Isonychia bicolor* exposed to the neonicotinoid insecticide imidacloprid. *Aquat. Toxicol.* **2016**, *178*, 49–57. [CrossRef]

75. Van den Brink, P.J.; Van Smeden, J.M.; Bekele, R.S.; Dierick, W.; De Gelder, D.M.; Noteboom, M.; Roessink, I. Acute and chronic toxicity of neonicotinoids to nymphs of a mayfly species and some notes on seasonal differences. *Environ. Toxicol. Chem.* **2016**, *35*, 128–133. [CrossRef]

76. Raby, M.; Zhao, X.M.; Hao, C.Y.; Poirier, D.G.; Sibley, P.K. Chronic effects of an environmentally–relevant, short–term neonicotinoid insecticide pulse on four aquatic invertebrates. *Sci. Total Environ.* **2018**, *639*, 1543–1552. [CrossRef]

77. Raby, M.; Nowierski, M.; Perlov, D.; Zhao, X.; Hao, C.; Poirier, D.G.; Sibley, P.K. Acute toxicity of 6 neonicotinoid insecticides to freshwater invertebrates. *Environ. Toxicol. Chem.* **2018**, *37*, 1430–1445. [CrossRef]

78. Raby, M.; Maloney, E.; Poirier, D.G.; Sibley, P.K. Acute effects of binary mixtures of imidacloprid and tebuconazole on 4 freshwater invertebrates. *Environ. Toxicol. Chem.* **2019**, *38*, 1093–1103. [CrossRef]

79. Macaulay, S.J.; Hageman, K.J.; Alumbaugh, R.E.; Lyons, S.M.; Piggott, J.J.; Matthaei, C.D. Chronic toxicities of neonicotinoids to nymphs of the common New Zealand mayfly *Deleatidium* spp. *Environ Toxicol. Chem.* **2019**. [CrossRef] [PubMed]

80. Diamond, J.M.; Winchester, E.L.; Mackler, D.G.; Gruber, D. Use of the mayfly *Stenonema modestum* (Heptageniidae) in subacute toxicity assessments. *Environ. Toxicol. Chem.* **1992**, *11*, 415–425. [CrossRef]

81. Camp, A.; Funk, D.; Buchwalter, D.B. A stressful shortness of breath: Molting disrupts breathing in the mayfly *Cloeon dipterum*. *Freshw. Sci.* **2014**, *33*, 695–699. [CrossRef]

82. Magbanua, F.S.; Townsend, C.R.; Hageman, K.J.; Lange, K.; Lear, G.; Lewis, G.D.; Matthaei, C.D. Understanding the combined influence of fine sediment and glyphosate herbicide on stream periphyton communities. *Water Res.* **2013**, *47*, 5110–5120. [CrossRef]

83. Magbanua, F.S.; Townsend, C.R.; Hageman, K.J.; Matthaei, C.D. Individual and combined effects of fine sediment and the herbicide glyphosate on benthic macroinvertebrates and stream ecosystem function. *Freshw. Biol.* **2013**, *58*, 1729–1744. [CrossRef]

84. Chapman, R.B. A Review of Insecticide Use on Pastures and Forage Crops in New Zealand. Available online: http://agpest.co.nz/wp-content/uploads/2013/06/A-review-of-insecticide-use-on-pastures-and-forage-crops-in-New-Zealand.pdf (accessed on 2 October 2019).

85. Hageman, K.J.; Aebig, C.H.F.; Luong, K.H.; Kaserzon, S.L.; Wong, C.S.; Reeks, T.; Greenwood, M.; Macaulay, S.; Matthaei, C.D. Current–use pesticides in New Zealand streams: Comparing results from grab samples and three types of passive samplers. *Environ. Pollut.* **2019**, *254*, 112973. [CrossRef]

86. Moreau, M.; Hadfield, J.; Hughey, J.; Sanders, F.; Lapworth, D.J.; White, D.; Civil, W. A baseline assessment of emerging organic contaminants in New Zealand groundwater. *Sci. Total Environ.* **2019**, *686*, 425–439. [CrossRef]

87. Magbanua, F.S.; Townsend, C.R.; Blackwell, G.L.; Matthaei, C.D. Responses of stream macroinvertebrates and ecosystem function to conventional, integrated and organic farming. *J. Appl. Ecol.* **2010**, *47*, 1014–1025. [CrossRef]

88. Shahpoury, P.; Hageman, K.J.; Matthaei, C.D.; Magbanua, F.S. Chlorinated pesticides in stream sediments from organic, integrated and conventional farms. *Environ. Pollut.* **2013**, *181*, 219–225. [CrossRef]

Article

Mapping Micro-Pollutants and Their Impacts on the Size Structure of Streambed Communities

Ignacio Peralta-Maraver [1], Malte Posselt [2], Daniel M. Perkins [1] and Anne L. Robertson [1,*]

[1] Life Sciences Department, University of Roehampton, London SW15 4JD, UK;
    nacho.peralta@roehampton.ac.uk (I.P.-M.); daniel.perkins@roehampton.ac.uk (D.M.P.)
[2] Department of Environmental Science and Analytical Chemistry (ACES), Stockholm University,
    11418 Stockholm, Sweden; malte.posselt@aces.su.se
* Correspondence: a.robertson@roehampton.ac.uk

Received: 22 November 2019; Accepted: 7 December 2019; Published: 11 December 2019

Abstract: Recently there has been increasing concern over the vast array of emerging organic contaminants (EOCs) detected in streams and rivers worldwide. Understanding of the ecological implications of these compounds is limited to local scale case studies, partly as a result of technical limitations and a lack of integrative analyses. Here, we apply state-of-the-art instrumentation to analyze a complex suite of EOCs in the streambed of 30 UK streams and their effect on streambed communities. We apply the abundance–body mass (N–M) relationship approach as an integrative metric of the deviation of natural communities from reference status as a result of EOC pollution. Our analysis includes information regarding the N and M for individual prokaryotes, unicellular flagellates and ciliates, meiofauna, and macroinvertebrates. We detect a strong significant dependence of the N–M relationship coefficients with the presence of EOCs in the system, to the point of shielding the effect of other important environmental factors such as temperature, pH, and productivity. However, contrary to other stressors, EOC pollution showed a positive effect on the N–M coefficient in our work. This phenomenon can be largely explained by the increase in large-size tolerant taxa under polluted conditions. We discuss the potential implications of these results in relation to bioaccumulation and biomagnification processes. Our findings shed light on the impact of EOCs on the organization and ecology of the whole streambed community for the first time.

Keywords: abundance-size scaling theory; benthos; hyporheos; freshwater communities; pharmaceuticals; large scale survey

1. Introduction

Today, most of the world's rivers transport contaminants derived from anthropogenic activities [1,2] in a concomitant reduction of important ecosystem services such as clean drinking water [3], leading to global public alarm [1,4]. This problem is predicted to become more acute in the coming decades as a result of increasing concentrations of emerging organic compounds (EOCs), as well as their transformation products, detected in surface and groundwater systems globally [5,6]. EOCs, also known as trace organic compounds or micro-pollutants, are compounds of anthropogenic origin that contaminate natural systems (by up to several micrograms per liter) and which may lead to adverse effects in wildlife, including endocrine disruption, behavioral alterations, and developmental inhibition [7–10]. EOCs comprise a vast set of synthetic chemicals, ranging from daily-use pharmaceuticals and personal care products to pesticides and agricultural chemicals [6]. EOC pollution occurs when these chemicals enter natural systems in many different ways, including as a part of wastewater treatment plant release and percolation from agricultural areas in floodplains [11], resulting in a widespread and constant source of pollution (EOC pollution) with the potential to affect all levels of biological organization [9]. However, characterization and quantification of EOCs in riverine systems is still limited, as is our understanding of their relationship with environmental gradients, especially at large spatial scales [4]. This is largely because determining low concentrations of these substances can be a challenge using existing analytical methodologies [12]. Consequently, most EOCs have been determined to be low risk due to low environmental concentrations [13], and their effect on natural systems is largely unknown.

Most anthropogenic effluents are discharged to surface streams and rivers where water is exchanged between the open channel and the saturated permeable streambed sediments [14]. Consequently, dissolved EOCs penetrate the sediments, and organisms located here may be exposed to the effects of EOCs for longer periods of time because of the extended residence times of water in the pore-spaces [15,16]. Streambed sediments harbor diverse and productive biological consortia, whose components range from prokaryotes and microscopic single cell eukaryotes (e.g., flagellates and ciliates) through to meio- and macrofauna (rotifers, copepods, and insect larvae). This translates into a great diversity of life strategies and adaptation capabilities, and, consequently, understanding the effect of EOCs on the whole streambed assemblage is challenging. For example, it might be expected that the rapid population growth and adaptation capacity of prokaryotes might result in a less detrimental effect of chronic low levels of pollution compared to larger size fractions with longer life cycles and lower recruitment rates. However, current insights on the effect of EOCs in streambed communities are limited and target single groups such as prokaryotes [17,18], protists [18], and invertebrates [19,20]. Previous studies have focused on reach or local scales with low replication power. However, the functions and services provided by the streambed are mediated by all the different groups of organisms and the complex interactions between them and the environment [21]. Thus, a more integrative analysis ensuring the representation of the whole streambed assemblage, as well as large-scale approaches, must be undertaken to fully understand how riverine ecosystems will respond to the input of EOCs.

Body size scaling represents a synthetic analytical framework [22] which can be used to integrate the whole streambed assemblage. In particular, the allometric relationship between abundance (N) and body mass (M) is a widely studied pattern in ecological research [23,24]. When individual organisms are grouped into body mass classes, regardless of their taxonomy, the slope of the N–M relationship, or 'size spectra', provides an integrated measure of the community structure and the energy flow across trophic levels [25]. This is especially true in freshwater ecosystems, where communities are strongly size-structured [26,27] and gape-limited predation predominates [25]. The intercept of the N–M relationship provides a proxy for the community carrying capacity while the slope represents the energy flow and trophic transference efficiency in the system. Thus, N–M coefficients can be used as a quantitative measure of deviation of a natural community in relation to a reference status due to anthropic stressors [28]. Typically, intercepts decrease and size spectra become steeper under stress conditions [29]. To date, we do not know how EOCs affect the N–M relationship despite the insights that this approach could afford with regard to understanding how EOCs compromise the natural functioning of riverine systems.

Here, we report for the first time the effect of EOC pollution on community composition and N–M relationship coefficients in streambed communities following a regional scale approach. For this purpose, we make use of a large dataset from an existing survey study characterizing streambed communities and environmental features across 30 UK streams [21]. We complement that dataset with unpublished concentrations of 24 model EOCs measured in the pore-space of the streambeds collected at the same temporal and spatial resolution as the community samples. Our study delivers valuable information regarding the presence and concentration of EOCs in natural freshwater systems and their relationship with environmental gradients at a regional scale. Additionally, and exceptionally, these datasets include information of the N and M for individual prokaryotes, unicellular flagellates and ciliates, meiofauna (with body lengths between 0.45 and 500 μm), and macroinvertebrates (body length > 500 μm). Thus, the data comprise more than 10 orders of magnitude in terms of M, ensuring very good representation of the streambed assemblages. We hypothesize that EOC pollution results in detrimental effects to streambed communities, especially for the large-size fractions with lower adaptation capacity to constant exposure to EOCs. Hence, we predict N–M coefficients (intercept and slope) to decrease compared with reference systems. Our findings help understand how EOCs shape the structure, metabolic capacity, and energy flow through components of the streambed assemblage.

2. Methods

2.1. Data Acquisition

Here, we complemented open-access available data from a large survey project (Peralta-Maraver et al. 2019a) with data of EOC concentration, and calculation of N–M relationship coefficients. This dataset comprises 30 streams covering 10 different catchments across England and Wales (UK, Figure 1). Streams varied from small upland, acidic headwaters to large lowland, base-rich chalk streams, covering a large productivity and pollution gradient. Original datasets included information on a large set of environmental variables by study site: canopy cover, sediment morphology (cobbles, gravel, sand, and silt), leaf litter, depth and width of channel, submerged plants and submerged wood, temperature, pH, altitude, latitude, longitude, dissolved organic carbon, ammonium, nitrate, and phosphate [21].

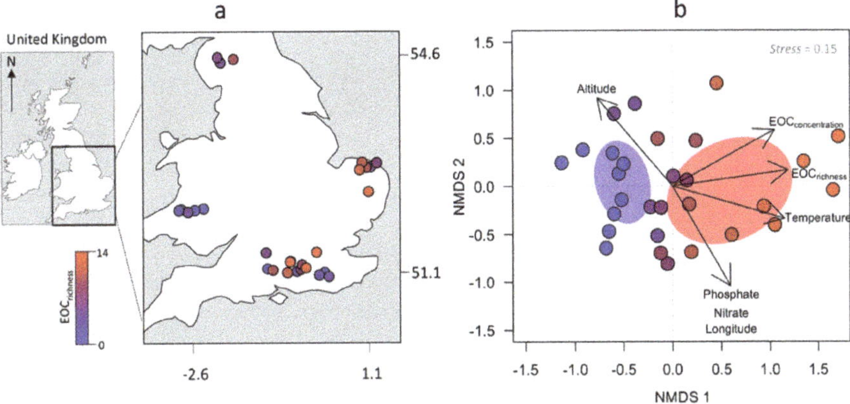

Figure 1. (**a**) Locations of the study systems in the United Kingdom, including number of emerging organic contaminants (EOCrichness). (**b**) Non-metric multidimensional scaling (NMDS) ordination model based on Canberra index comparing the dissimilarities in profile and concentration (μg L^{-1}) of dissolved compounds (macronutrients and EOCs) across the 30 studied rivers. Environmental gradients that were significantly correlated ($p < 0.05$) with the ordination are overlapped with the ordination. The arrows depict the relationship of fitted variables with the ordination.

Streambed communities were originally sampled using colonization traps (mesh = 0.5 cm, volume = 38–45 mL) containing three different organic substrates [21]. At each study site, six colonization traps were installed in pairs in the streambed at 0–2 and 15 cm depth for between 29–61 days. After incubation, colonization traps were collected and streambed communities were processed in the laboratory. Sampled organisms were identified and counted (N) and their body dimensions measured. Then, body dimensions (length and width in μm) of all collected individuals were converted into dry carbon content (M) using allometric relationships (further details on sampling design and sample processing are available in Peralta-Maraver et al. [21]).

2.2. Sampling and Processing of EOCs

Streambed pore-water samples were collected from the studied sites using a dive point piezometer during removal of colonization traps. One sampler per stream (n = 30) was pushed vertically into the streambed sediments to a depth of ~7.5 cm and 50 mL of water was pumped manually. Samples were stored in a coolbox at 4 °C and transported to the laboratory within 24 h, where they were frozen until analysis.

EOCs were analyzed using a previously developed direct-injection ultra-high-performance liquid chromatography–tandem mass spectrometry (UHPLC-MS/MS) method (Posselt et al. 2018). A total of 37 polar organic substances (mostly pharmaceuticals and their transformation products) were selected based on their concentration ranges, detection frequency, degradation behavior, and potential persistence, as well as their occurrence on priority lists [12,30–32]. Water samples were defrosted and vortexed and a sample volume of 800 µL was combined with 195 µL methanol and 5 µL of an internal standard mix. Afterwards, samples were vortexed again, filtered (Filtropur S 0.45 µm, PES membrane, Sarstedt AG&Co, Nuembrecht, Germany) into LC vials (2 mL, Thermo Scientific, Dreieich, Germany) and analyzed within 12 h. The sample injection volume was 20 µL. Liquid chromatography was performed using a Thermo Scientific Ultimate 3000 UHPLC system equipped with a Waters (Manchester, UK) Acquity UPLC HSS T3 column (1.8 µm, 2.1 mm × 100 mm). The mobile phase consisted of 10 mM acetic acid in deionized water (A) and 10 mM acetic acid in methanol (B). The flow rate was 500 µL min^{-1} for the gradient and 1000 µL min^{-1} for column equilibration. Instrumental analysis was carried out using a Thermo Scientific Quantiva triple-quadrupole mass spectrometer equipped with a heated electrospray ionization source. Detailed information on the LC gradient and MS instrument settings can be found in Posselt et al. (2018). A series of calibration standards (in 80% LC/MS grade water/20% MeOH) containing the 37 target compounds and isotopically labelled internal standards was measured three times. Data were processed using the Thermo Scientific Xcalibur 3.1.66.10 instrument software and quantification was performed using the internal standard method. Precision was determined by injecting a quality control standard every 15 samples. The relative standard deviation was <1–12% for all detected compounds except for valsartan acid (21%) and 4-hydroxydiclofenac (20%).

Concentrations in both analyzed blank samples were always below the method limit of quantification for all targets. Method limits of quantification for the 37 targeted compounds are provided in Table S1 and further information regarding materials, chemicals, and standards, as well as additional quality control data can be found elsewhere [12].

2.3. Statistical Analysis

First, a non-metric multidimensional scaling (NMDS) ordination model was applied to compare quantitatively the similarities in profile and concentration (µg L^{-1}) of dissolved compounds (macronutrients and EOCs) across the 30 studied rivers. Excessively large differences between the smallest nonzero and largest concentration values were reduced using the Wisconsin double standardization of variables [33]. This approach improves the detection ability of the similarity index used in NMDS ordination [34]. The Canberra index was used to produce the ordination model and it was run iteratively to find the ordination with the best fit (lower stress value). Subsequently, we evaluated the degree to which the number and concentration of EOCs and dissolved compounds such as nutrients were associated with the ordination axis. For this, we fitted all environmental variables collected originally (Table S1) and new data on profiles and concentrations of EOCs onto the resulting two-dimensional ordination following Peralta-Maraver et al. [35]. Degree correlation and significance of the association between fitted variables and the ordination axis was then assessed after a 1000 randomized permutation test.

Secondly, we applied multiple regression and backward model selection approaches to build the N–M relationship models comparing reference systems (no EOCs detected) with polluted sites (EOCs detected). We pooled data from all colonization traps by study site to provide an integrated sample of the streambed community (n = 30 streambed communities). We constructed the N–M relationship for each site by applying the logarithmic size-binning method [36]. Size bins were determined from the (log10) body mass range for each sampled community and the abundances of organisms were then summed within each size bin [23]. We used a total of six bins to maximize the number of size bins while minimizing the number of empty size bins in the analysis [23,26]. Next, we built a saturated model comparing reference systems with EOC-polluted (two-level factor), and all environmental gradients significantly related with the NMDS ordination. Also, and independently of their relationship with the NMDS ordination, we included pH in the saturated model as a classical driver of the N–M relationship in freshwater systems [37,38]. Covariates were dropped sequentially, and the model re-fitted. Then, the Akaike information criterion (AIC) was applied to select the model with the best fit, and Akaike weights (w_i) were used to quantify the relative support of each model in comparison to all alternative models (and therefore $\Sigma\, w_i = 1$). In addition, we studied the potential collinear relationship between all covariates included in the candidate models. Model selection and collinearity testing allowed us to inspect potential confounding effects of EOCs with underlying gradients, such as productivity. Model validation was finally applied to verify the underlying assumptions following Zuur et al. [39]. This encompassed testing normality and homoscedasticity of model residuals and their potential dependence with those variables included and not included in the model (e.g., study site).

All statistical analyses were performed using R software (R Core Team, 2019). NMDS ordinations and subsequent variable fitting were carried out using the functions metaMDS and envfit of the R-package Vegan [40].

3. Results

From the 37 targeted compounds, a diverse set of 24 EOCs, including pharmaceuticals and other organic contaminants, were collected from the streambed of two thirds of the study sites (Figure 1, Table 1). The most EOC-polluted sites were mainly distributed in the east and southeast regions of England. The 10 streams unpolluted by EOCs, hereafter called reference sites, were mainly located in the west regions of Wales but were also represented in the southeast of England. The NMDS model based on the 24 EOCs and macronutrients [nitrate, phosphate, and dissolved organic carbon (DOC)] produced a two-dimensional ordination with a very high goodness of fit between the distances in the ordination against the original data (linear fit $R^2 = 0.995$, non-metric fit $R^2 = 0.990$). The resulting ordination (Figure 1b) showed a strong increasing gradient of number of EOCs and concentration positively related with axis 1 ($R^2 = 0.94$, $p < 0.01$), while dispersion along axis 2 was better explained by the presence and concentration of dissolved nitrate ($R^2 = 0.94$, $p < 0.01$) and phosphate ($R^2 = 0.21$, $p = 0.04$), but not DOC ($R^2 = 0.03$, $p = 0.62$). Environmental fitting onto the ordination showed that number and concentration of EOCs and macronutrients increased significantly along environmental gradients of longitude ($R^2 = 0.56$, $p < 0.04$) and temperature ($R^2 = 0.26$, $p = 0.02$), and in the lowland regions of the UK ($R^2 = 0.58$, $p < 0.001$; Figure 1b).

Table 1. Concentration ($\mu g\ L^{-1}$) of the target EOCs analyzed across the 30 studied streams: 1H-benzotriazole (1.H.B.), 2/4-chlorobenzoic acid (C.Acid), 4-hydroxydiclofenac (4.H.D.), acesulfame (Acesu), acetaminophen (Aceta), carbamazepine (Carba), clofibric acid (C.Acid2), diclofenac (Diclo), 11-dihydroxy carbamazepine (11.D.C), furosemide (Furo), gemfibrozil (Gem), guanylurea (Guan), ibuprofen (Ibu), metformin (Metf), metoprolol acid (M.Acid), metoprolol (Meto), naproxen (Napro), O-desmethylvenlafaxine (O.Des), oxazepam (Oxa), propranolol (Prop), sitagliptin (Sita), sotalol (Sota), tramadol (Trama), and venlafaxine (Venl). Table shows latitude (Lat) and longitude (Lon) and total amount of EOCs of studied streams.

River	Lat	Lon	Acesu	Aceta	Sita	O.Des	11.D.C.	Napro	Guan	Metf	Venl	M.Acid	Carba	Oxa	Prop	Sota	4.H.D.	Trama	Diclo	1.H.B.	Ibu	11.D.C.	C.Acid2	Furo	Meto	Gem	Tot EOCs
Beverly Brooks	51.44	0.25	13.8		0.1	0.4	1.1		44.8	1.5	0.6	0.6	0.2		0.6	0.1		0.2			2.9	0.6					14
Loddon	51.42	1.72	0.8		0.1	0.4	0.8		9.7	0.2	0.2	0.1	0.2			0.7		0.1	0.4			0.1					13
Wey	51.19	0.68	0.8		0.5	0.1	0.1	0.7	5.2	0.4	0.1	0.8	0.3			0.2		0.6									12
Waveney	52.42	1.36	0.5	0.1		0.6	0.2		1.7	0.4		0.2	0.3							0.1							9
Wensum	52.42	1.36	0.2	0.4	0.3	0.4	0.4		0.8	0.2										0.3							8
Deadwater	51.17	0.85	1.9		0.2									0.7	0.1						0.7			0.1			6
Stiffkey	52.92	0.89		0.3	0.2						0.2								0.2	0.1							5
Tat	52.82	0.75	0.5	0.9																			0.2		0.9		4
River Leith	54.61	−2.62	0.5					0.4									0.4										3
Nadder	51.12	0.90	1.7																				0.4	0.1			3
Test	51.14	1.47	1.0			0.2		0.4																			3
Glaven	52.93	1.63	0.2	0.7													0.4										3
Lamports	51.15	1.72	0.1					0.3						0.1													3
Lyde	51.29	1.72	0.2	0.8															0.2								3
GI1	52.14	−3.84			0.4							0.1															2
Howe Beck	54.68	−2.59						0.1						0.7													2
Bure	52.82	1.21		0.1													0.1										2
River Crowdundle	51.15	1.72	0.1																							0.1	2
Kennet	51.42	1.72	0.2													0.7											2
River Lyvennet	54.68	−2.61	0.1				0.5																				2
LI7	52.13	−3.75																									0
LI8	52.16	−3.75																									0
LI3	52.14	−3.73																									0
Old Lodge	54.65	−2.64																									0
Lone Oak	51.77	0.13																									0
LI6	51.44	0.25																									0
Broadstone Stream	51.89	0.57																									0
Oakhanger	51.45	0.79																									0
Anton	51.15	1.46																									0
Morland Beck	51.23	1.72																									0

After model selection routines, all studied variables were excluded from the N–M relationship except EOC pollution in the system (Table 2). The AIC model selection approach suggested a certain improvement of model fitting when adding pH and temperature. However, those variables were also excluded in favor of a model simply comparing reference and polluted sites. AIC and Akaike weight indicate a very strong support of the model including the interaction between M and the EOC pollution (presence/absence of EOCs). This specifies that presence of EOCs in the system strongly determines the intercept and slope of the N–M relationship model. The fitted N–M relationship model including information of the comparison between polluted or reference streams also had a high explanatory capacity (R = 0.67).

Table 2. Comparison of the regression models testing EOC pollution, pH, temperature (Temp), longitude (Lon), altitude (alt), nitrate (Nit), and phosphate (Phos) on the abundance–body mass (N–M) relationship (all models include an intercept, which has not been shown for simplicity). Legend: AIC, Akaike information criterion; LogLik, maximum likelihood estimator; wi, Akaike weight. Candidate model with the best fit is highlighted in bold.

Response	Predictors	N	AIC	ΔAIC	LogLik	wi
	log10(M) × EOCs + pH + Temp + Lon + Lat + Alt + Nit + Phos	12	403.82	9.23	0.01	0.00
	log10(M) × EOCs + pH + Temp + Lon + Lat + Alt + Nit	11	402.30	7.70	0.02	0.01
	log10(M) × EOCs + pH + Temp + Lon + Lat + Alt	10	402.35	7.75	0.02	0.01
	log10(M) × EOCs + pH + Temp + Lon + Lat	9	401.17	6.57	0.04	0.02
	log10(M) × EOCs + pH + Temp + Lon	8	399.18	4.59	0.10	0.05
Log 10 (N)	log10(M) × EOCs + pH + Temp	7	397.32	2.72	0.26	0.12
	log10(M) × EOCs + pH	6	396.15	1.55	0.46	0.22
	log10(M) × EOCs	**5**	**394.60**	**0.00**	**1.00**	**0.47**
	log10(M) + EOCs	4	397.88	3.28	0.19	0.09
	log10(M)	3	401.22	6.62	0.04	0.02

When over more than 10 orders of magnitude in body mass, from flagellates to macroinvertebrates, abundance declined linearly with body mass with an average N–M slope of −0.37 (95% CI = −0.42, −0.31). However, N–M relationship coefficients varied significantly between reference and polluted streams (Table 3). We found that N–M intercepts (as a proxy for community carrying capacity) were higher and size spectra slopes shallower in polluted compared to reference streams. The difference in the 95% CI between fitted regression in the reference and polluted streams became visible in the large size fraction of the N–M relationship, and we attribute this variation to the increase in abundance of the large-size fractions of organisms in polluted sites (Figure 2) which was chiefly associated with increases in pollutant-tolerant groups such as Asellidae and Oligochaeta (Figure 2b).

Table 3. Summary table of the fitted abundance–body mass regression (fixed coefficients). Fixed coefficients (*Coef*), standard errors (*SE*), degrees of freedom (*DF*), *t* values, and *p* values (*p*) are given. Significance codes (*Sig*): 0 (***), 0.01 (*).

Fixed Equation Terms	Coef	SE	*t* Value	*p* Value	Sig
Intercept	2.95	0.12	25.20	> 0.001	***
Log_{10} body mass	−0.37	0.03	−13.87	> 0.001	***
Presence/absence of EOCs	−0.77	0.24	−3.22	0.001	***
Log_{10} body mass * EOC pollution	−0.12	0.05	0.05	0.02	*

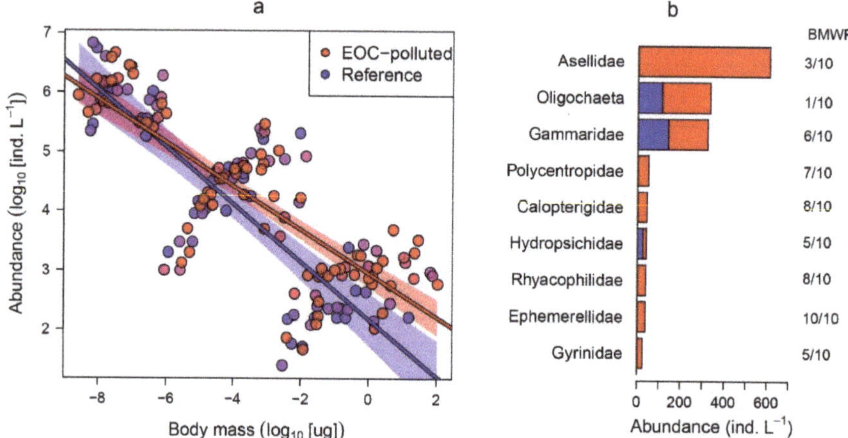

Figure 2. (**a**) Results from the linear regression of log10 total abundance on log10 body mass bin (N–M relationship) comparing EOC-polluted and reference streams. Each data point denotes the abundance of a given size class for each sampling unit. The fitted lines represent the average N–M slope for EOC-polluted and reference streams. (**b**) Abundance and biological monitoring working party (BMWP) score of the macroinvertebrate families belonging to the largest size bin across the studied EOC-polluted and reference streams.

4. Discussion

In this work, we identified a diverse set of EOCs in streambed sediments across a large regional scale and examined their effect on the size structure of streambed assemblages. We found complex mixtures of EOCs in two thirds of all studied streams, yet we screened our water samples for only 37 target compounds as a small subset of the possible compounds found in natural systems. More than 70,000 daily-use chemicals are registered in the European Union with the potential to enter surface and subsurface water systems [5]. Nevertheless, technical limitations on the detection of EOCs make quantifying and regulating such a variety of chemicals in natural systems unrealistic. Therefore, centering efforts on analyzing carefully selected target chemicals as markers seems to be a suitable strategy to assess the degree of EOC pollution at large spatial scales.

Our findings evidence the strong effect that the presence of EOCs has over the organization of streambed communities, to the point of shielding the effect of other important environmental factors. Temperature, pH, and productivity are considered major drivers in freshwater ecology and determining factors of the size structure and metabolic capacity of streambed communities [26,37,38,41,42]. Even so, both the intercept and size spectra slope of the N–M relationship exhibited higher sensitivity in this work to EOC pollution than to those environmental variables. Hence, the N–M relationship approach seems to be an appropriate analytical and integrative statistical tool for testing deviation of natural communities from a reference status as a consequence of EOC pollution. However, and contrary to our predictions, N–M relationship coefficients increased under polluted conditions. That is, the intercept (i.e., the carrying capacity of the community) increased and the size spectrum slope became shallower. As predicted, the smallest size fractions showed low variation in their abundance when comparing polluted and reference streams. Thus, this pattern is mainly due to the notably higher abundance of certain groups of macroinvertebrates that push up the N–M relationship from the extreme of the mean and largest size fraction. A closer look at the taxa that most influence variation in the N–M relationship (Figure 2) reveals that they mainly belong to tolerant groups of organisms with medium to very low biological monitoring working party scores (BMWP) [43].

The size spectra slope of the M–N relationship describes the rate of biomass depletion through different levels of the food web in freshwater systems, typically becoming shallower as this rate increases [25,26]. Consequently, the existence of more abundant tolerant invertebrates in our study might imply a better transference of biomass and energy between trophic levels and potentially longer food chains. Considering these predictions, we are particularly concerned that several EOCs bioaccumulate untransformed in the cells and tissues of aquatic organisms and even biomagnify through different trophic levels [44–46]. Several of the measured EOCs in our survey study, such as diclofenac, ibuprofen, carbamazepine, metoprolol, gemfibrozil, oxazepam, tramadol, and venlafaxine are already known to bioaccumulate in biofilms, invertebrates, and fish in the food webs of stream ecosystems [45,47,48]. In addition, some of these EOCs in larval tissues can be conserved through metamorphosis and adults play the role of a biological vector transporting pharmaceuticals to terrestrial predators [46]. Under our theoretical framework, bioaccumulation is expected to be more acute under a scenario of EOC pollution and the concomitant population increase of tolerant macroinvertebrates. In either case, future experimental approaches are needed to test our prediction. Hence, controlled mesocosm experiments analyzing uptake of EOCs by model organisms and the transference of these EOCs through trophic levels represent a fruitful strategy to assessing these ecological mechanisms. Our metrics involve a broad range of body sizes; however, species-specific responses from different size groups are expected. Therefore, controlled experiments will also shed light on how different populations of taxa will respond to the presence and concentration of different EOCs.

In addition, from the total set of targeted compounds selected as model EOCs, the artificial low-calorie sweetener acesulfame was the most widespread chemical and was detected almost constantly in all polluted streams in our study. This pattern is consistent with previous studies in which acesulfame has been reported as the most ubiquitous EOC in streambed sediments and groundwater across Europe [12,30–32,49–51]. Thus, our findings strongly support the use of acesulfame as a marker of EOC pollution for application in management and regulation of surface and subsurface waters [49]. Acesulfame is an anthropic source-specific compound released to the environment in high quantities (in quantities of up to 13.8 μg L^{-1} in this study) and is amenable to rapid and sensitive analysis. It is sufficiently persistent and hydrophilic enough to penetrate into streambed and groundwater systems [49]. In fact, combining analysis of acesulfame with that of acetaminophen (an analgesic pharmaceutical) and sitagliptin (an antihyperglycemic pharmaceutical) allowed us to distinguish between polluted and reference streams in our study.

In summary, in this work we detected strong variation from reference status when comparing N–M relationship coefficients between polluted and reference streams. Even though the ecological mechanisms remain unclear, EOC pollution was associated with an increase in abundance of large-size tolerant macroinvertebrates. This resulted in less size-structured assemblages under polluted conditions with direct implications for the structure of the streambed community and potential biomagnification processes along food webs. With this in mind, future research which characterizes food web structures (feeding links) and quantifies concentrations of EOCs at different trophic levels would be particularly instructive.

Supplementary Materials: The following is available online at http://www.mdpi.com/2073-4441/11/12/2610/s1, Table S1: LC-MS/MS method limits of quantification in μg L^{-1}.

Author Contributions: Conceptualization, I.P.-M. and A.L.R.; methodology, I.P.-M., M.P., D.M.P., A.L.R.; software, I.P.-M.; validation, I.P.-M., D.M.P.; formal analysis, I.P.-M., M.P.; investigation, I.P.-M., M.P., D.M.P., A.L.R.; resources, I.P.-M., M.P., D.M.P., A.L.R.; data curation, I.P.-M., M.P.; writing—original draft preparation, I.P.-M., D.M.P., A.L.R.; writing—review and editing, I.P.-M., M.P., D.M.P., A.L.R.; visualization, I.P.-M., A.L.R.; supervision, A.L.R.; project administration, I.P.-M., A.L.R.; funding acquisition, D.M.P., A.L.R.

Funding: This project was funded by the European Union Horizon 2020 research and innovation programme under Marie Skłodowska-Curie grant agreement No. 641939.

Acknowledgments: The authors thank Jon Benskin and the two anonymous reviewers, who provided useful suggestions that improved the original manuscript.

Conflicts of Interest: No authors have any conflicting interest.

References

1. Vörösmarty, C.J.; McIntyre, P.B.; Gessner, M.O.; Dudgeon, D.; Prusevich, A.; Green, P.; Glidden, S.; Bunn, S.E.; Sullivan, C.A.; Liermann, C.R.; et al. Global threats to human water security and river biodiversity. *Nature* **2010**, *467*, 555–561. [CrossRef] [PubMed]
2. Wen, Y.; Schoups, G.; Van De Giesen, N. Organic pollution of rivers: Combined threats of urbanization, livestock farming and global climate change. *Sci. Rep.* **2017**, *7*, 43289. [CrossRef] [PubMed]
3. Malaj, E.; Peter, C.; Grote, M.; Kühne, R.; Mondy, C.P.; Usseglio-Polatera, P. Organic chemicals jeopardize the health of freshwater ecosystems on the continental scale. *Proc. Natl. Acad. Sci. USA* **2014**, *111*, 9549–9554. [CrossRef] [PubMed]
4. Schwarzenbach, R.P.; Escher, B.I.; Fenner, K.; Hofstetter, T.B.; Johnson, C.A.; Von Gunten, U.; Wehrli, B. The challenge of micropollutants in aquatic systems. *Science* **2006**, *313*, 1072–1077. [CrossRef] [PubMed]
5. Reemtsma, T.; Weiss, S.; Mueller, J.; Petrovic, M.; González, S.; Barcelo, D.; Ventura, F.; Knepper, T.P. Polar pollutants entry into the water cycle by municipal wastewater: A European perspective. *Environ. Sci. Technol.* **2006**, *40*, 5451–5458. [CrossRef] [PubMed]
6. Pal, A.; Gin, K.Y.H.; Lin, A.Y.C.; Reinhard, M. Impacts of emerging organic contaminants on freshwater resources: Review of recent occurrences, sources, fate and effects. *Sci. Total Environ.* **2010**, *408*, 6062–6069. [CrossRef] [PubMed]
7. Kidd, K.A.; Blanchfield, P.J.; Mills, K.H.; Palace, V.P.; Evans, R.E.; Lazorchak, J.M.; Flick, R.W. Collapse of a fish population after exposure to a synthetic estrogen. *Proc. Natl. Acad. Sci. USA* **2007**, *104*, 8897–8901. [CrossRef]
8. Galus, M.; Rangarajan, S.; Lai, A.; Shaya, L.; Balshine, S.; Wilson, J.Y. Effects of chronic, parental pharmaceutical exposure on zebrafish (Danio rerio) offspring. *Aquat. Toxicol.* **2014**, *151*, 124–134. [CrossRef]
9. Stamm, C.; Räsänen, K.; Burdon, F.J.; Altermatt, F.; Jokela, J.; Joss, A.; Ackermann, M.; Eggen, R.I. Unravelling the impacts of micropollutants in aquatic ecosystems: Interdisciplinary studies at the interface of large-scale ecology. *Adv. Ecol. Res.* **2016**, *55*, 183–223.
10. Kubec, J.; Hossain, S.; Grabicová, K.; Randák, T.; Kouba, A.; Grabic, R.; Roje, S.; Buřič, M. Oxazepam alters the behavior of crayfish at diluted concentrations, venlafaxine does not. *Water* **2019**, *11*, 196. [CrossRef]
11. Lewandowski, J.; Arnon, S.; Banks, E.; Batelaan, O.; Betterle, A.; Broecker, T.; Coll, C.; Drummond, J.D.; Garcia, J.G.; Galloway, J.; et al. Is the Hyporheic Zone Relevant beyond the Scientific Community? *Water* **2019**, *11*, 2230. [CrossRef]
12. Posselt, M.; Jaeger, A.; Schaper, J.L.; Radke, M.; Benskin, J.P. Determination of polar organic micropollutants in surface and pore water by high-resolution sampling-direct injection-ultra high performance liquid chromatography-tandem mass spectrometry. *Environ. Sci. Process. Impacts* **2018**, *20*, 1716–1727. [CrossRef] [PubMed]
13. Richmond, E.K.; Grace, M.R.; Kelly, J.J.; Reisinger, A.J.; Rosi, E.J.; Walters, D.M. Pharmaceuticals and personal care products (PPCPs) are ecological disrupting compounds (EcoDC). *Elem. Sci. Anth.* **2017**, *5*, 66. [CrossRef]
14. Findlay, S. Importance of surface-subsurface exchange in stream ecosystems: The hyporheic zone. *Limnol. Oceanogr.* **1995**, *40*, 159–164. [CrossRef]
15. Hester, E.T.; Young, K.I.; Widdowson, M.A. Mixing of surface and groundwater induced by riverbed dunes: Implications for hyporheic zone definitions and pollutant reactions. *Water Resour. Res.* **2013**, *49*, 5221–5237. [CrossRef]
16. Subirats, J.; Triadó-Margarit, X.; Mandaric, L.; Acuña, V.; Balcázar, J.L.; Sabater, S.; Borrego, C.M. Wastewater pollution differently affects the antibiotic resistance gene pool and biofilm bacterial communities across streambed compartments. *Mol. Ecol.* **2017**, *26*, 5567–5581. [CrossRef]
17. Subirats, J.; Timoner, X.; Sànchez-Melsió, A.; Balcázar, J.L.; Acuña, V.; Sabater, S.; Borrego, C.M. Emerging contaminants and nutrients synergistically affect the spread of class 1 integron-integrase (intI1) and sul1 genes within stable streambed bacterial communities. *Water Res.* **2018**, *138*, 77–85. [CrossRef]
18. Láng, J.; Kőhidai, L. Effects of the aquatic contaminant human pharmaceuticals and their mixtures on the proliferation and migratory responses of the bioindicator freshwater ciliate Tetrahymena. *Chemosphere* **2012**, *89*, 592–601. [CrossRef]

19. Althakafy, J.T.; Kulsing, C.; Grace, M.R.; Marriott, P.J. Determination of selected emerging contaminants in freshwater invertebrates using a universal extraction technique and liquid chromatography accurate mass spectrometry. *J. Sep. Sci.* **2018**, *41*, 3706–3715. [CrossRef]

20. Miller, T.H.; Ng, K.T.; Bury, S.T.; Bury, S.E.; Bury, N.R.; Barron, L.P. Biomonitoring of pesticides, pharmaceuticals and illicit drugs in a freshwater invertebrate to estimate toxic or effect pressure. *Environ. Int.* **2019**, *129*, 595–606. [CrossRef]

21. Peralta-Maraver, I.; Perkins, D.M.; Thompson, M.S.; Fussmann, K.; Reiss, J.; Robertson, A.L. Comparing biotic drivers of litter breakdown across stream compartments. *J. An. Ecol.* **2019**, *88*, 1146–1157. [CrossRef] [PubMed]

22. Brown, J.H.; Gillooly, J.F.; Allen, A.P.; Savage, V.M.; West, G.B. Toward a metabolic theory of ecology. *Ecology* **2004**, *85*, 1771–1789. [CrossRef]

23. White, E.P.; Ernest, S.M.; Kerkhoff, A.J.; Enquist, B.J. Relationships between body size and abundance in ecology. *Trends Ecol. Evol.* **2007**, *22*, 323–330. [CrossRef] [PubMed]

24. Trebilco, R.; Baum, J.K.; Salomon, A.K.; Dulvy, N.K. Ecosystem ecology: Size-based constraints on the pyramids of life. *Trends Ecol. Evol.* **2013**, *28*, 423–431. [CrossRef]

25. Kerr, S.R.; Dickie, L.M. *The Biomass Spectrum: A PredatorPrey Theory of Aquatic Production*; Columbia University Press: Chichester, NY, USA, 2001.

26. Perkins, D.M.; Durance, I.; Edwards, F.K.; Grey, J.; Hildrew, A.G.; Jackson, M.; Jones, J.I.; Lauridsen, R.B.; Layer-Dobra, K.; Thompson, M.S.; et al. Bending the rules: Exploitation of allochthonous resources by a top-predator modifies size-abundance scaling in stream food webs. *Ecol. Lett.* **2018**, *21*, 1771–1780. [CrossRef]

27. Peralta-Maraver, I.; Robertson, A.L.; Perkins, D.M. Depth and vertical hydrodynamics constrain the size structure of a lowland streambed community. *Biol. Lett.* **2019**, *15*, 20190317. [CrossRef]

28. Petchey, O.L.; Morin, P.J.; Hulot, F.D. Contributions of aquatic model systems to our understanding of biodiversity and ecosystem functioning. In *Biodiversity and Ecosystem Functioning—Synthesis and Perspectives*; Loreau, M., Naeem, S., Inchausti, P., Eds.; Oxford University Press: Oxford, UK, 2002; pp. 127–138.

29. Petchey, O.L.; Belgrano, A. Body-size distributions and size-spectra: Universal indicators of ecological status? *Biol. Lett.* **2010**, *6*, 434–437. [CrossRef]

30. Li, Z.; Sobek, A.; Radke, M. Fate of pharmaceuticals and their transformation products in four small European rivers receiving treated wastewater. *Environ. Sci. Technol.* **2016**, *50*, 5614–5621. [CrossRef]

31. Schaper, J.L.; Posselt, M.; Bouchez, C.; Jaeger, A.; Nuetzmann, G.; Putschew, A.; Singer, G.; Lewandowski, J. Fate of Trace Organic Compounds in the Hyporheic Zone: Influence of Retardation, the Benthic Biolayer, and Organic Carbon. *Environ. Sci. Technol.* **2019**, *53*, 4224–4234. [CrossRef]

32. Mechelke, J.; Vermeirssen, E.L.; Hollender, J. Passive sampling of organic contaminants across the water-sediment interface of an urban stream. *Water Res.* **2019**, *165*, 114966. [CrossRef]

33. Bray, J.R.; Curtis, J.T. An ordination of the upland forest communities of southern Wisconsin. *Ecol. Monogr.* **1957**, *27*, 325–349. [CrossRef]

34. Oksanen, J. Multivariate Analysis of Ecological Communities in R: Vegan Tutorial. 2015. Available online: http://cc.oulu.fi/~{}jarioksa/opetus/metodi/veg-antutor.pdf (accessed on 19 November 2019).

35. Peralta-Maraver, I.; Robertson, A.L.; Rezende, E.L.; Lemes da Silva, A.L.; Tonetta, D.; Lopes, M.; Schmitt, R.; Leite, N.K.; Nuñer, A.; Petrucio, M.M. Winter is coming: Food web structure and seasonality in a subtropical freshwater coastal lake. *Ecol. Evol.* **2017**, *7*, 4534–4542. [CrossRef] [PubMed]

36. Edwards, A.M.; Robinson, J.P.; Plank, M.J.; Baum, J.K.; Blanchard, J.L. Testing and recommending methods for fitting size spectra to data. *Methods Ecol. Evol.* **2017**, *8*, 57–67. [CrossRef]

37. Mulder, C.; Elser, J.J. Soil acidity, ecological stoichiometry and allometric scaling in grassland food webs. *Glob. Chang. Biol.* **2009**, *15*, 2730–2738. [CrossRef]

38. Layer, K.; Riede, J.O.; Hildrew, A.G.; Woodward, G. Food web structure and stability in 20 streams across a wide ph gradient. *Adv. Ecol. Res.* **2010**, *42*, 265–299.

39. Zuur, A.F.; Ieno, E.N.; Walker, N.J.; Savaliev, A.A.; Smith, G.M. *Mixed Effects Models and Extensions in Ecology with R*; Springer: New York, NY, USA, 2009; p. 57.

40. Oksanen, J.; Blanchet, F.G.; Kindt, R.; Legendre, P.; Minchin, P.R.; O'Hara, R.B.; Simpson, G.L.; Solymos, P.; Stevens, M.H.H.; Wagner, H. Vegan: Community Ecology Package. R Package Version 2.0–10. 2013. Available online: http://CRAN.R--project.org/package=vegan (accessed on 7 November 2019).

41. Dossena, M.; Yvon-Durocher, G.; Grey, J.; Montoya, J.M.; Perkins, D.M.; Trimmer, M.; Woodward, G. Warming alters community size structure and ecosystem functioning. *Proc. Biol. Sci. Replaces Proc. R. Soc.* **2012**, *279*, 3011–3019. [CrossRef]

42. O'Gorman, E.J.; Pichler, D.E.; Adams, G.; Benstead, J.P.; Cohen, H.; Craig, N.; Cross, W.F.; Demars, B.O.; Friberg, N.; Gislason, G.M.; et al. Impacts of warming on the structure and functioning of aquatic communities: Individual-to ecosystem-level responses. *Adv. Ecol. Res.* **2012**, *47*, 81–176.

43. Hawkes, H.A. Origin and development of the biological monitoring working party score system. *Water Res.* **1998**, *32*, 964–968. [CrossRef]

44. Zenker, A.; Cicero, M.R.; Prestinaci, F.; Bottoni, P.; Carere, M. Bioaccumulation and biomagnification potential of pharmaceuticals with a focus to the aquatic environment. *J. Environ. Manag.* **2014**, *133*, 378–387. [CrossRef]

45. Ruhí, A.; Acuña, V.; Barceló, D.; Huerta, B.; Mor, J.R.; Rodríguez-Mozaz, S.; Sabater, S. Bioaccumulation and trophic magnification of pharmaceuticals and endocrine disruptors in a Mediterranean river food web. *Sci. Total Environ.* **2016**, *540*, 250–259. [CrossRef]

46. Richmond, E.K.; Rosi, E.J.; Walters, D.M.; Fick, J.; Hamilton, S.K.; Brodin, T.; Sundelin, A.; Grace, M.R. A diverse suite of pharmaceuticals contaminates stream and riparian food webs. *Nat. Commun.* **2019**, *9*, 4491. [CrossRef] [PubMed]

47. Brown, J.N.; Paxéus, N.; Förlin, L.; Larsson, D.J. Variations in bioconcentration of human pharmaceuticals from sewage effluents into fish blood plasma. *Environ. Toxicol. Pharmacol.* **2007**, *24*, 267–274. [CrossRef] [PubMed]

48. Fick, J.; Lindberg, R.H.; Tysklind, M.; Larsson, D.J. Predicted critical environmental concentrations for 500 pharmaceuticals. *Regul. Toxicol. Pharmacol.* **2010**, *58*, 516–523. [CrossRef] [PubMed]

49. Buerge, I.J.; Buser, H.-R.; Kahle, M.; Müller, M.D.; Poiger, T. Ubiquitous occurrence of the artificial sweetener acesulfame in the aquatic environment: An ideal chemical marker of domestic wastewater in groundwater. *Environ. Sci. Technol.* **2009**, *43*, 4381–4385. [CrossRef] [PubMed]

50. Jaeger, A.; Posselt, M.; Betterle, A.; Schaper, J.; Mechelke, J.; Coll, C.; Lewandowski, J. Spatial and Temporal Variability in Attenuation of Polar Organic Micropollutants in an Urban Lowland Stream. *Environ. Sci. Technol.* **2019**, *53*, 2383–2395. [CrossRef]

51. Jaeger, A.; Coll, C.; Posselt, M.; Mechelke, J.; Rutere, C.; Betterle, A.; Raza, M.; Mehrtens, A.; Meinikmann, K.; Portmann, A.; et al. Using recirculating flumes and a response surface model to investigate the role of hyporheic exchange and bacterial diversity on micropollutant. *Environ. Sci. Process. Impacts* **2019**, *21*. [CrossRef]

Article

The Application of Artificial Mussels in Conjunction with Transplanted Bivalves to Assess Elemental Exposure in a Platinum Mining Area

Marelize Labuschagne [1,*], Victor Wepener [1], Milen Nachev [2], Sonja Zimmermann [1,2], Bernd Sures [2] and Nico J. Smit [1]

[1]   Water Research Group, Unit for Environmental Sciences and Management, North-West University, 2520 Potchefstroom, South Africa; victor.wepener@nwu.ac.za (V.W.); sonja.zimmermann@uni-due.de (S.Z.); nico.smit@nwu.ac.za (N.J.S.)

[2]   Department of Aquatic Ecology and Centre for Water and Environmental Research, University of Duisburg-Essen, 45141 Essen, Germany; milen.nachev@uni-due.de (M.N.); bernd.sures@uni-due.de (B.S.)

*   Correspondence: marelizelabuschagne94@gmail.com; Tel.: +27-018-285-2279

Received: 31 October 2019; Accepted: 17 December 2019; Published: 20 December 2019

Abstract: There is increasing evidence that platinum group elements (PGE) are pollutants of emerging concern worldwide. Limited information exists on levels, particularly in regions where PGEs are mined. A passive sampling device (i.e., the artificial mussel (AM)) and transplanted indicator organisms (i.e., the freshwater clam *Corbicula fluminalis africana*) were deployed along a PGE mining gradient in the Hex River, South Africa, and concentrations of As, Cd, Co, Cr, Ni, Pb, Pt, V, and Zn were determined after six weeks of exposure. Results showed differential uptake patterns for Pt, Cr, and Ni between the AMs and clams indicating availability differences. For monitoring purposes, a combination of AMs and indicator organisms provides a more holistic assessment of element exposure in aquatic environments.

Keywords: platinum; bioaccumulation; passive sample; freshwater clam; *Corbicula fluminalis africana*

1. Introduction

In recent decades, the environmental concentrations of platinum group elements (PGE) have increased in different environmental matrices worldwide since these precious metals have been used more often in a number of applications [1]. The biggest contributors to PGEs in the environment are the use of automotive catalytic converters and the mining activities. Mining activities in South Africa greatly contribute to PGE emissions in the environment [1]. It is particularly in the platinum mining regions of the North-West Province of South Africa where PGE and associated metals show increased levels in the aquatic environment [2,3].

Bioaccumulation is an important process where living organisms take up toxicants at a greater rate than the rate at which they eliminate these substances [4]. Many factors can influence the bioaccumulation of metals in the environment such as pH, conductivity, temperature, and salinity, as well as biotic factors such as age, body size, and reproductive status of the bioaccumulating organism [4].

It is this process of bioaccumulation that is applied in monitoring studies where organisms such as bivalves are used as indicators of metal exposure [5]. Traditionally, resident bivalve populations are used as indicator species (passive biomonitoring) [6]. However, in instances where no resident species are available, organisms that were collected from an unstressed or otherwise unpolluted population can be translocated to a selected site to determine its degree of pollution (active biomonitoring) [7].

Suitable transplanted organisms are selected based on biological characteristics, such as tolerance and abundance. Transplanted organisms facilitate the investigation of areas where native specimens are absent or compensatory adaptive mechanisms occur in the native populations from contaminated areas [8]. For this study, the freshwater clam *Corbicula fluminalis africana* was selected as a bioindicator organism. Bivalves are excellent bioindicators because of their sensitivity to contaminants, sedentary lifestyle, wide geographical distribution, high abundance, high bioaccumulation capacity, and their sturdiness for both laboratory and field studies [4].

Despite the widespread use of living organisms as indicators of metal bioaccumulation, they also have disadvantages. The kinetics of metal uptake and elimination by these organisms are sometimes not well understood [9] and are affected by factors such as temperature, life stage of the organism (e.g., larval versus adult), size, depth, and reproduction [7]. Different indicator species have different natural distribution patterns. Therefore, they are not always present in all ecosystems that need to be monitored to prevent the comparison between different sites [7].

To overcome these disadvantages, passive sampling devices such as the artificial mussel (AM) have been utilised. The advantages of using AMs for monitoring metals are the provision of time-integrated estimates of metal concentrations within the aquatic ecosystems, the possibility of monitoring water bodies with unfavourable conditions for living indicator species, and the avoidance of killing organisms [10,11]. These devices are not affected by biotic and abiotic factors and, thus, can be deployed in marine and freshwater ecosystems [12,13].

Several studies included both AMs and transplanted organisms to monitor metal exposure. The results of these studies indicate that AMs are less affected by salinity and temperature than transplanted organisms [10,14,15] and that AMs and bivalves accumulate metals at different levels but in a similar pattern [13,15].

The present study aims to determine if the AM can be used as a tool to determine the biologically available concentrations of Pt and other elements in freshwater ecosystems. The AM was tested in a field study alongside an established bivalve bioaccumulation indicator to determine the degree of similarity in element bioaccumulation. Additionally, the concentrations measured in the AMs and bivalves were related to the environmental concentrations in the water at each of the study sites.

2. Materials and Methods

2.1. Description of the Study Area and Sampling Sites

The Hex River flows through the city of Rustenburg, which is the most populated municipality in the North West Province of South Africa. Rustenburg is located on the western limb of the Bushveld Igneous Complex where the main industrial activity is mining [16]. These mining activities include mining for platinum and chromium. Along the course of the Hex River, there are impoundments, which can be affected by mining activities [17].

Olifantsnek Dam is situated to the south of Rustenburg near the origin of the Hex River and is separated from the mining area by the Magaliesberg Mountain range. It was, therefore, chosen as the reference site [17]. According to Roux [18], the part of the Hex River that joins with Olifantsnek Dam is in a good to fair ecological state. The main impacts on this impoundment include water abstraction and farming activities. The ecological category of this stretch of the river classifies it as largely natural/moderately modified. Olifantsnek Dam is managed by private owners, where the main activity is recreation, e.g., fishing and sailing [18].

Bospoort Dam is a small state-owned impoundment situated along the course of the Hex River in Rustenburg [17]. The land-uses in the catchment include urban developments, intensive mining activities, and agricultural activities [19]. The water of Bospoort Dam is also used for irrigation and, in the past, for domestic water supply. However, due to a decrease in the water quality and an increase in algal blooms that influenced the taste and odor of the water, Bospoort Water Treatment Works (WTW) ceased to operate [19]. The Bospoort Dam is classified as being in a hypereutrophic state, where the possible influences include sewage treatment works, agricultural and urban run-offs, as well as recirculation of nutrients that can be found in the sediments [19]. In 2008, a report was published on the water quality of Bospoort Dam stating that the Cd, As, and Pb concentrations in the water are of concern [19]. The system also shows a high salt content, which is generally indicated by high conductivity values [19]. Urban surface runoff and mining activities are possible sources of ions that contributed to the high conductivity in the Bospoort dam, while the possible sources of elevated metal concentrations include mining activities and runoff from agricultural fields [19]. According to Roux [18], the Hex River upstream of Bospoort Dam is classified as seriously to critically modified, whereas the river below the dam is in a moderate to largely modified category.

Two sampling surveys during the low flow periods (dry seasons—May to June) in 2017 and 2018 were undertaken at the selected sites (Figure 1). In the second survey, a third site was added, which was located in a pollution control dam on the premises of a platinum mine to quantify the release of Pt and other elements by that mine. At each of these sampling sites, water samples were taken and AMs and clams were collected after exposure.

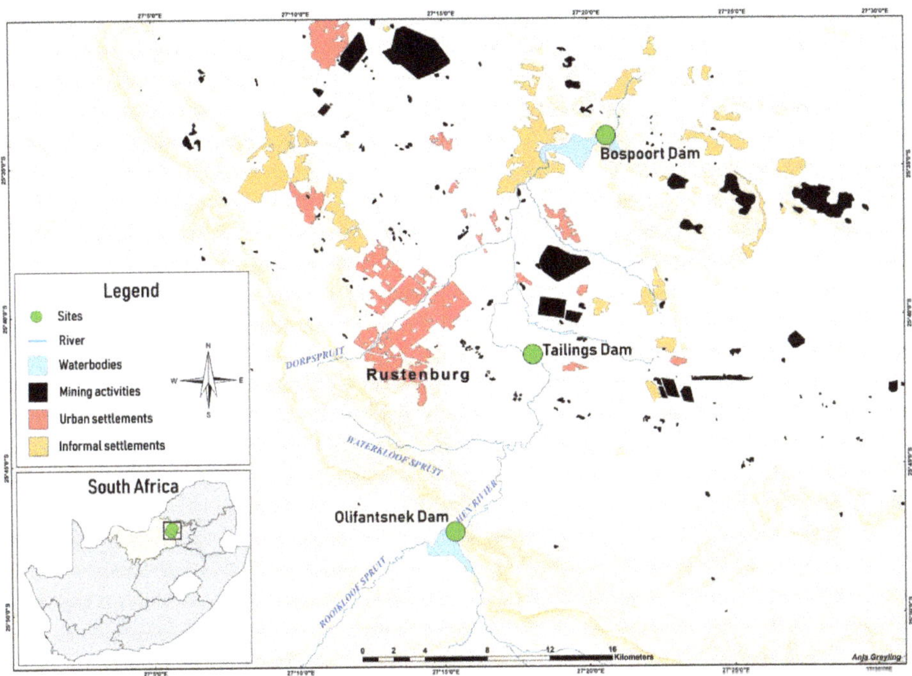

Figure 1. Map of the Hex River System in Rustenburg, North West Province, South Africa indicating the position of the sampling sites.

2.2. Passive Sampling Device—Artificial Mussel

The artificial mussel design was adapted for use in freshwater from the original design developed for marine environments by Wu, Lau, Fung, Ko, and Leung [10]. The artificial mussel consists of a Perspex tubing, containing 200 mg Chelex beads suspended in ultrapure water inside a 1 cm Perspex spacer (Figure 2). These components are enclosed between two polyacrylamide gel layers that consist of acrylamide (Acrylamide for electrophoresis, ≥99% (HPLC) powder, Sigma, Germany), N,N methylenebis-acrylamide (BioReagent, suitable for electrophoresis, 99%, Sigma, Germany), ammonium peroxidisulfate (Reagent grade, 98%, Sigma, Germany), and N,N,N′,N′-Tetramethylethylenediamine (BioReagent, for molecular biology, ≥99% (GC), Sigma, Germany) [20].

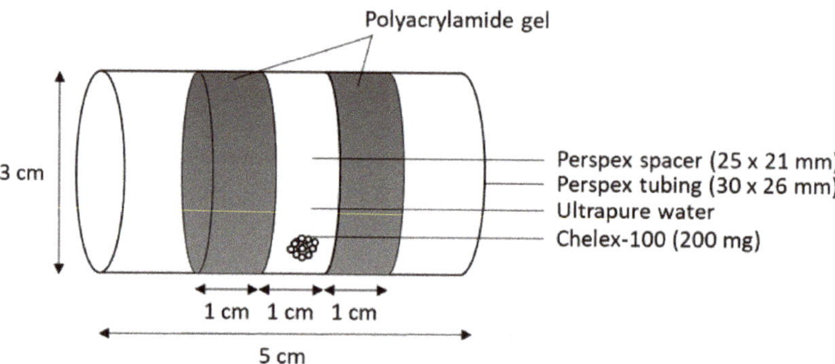

Figure 2. The artificial mussel design (Wu et al., 2007).

Once the AMs were assembled, the devices were placed in containers filled with ultrapure water and stored in the laboratory until being deployed. For the transportation of the AMs to the field, the devices were removed from the water and both ends of the tube were plugged with cotton soaked in ultrapure water. This prevents any damage that may occur during transportation. One set of 10 AMs was kept in the laboratory to serve as a non-exposed control.

2.3. Corbicula Fluminalis Africana

In addition to the deployment of the AM, an established bioaccumulation indicator species was concomitantly deployed. Freshwater clams, *Corbicula fluminalis africana*, with an average shell length of 28 ± 5 mm were collected from the Mooi River, Potchefstroom, South Africa. The clams were transported to the laboratory, where they were kept in reconstituted freshwater for mussels [21] for a depuration period of two weeks. During laboratory maintenance, the clams were fed with *Spirulina* cultured in the laboratory at the North-West University. Water was changed regularly throughout the two-week depuration period. For active biomonitoring, the clams were transported to the field in plastic containers filled with reconstituted freshwater and supplied with oxygen using a portable pump. A control group of clams was prepared for metal analyses to determine the initial concentrations before the clams were deployed at the sampling sites.

2.4. Deployment of the Artificial Mussels and Clams

The AMs and clams were deployed at each sampling site in two plastic baskets (Figure 3). Each basket was fitted with a mesh bag in the bottom of the basket. In addition, a mesh covered the top of the basket. Once arriving at the site, the mesh bags were filled with sediment from that site. This was done to create a natural habitat for the clams for the exposure period. Thereafter, 10 clams were transferred to each basket. In total, 20 transplanted clams were deployed at each site. After transferring the clams, the mesh bags were closed to prevent the organisms from escaping. The top half of each basket was then used to attach the AMs. Fifteen AMs were deployed at each site by fastening them to the side of the plastic baskets with cable ties. The baskets were covered with mesh to prevent objects from damaging the AMs during the survey. Once the AMs and transplanted clams were secured, the baskets were tied together with rope and cable ties and secured to a weight. The weights ensured the submersion of the baskets within the water body. The baskets were secured to buoys and permanent structures within the impoundments and the pollution control dam on the mine site.

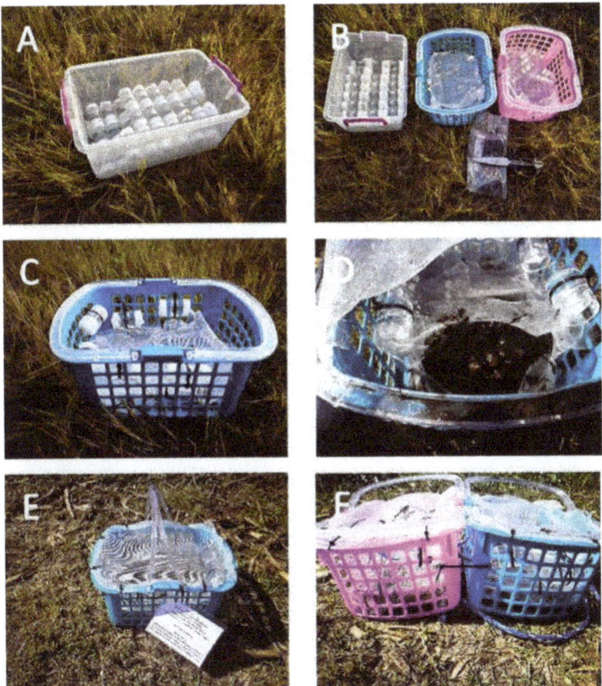

Figure 3. Steps taken to assemble the plastic baskets for the artificial mussels and transplanted clams. (**A**) Storage of artificial mussels before deployment. (**B**) Plastic baskets in which the artificial mussels and transplanted clams were deployed. (**C**) The AMs were secured to the plastic basket with cable ties. (**D**) Sediment from each site was placed in a mesh bag in the bottom of the basket for the transplanted clams. (**E**,**F**) Two baskets were deployed at each site. These baskets were secured together with rope and cable ties before submerging them in the water.

2.5. Retrieval of Artificial Mussels and Clams

Each batch of AMs and transplanted clams were retrieved after six weeks of exposure. Each AM was then rinsed with water from the site to remove any silt or algae that might have accumulated at the surface during the exposure period [20]. Cotton pads were soaked in water from the site, and placed in both ends of the AM to form a plug. This ensured that the gel would not get damaged during transportation in order to prevent the water and Chelex beads from seeping from the AM. At the laboratory, the containers with the AMs were placed at room temperature until further analysis.

The transplanted clams were retrieved at the same time as the AMs. These organisms were removed from the mesh bags and placed in containers that were fitted with an oxygen pump and water from the site. They were transported back to the laboratory and were killed by deep-freezing after they were measured and weighed. For each sampling site, soft tissues were dissected from the shells, placed in acid pre-washed polypropylene tubes, and frozen at $-4\ ^{\circ}\text{C}$ until further preparation. The clam tissue samples were freeze-dried before elemental analysis.

2.6. Water Quality Parameters and Sample Collection

In situ water quality parameters, i.e., pH, electrical conductivity (EC), total dissolved solids (TDS), temperature, and dissolved oxygen (DO) concentrations, were determined at each sampling site at the day of deployment as well as at the day of retrieval of the AMs and clams. At the same time, three water samples were collected at each site. The sampling procedure was as follows: 10 mL water was taken and acidified with 10 µL HNO$_3$ (sub-boiled from 65%; p.a. quality, Merck, Darmstadt, Germany) and stored in polypropylene tubes at room temperature until elemental analysis.

2.7. Element Analysis

For elemental analysis, the content of each control (n = 10) and field-deployed (n = 15) AM was emptied into a 15-mL polypropylene tube. After centrifugation (2 min at 1000× g), the supernatant was removed and the beads were rinsed with 5 mL of ultrapure water. The supernatant was removed and the beads were eluted with a mixture of 4.5 mL 6 M HNO$_3$ (sub-boiled from 65%, Merck, Darmstadt, Germany) and 0.5 mL HCl (37%, suprapure, Merck, Darmstadt, Germany) for approximately 2 hours to ensure that all bound elements were released from the beads. After centrifugation, the supernatant was removed and placed in polypropylene tubes at room temperature for further analysis.

For digestion of the freeze-dried clam soft tissue, five replicate samples of at least 60 mg (dry weight) were weighed into 20-mL TFM® vessels (MarsXpress, CEM, Kamp-Lintfort, Germany). The digestion was carried out in a microwave digestion system (CEM, Mars 6) with a mixture of 2.5 mL H$_2$O$_2$ (30%, Suprapur®, Merck, Darmstadt, Germany) and 1.3 mL HNO$_3$ (sub-boiled from 65%, p.a. quality, Merck, Darmstadt, Germany), according to Zimmermann et al. [22]. The clear digest solutions were then transferred to 5-mL glass flasks and brought to volume with 1% HNO$_3$. The digested solutions were stored at room temperature in polypropylene tubes until elemental detection.

Concentrations of As, Cd, Co, Ni, Pb, Pt, V, and Zn in the water samples, AMs, and clam tissue samples were determined using a quadrupole Inductively Coupled Plasma Mass Spectrometry (ICP-MS) system (Elan 6000, Perkin Elmer, Rodgau, Germany) equipped with an autosampler system (AS-90, Perkin Elmer, Rodgau, Germany). Cr concentrations were analyzed by atomic absorption spectrometry (AAS) as described by Erasmus et al. [3]. Hafnium oxide interference rates on Pt-194 were below 2%. Therefore, no Hf correction was performed. For the ICP-MS analysis, the wash time was set to 30 s with 2% HNO$_3$ to avoid contamination. After every 10 samples, a standard solution (10 µg/L), for all elements measured, was used to control the accuracy and stability of the measurements. Before measuring, samples were diluted 1:10 with an internal standard solution, which consisted of 1% HNO$_3$ and 10 µg/L thulium and yttrium (Certipur®, Merck, Darmstadt, Germany). Calibration of the instrument was performed using a series of 11 dilutions of the standard solution. With this, the concentrations of the sample analytes were calculated using regression lines with a correlation factor of ≥0.999. Detection limits were calculated as three times the standard deviation of the concentrations in 10 analytical blanks (Table 1).

Table 1. Detection limits for metal concentrations analyzed in water, artificial mussels, and transplanted clams.

Metal Analysed	Detection Limit for Water (µg/L)	Detection Limit for Artificial Mussels (µg/g)	Detection Limit for Transplanted Clams (µg/g)
As	0.203	0.0056	0.0169
Cd	0.022	0.0004	0.0018
Co	0.112	0.003	0.0099
Cr	0.151	0.039	0.130
Ni	0.463	0.011	0.0386
Pb	0.0237	0.0003	0.0020
Pt	0.0107	0.00015	0.0009
V	0.96	0.025	0.0798
Zn	17.7	0.053	1.48

The AM data was converted to µg/g units by referring the total eluted analyte mass (in µg) to the mass of Chelex beads (200 mg) in an AM. Thus, the AM data of the present study is comparable with the concentrations measured in the transplanted clams and other studies [20,23].

2.8. Statistical Analysis

Normal distribution of data was checked using the Kolmogorov-Smirnov test with a Dallal-Wilkinson-Lilliefor P-value. For element analyses in water samples of different sites, two-way ANOVAs were performed, which were followed by Tukey's multiple comparison test. For element analyses of both the AMs and transplanted clams, the data were log-transformed to compare all data sets since not all data passed normality. For statistical analyses, all samples that were below the detection limit were set to the detection limit divided by two [24]. A two-way ANOVA was performed to determine significant differences between surveys and sites for the AMs and transplanted clams. Thereafter, Tukey's multi-comparisons test was performed. Statistical significance was set at $p < 0.05$ for all comparisons. Principal component analyses (PCA) were conducted using Canoco 5 showing the grouping of AMs and transplanted clams based on the similarity of element concentrations at the different sites and surveys.

3. Results

The water quality parameters, i.e., pH, electrical conductivity (EC), total dissolved solids (TDS), temperature (Temp), and dissolved oxygen (DO) concentrations showed differences between sampling sites and survey years, respectively (Table 2). The temperature for the first survey was cooler than during the second survey, which influences the DO concentrations. During the second survey, the DO concentration was less than during the first survey. In both surveys, the Bospoort Dam had lower DO concentrations than the Olifantsnek Dam. The EC and TDS values differed significantly between the three sites. Bospoort Dam had significantly higher EC and TDS values than Olifantsnek Dam, whereas the pollution control dam showed the highest values. The pH was relatively stable throughout both surveys.

Table 2. Water quality parameters were measured at each site during the first survey (2017) and the second survey (2018). The results are the average of parameters measured during a survey. For abbreviations, see text.

Site	pH	Temp (°C)	EC (μS/cm)	TDS (mg/L)	DO (mg/L)	DO (%)
			Survey 2017			
Olifantsnek Dam	8.21	19.3	173.8	119.8	10.12	103.3
Bospoort Dam	8.46	19.45	910.5	570	9.69	81.4
			Survey 2018			
Olifantsnek Dam	8.20	24.3	210	147	5.5	62.1
Bospoort Dam	7.65	28.3	1160	819	2.94	34.3
Pollution control dam	8.00	25.6	1775	1244	-	

The concentrations of the different elements in the water of the three sampling sites showed different patterns (Figures 4–6, left). The element concentrations in the water of the two impoundments, Olifantsnek Dam and Bospoort Dam, revealed no significant differences in a specific survey except for Ni in the first survey, where the water of Bospoort Dam contained significantly higher concentrations. However, in the second survey, the water of the pollution control dam contained significantly higher As, Co, Ni, V, and Zn concentrations than the other impoundments and significantly higher Cr and Pt concentrations when compared to only the Bospoort Dam.

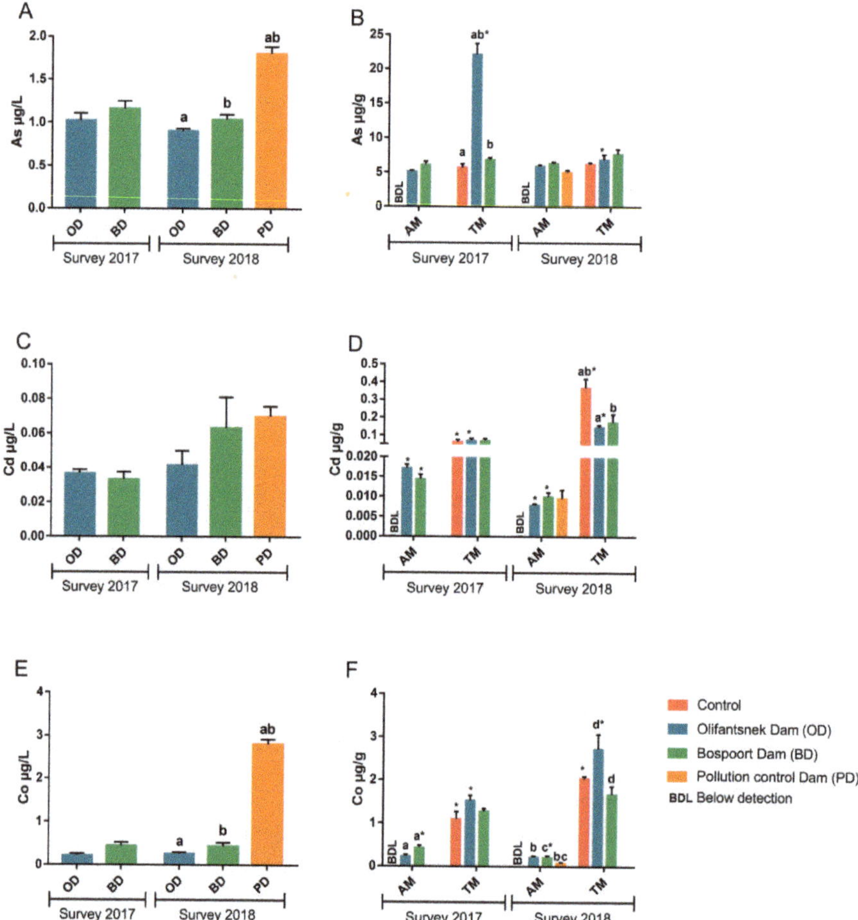

Figure 4. Arsenic (**A,B**), cadmium (**C,D**), and cobalt (**E,F**) concentrations (mean ± SEM) in water (**left**), artificial mussels (AM) and transplanted clams (TM) (**right**) during the two surveys in 2017 and 2018. Concentrations are presented in µg/L for water (n = 3), µg/g Chelex for the AMs (n = 15), and µg/g dry weight for transplanted clams (n = 5). Within a survey period, bars with a common letter superscript represent significant differences between different sites for the particular indicator group. Between surveys, an asterisk (*) indicates a significant temporal difference between the specific indicator at the particular site. Significance was regarded as $p < 0.05$. BDL = below detection limit.

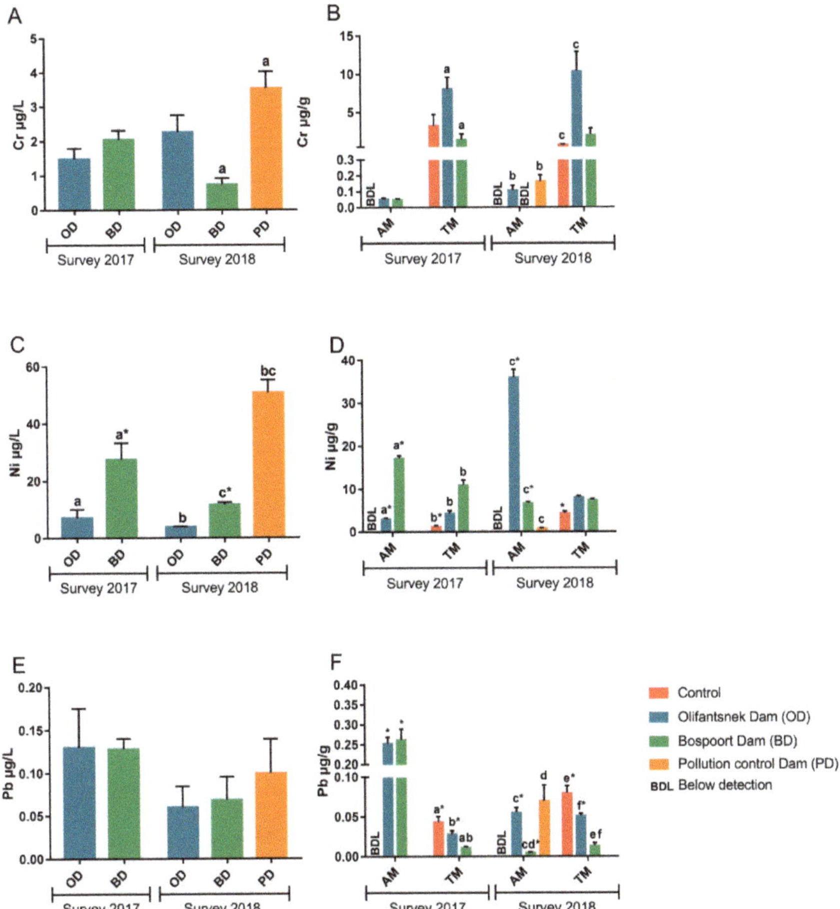

Figure 5. Chromium (**A,B**), nickel (**C,D**), and lead (**E,F**) concentrations (mean ± SEM) in water (**left**), and artificial mussels (AM) and transplanted clams (TM) (**right**) during the two surveys in 2017 and 2018. Concentrations are presented in µg/L for water (n = 3), µg/g Chelex for the AMs (n = 15), and µg/g dry weight for transplanted clams (n = 5). Within a survey, period bars with a common letter superscript represent significant differences between different sites for the particular indicator group. Between surveys, an asterisk (*) indicates a significant temporal difference between the specific indicator at the particular site. Significance was regarded as $p < 0.05$. BDL = below detection limit.

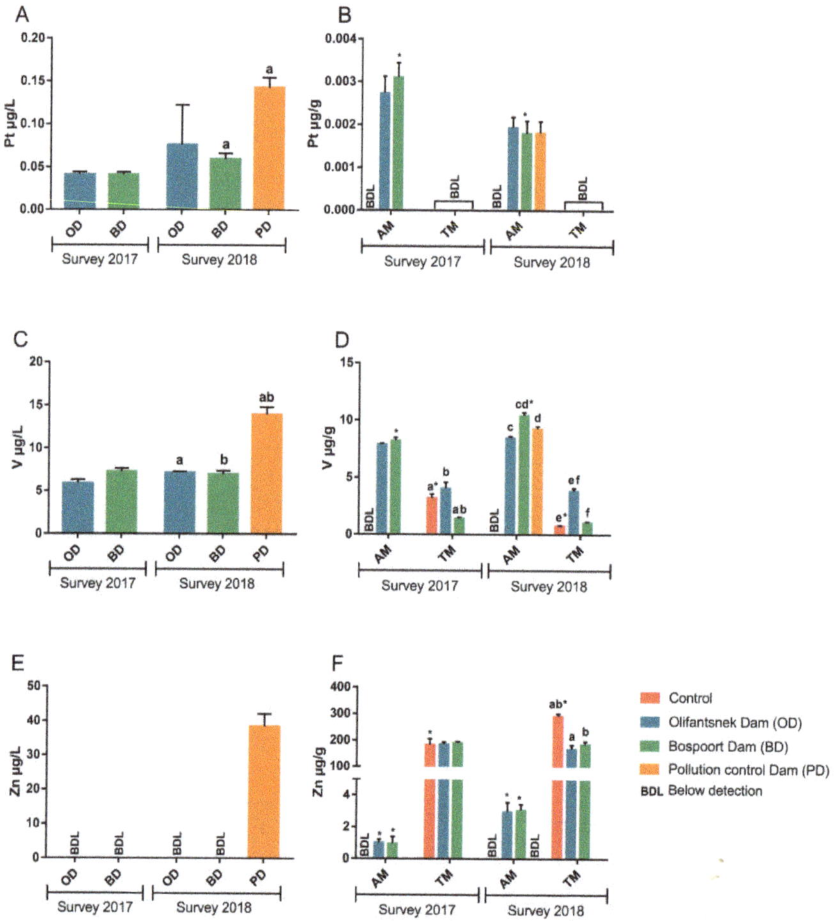

Figure 6. Platinum (**A,B**), vanadium (**C,D**), and zinc (**E,F**) concentrations (mean ± SEM) in water (**left**), and artificial mussels (AM) and transplanted clams (TM) (**right**) during the two surveys in 2017 and 2018. Concentrations are presented in µg/L for water (n = 3), µg/g Chelex for the AMs (n = 15), and µg/g dry weight for transplanted clams (n = 5). Within a survey, period bars with common letter superscript represent significant differences between different sites for the particular indicator group. Between surveys, an asterisk (*) indicates a significant temporal difference between the specific indicator at the particular site. Significance was regarded as $p < 0.05$. BDL = below detection limit.

Concentrations measured in the control AMs were below detection limits (Table 1) for all elements. Therefore, the element concentrations measured in the field-deployed AMs are attributed to accumulation from the ambient environment during the exposure period. In the first survey, the Co and Ni concentrations of the AMs from Bospoort Dam were significantly higher when compared to the Olifantsnek Dam. In the second survey, the AMs of the Olifantsnek Dam contained significantly higher Cr, Ni, and Pb concentrations, but significantly lower V concentrations when compared to Bospoort Dam. The AMs from the pollution control dam in the second survey contained similar As, Cd, and Pt concentrations, but significantly lower Co, Ni, and Zn concentrations when compared to the other impoundments. There were significant differences in the Cr concentrations of the AMs decreasing in the following order: pollution control dam > Olifantsnek Dam > Bospoort Dam.

The element concentrations in the transplanted clams showed a heterogeneous picture (Figures 4–6, right). The concentrations of Cd, Pb, Pt, and Zn in the transplanted clams of all three sampling sites and both surveys did not exceed the respective concentrations of the controls. Significantly higher concentrations occurred only for As and Ni (survey 2017), when compared to the control, as well as for Cr and V (survey 2018) in the clams of Olifantsnek Dam. However, for some elements, the transplanted clams contained lower concentrations as compared to the control, as seen for Cd and Zn (survey 2018) in both impoundments as well as for Pb (both surveys) and V (survey 2017) in Bospoort Dam. The Pt concentrations in all of the clams were below the detection limit. Unfortunately, no comparison can be made between the transplanted clams of the impoundments and the pollution control dam since all of the organisms died during the exposure period.

Thus, the AMs and transplanted clams accumulated different elements from the two impoundments, including Olifantsnek Dam and Bospoort Dam, in different patterns. This can be seen by the clear dissimilarity between the AMs and transplanted clams on the PCA (Figure 7). The AMs correlate strongly with Pb and V, while the transplanted clams correlate with Cd, Cr, Co, Zn, and As. The groupings of element accumulation in the AMs indicated site-specific and survey-specific patterns, while the temporal and spatial groups were less clear for the clams.

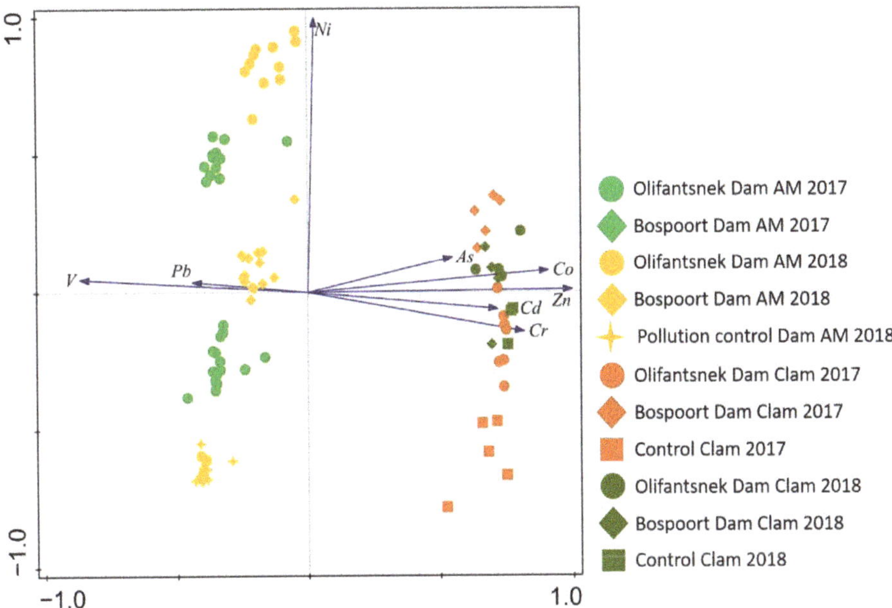

Figure 7. Principal component analysis (PCA) showing the grouping of AMs and transplanted clams based on the similarity of element bioaccumulation at the different sites and surveys. The ordination explains 93.7% of the total variation in the data with 79.5% on axis 1 and 14.2% on axis 2.

The PCA also indicated some groupings and correlations within the AMs (Figure 8). The AMs from the Olifantsnek Dam in 2017 and the pollution control dam in 2018 correlated strongly with Cd, while there was a strong correlation of the AMs with Pb and Pt in both impoundments in 2017. Olifantsnek Dam correlated with As, Co, and Ni, Bospoort Dam correlated with V, and the pollution control dam correlated with Cd in 2018. The grouping of the AMs of the three impoundments of the survey in 2018 indicated spatial differences.

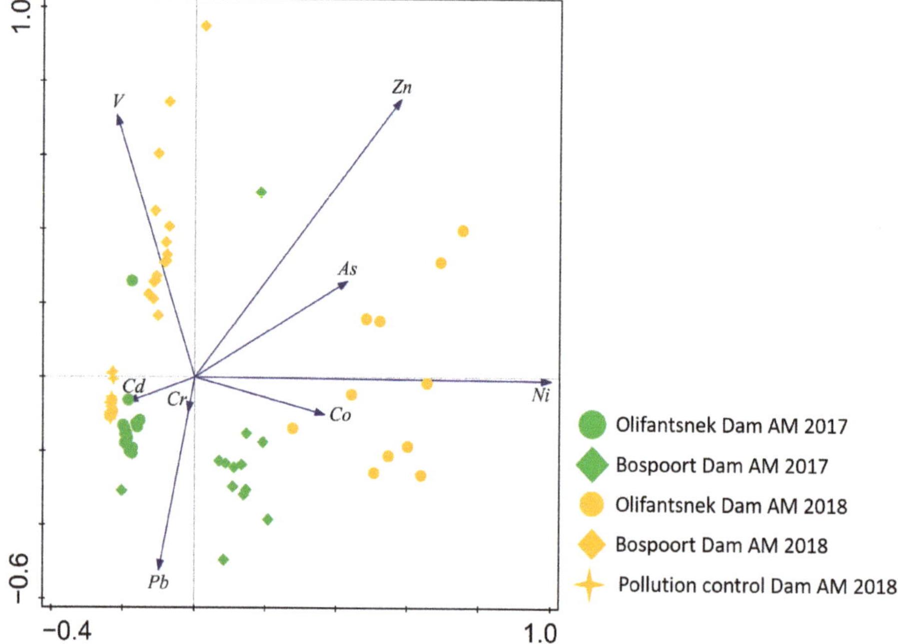

Figure 8. PCA illustrating the grouping of AMs within the three impoundments during the two surveys based on the similarity of metal bioaccumulation at the different sites and surveys. The ordination explains 98.9% of the total variation in the data with 97.3% on axis 1 and 1.6% on axis 2.

Within the transplanted clams, the PCA revealed groupings and correlations, which were different from the AM results (Figure 9). The bioaccumulation patterns of the clams from the Olifantsnek Dam in 2017 were distinctly different from all other clams and correlated strongly with As, Cr, and V. Additionally, the clams from Bospoort Dam correlated strongly with Ni. Furthermore, the control clams in 2018 showed a strong correlation with Cd, Pb, and Zn.

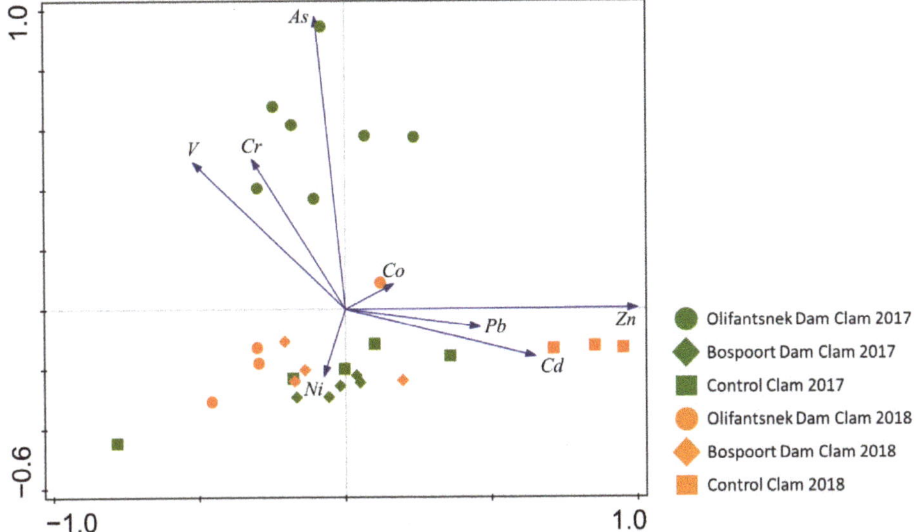

Figure 9. PCA illustrating the grouping of transplanted clams within the two impoundments during the two surveys based on the similarity of metal bioaccumulation at the different sites and surveys. The ordination explains 98.5% of the total variation in the data with 95.7% on axis 1 and 2.8% on axis 2.

4. Discussion

Point and diffuse source pollution from mining and production activities are recognized as an important source of contamination in the environment [16]. Platinum group elements are mined from primary deposits where they are associated with other elements such as Cu and Ni [16]. Elevated concentrations of PGE and other mining-associated elements have been reported in the vicinity of mining areas [2]. Analyses of road dust samples from four highly populated cities in South Africa demonstrated that Rustenburg had the most elevated Pt concentrations associated with mining and production activities [16]. The Hex River, which flows through Rustenburg, drains one of the most important mining areas in the western limb of the Bushveld Igneous Complex [2]. Rustenburg is located in a large basin, which forms part of the Bushveld Igneous Complex, where the ore mainly consists of PGEs. In addition to the PGEs and chromite, the ore contains deposits of gold and sulphides of nickel, copper, and iron [17].

4.1. Metals in Water

Olifantsnek Dam is geographically separated from the mining area in Rustenburg, while the Bospoort Dam is located in the vicinity of many informal settlements as well as many of the mines found in this area [17]. The element concentrations in the water showed that Bospoort Dam was slightly more polluted than the Olifantsnek Dam. The Hex River flows in a northerly direction through all the industrial and mining areas [2]. Consequently, wastes from these activities, which is discharged into the environment, will end up in the Hex River, where Bospoort Dam acts as a sink by accumulating the discharged elements. Since both of these impoundments are located in an area that contains geologically-rich deposits, it is assumed that some elements will occur naturally in high concentrations in both impoundments [2,25].

According to the weather data of the Rustenburg region, there was almost no rain during the first survey [26]. This resulted in very low flow to no flow in the river between the two impoundments during the first survey. The rainfall and ensuing run-off were much greater and more variable during the second survey in 2018. This is also reflected in the far greater variation in physic-chemical parameters. This is supported by Prathumratana et al. [27] who showed that rainfall and associated run-off influences variations in site-specific water quality.

The increase in element concentrations measured in Bospoort Dam compared to Olifantsnek Dam can be linked to a local point source located along the course of the Hex River and not from run-off from the greater catchment. Bospoort Dam contained slightly higher Cd, Co, and Ni concentrations than the Olifantsnek Dam. In general, element concentrations in the water of the Bospoort Dam were higher when compared to the Olifantsnek Dam. The significant difference in metal concentrations from the two impoundments at the end of the first survey can be attributed to the change in season, where there was almost no rainfall during this period. During this period, metal concentrations within the water became more concentrated due to low inflow and outflow of the impoundment and water evaporation.

The water data indicated that almost all of the analysed elements were present in concentrations below South Africa's target water quality range (see supplement data Table S2), for both aquatic ecosystems and domestic use [28,29]. However, the Zn concentrations in the pollution control dam were above the Target Water Quality Range (TWQR) (2 µg/L) for aquatic ecosystems. The concentrations found in the pollution control dam were above the acute effect value (36 µg/L) [29].

The pollution control dam from the mine contained significantly higher As, Co, Cr, Ni, Pt, V, and Zn concentrations than the other two impoundments. Pollution control dams are generally constructed to collect the sludge containing the residues rejected from the ore, to allow solids to settle and to let the liquid phase evaporate [17]. The main environmental risk of such pollution control dams is the possibility that the wall of the dam comes apart and the sludge flows over the mining areas, farmlands, and roads and into aquatic ecosystems [17]. Furthermore, the pollution control dam can be overflown during heavy rain so that over-spilling water containing soluble contaminants and re-mobilized particles can enter the nearby aquatic system. Therefore, it is important to control the element concentrations within these dams. The pollution control dam of the present study is located near the river, which facilitates an entry of contaminants into the Hex River system.

South Africa is a mineral resource-rich region and, consequently, metal and metalloid concentrations are elevated in rivers that drain through mining regions. The Olifants River is generally regarded as the most metal-enriched system in South Africa. Studies by Gerber et al. recorded average concentrations of 0.75 µg/L As, 7.41 µg/L Cd, 5.26 µg/L Co, 4.24 µg/L Cr, 1.43 µg/L Ni, 4.74 µg/L Pb, and 2.68 µg/L Zn [30]. In this case, Cd concentrations were orders of magnitude higher than the values in the present study. Additionally, the highest Cr concentrations measured in the Olifants River were approximately double than those measured in the Olifantsnek Dam and the Bospoort Dam. However, the Ni concentrations were 3, 13, and 35 times higher in the Olifantsnek Dam, the Bospoort Dam, and the pollution control dam, respectively. In addition, the Zn concentrations in the water of the pollution control dam of the present study were 14 times higher than in the study by Gerber et al. [30].

For the water of the Marico River system, which is one of the least impacted systems in a mining region in South Africa [31], levels of 0.858 ± 0.1912 µg/L As, <0.0001 µg/L Cd, 0.693 ± 0.056 µg/L Co, 5.9 ± 0.154 µg/L Cr, 7.158 ± 0.586 µg/L Ni, 0.009 ± 0.007 µg/L Pb, and 2.275 ± 1.33 µg/L Zn were reported. The aqueous As concentrations in the Olifantsnek Dam and the Bospoort Dam were similar. The Co concentration was even lower when compared to the Marico River. The pollution control dam, however, contained 2-times and 4-times the concentrations of As and Co, respectively, which was found within the Marico River. However, the As, Co, Ni, and Zn concentrations in the water of the pollution control dam significantly exceeded the respective concentrations in the two other impoundments as well as in the Marico River system, which indicated that these elements were derived from the nearby mines. In contrast, although the mining activities near Rustenburg include Cr mining, the Cr concentrations within the impoundments and the pollution control dam were lower than those found in the Marico River. A different picture shows the aqueous Cd and Pb concentrations, which were higher at all three sites of the present study. Higher Cd and Pb concentrations observed upstream of the mining and other human activities may be the result of the geogenic input from the natural geology [2].

In recent decades, the demand and consumption of platinum group elements, especially in the automobile catalytic converter production, has continuously increased, which resulted in increasing anthropogenic emissions of these metals worldwide [32]. In general, the water solubility of PGE from road dust as well as their biological availability decreased in the order Pd > Pt > Rh [32,33]. Studies on Pt concentrations in the water of rivers without a PGE mining impact recorded levels between 0.006–2.6 ng/L [34–37]. In the present study, the concentrations found within the two impoundments were considerably higher and ranged between 41.5–76.2 ng/L. The water of the pollution control dam contained Pt concentrations (145 ng/L), which were three times higher than the highest concentration found in the two impoundments. This clearly indicates the entry of Pt from the mining activities into the aquatic ecosystem.

4.2. Transplanted Organisms

In the present study, *C. fluminalis africana* was used for active monitoring of metal and metalloid contaminations. Significantly higher concentrations when compared to the control occurred only for As, Cr, Ni, and V, but not for all sampling sites and surveys, respectively. The pre-exposure of the transplanted organisms from a presumed non-contaminated reference site could have resulted in high background levels so that additional bioaccumulation due to the active biomonitoring is not detectable or depressed. In a study by Ruchter and Sures, *Corbicula* sp. were analysed to determine the influence of road runoff near Karlsruhe, Germany [38]. Reference *Corbicula* sp. were analysed to determine background levels of Cd, Cr, Ni, Pb, Pt, and Zn in the organisms before the inlet [38]. These concentrations were much lower than the concentrations measured in the control group in this study except Pb. They were able to detect Pt concentrations. Furthermore, the effects of upswelling and the change in habitat of the transplanted clams can affect the outcome of active biomonitoring studies [7]. The transplanted clams contained higher As, Cd, Co, Cr, and Zn concentrations. The Olifantsnek Dam in 2017 was the only group that was distinctly different.

In studies that make use of transplanted organisms, it is difficult to compare the results to data found in the literature. Nevertheless, when comparing the bioaccumulation by the clams with aquatic snails that were transplanted to a gold mining area [13], both transplanted organisms show similar Co concentrations. However, the As, Cr, Ni, V, and Zn concentrations of the clams were lower when compared with the snails, while the Cd and Pb concentrations were much higher [13].

The higher water temperatures recorded in Bospoort Dam may have resulted in the metabolism of the transplanted clams being higher than in the Olifantsnek Dam since increased temperature and the subsequent metabolism results in aquatic organisms accumulating more metals [39].

When PGEs from road dust accumulated in the soft tissue of bivalves, e.g., *Dreissena polymorpha*, the bioaccumulation of metals decreased in the following order: Cu > Cd > Ag > Pd > Pb > Sb > Fe > Pt > Rh [33]. Platinum concentrations in bivalves sampled from different freshwater ecosystems ranged between 0.00001–0.0013 µg/g [33,38,40], whereas the Pt levels of the transplanted clams from the present study were all under the detection limit of 0.0009 µg/g.

The bioaccumulation in the transplanted clams did not correlate with the exposure concentration in the water column. Claassens et al. [13] made use of transplanted organisms at different localities within the Koekemoer Spruit, South Africa, and found no correlation between the concentrations in the transplanted organisms and the ambient water. This may be explained by the fact that the indicator organisms integrate the pollutants over the whole exposure period, while water samples only indicate the exposure concentration at the time of sampling. Many factors can influence the bioaccumulation of metals in field studies, i.e., among others' seasonality, the bioavailability of metals, and toxicokinetic processes in the indicator organism [41]. Bivalves are filter feeders. Therefore, both the dissolved metal fraction and the particle-bound metals play a crucial role in the metal bioaccumulation.

4.3. Artificial Mussels

The present study reported the uptake of Pt in AMs for the first time. Although the AMs are only able to take up dissolved metals, the availability of Pt could only be shown by the use of the AMs and not by the transplanted clams, where the Pt levels were below the detection limit.

Artificial mussels have successfully been used as monitoring tools for metal bioaccumulation in the aquatic environment [11,13,20,23,42]. Many of these studies endorse the use of AMs during biomonitoring studies since they are not affected by water quality conditions that would greatly effect a bioindicator. Artificial mussels that were used for monitoring aquatic systems in a gold mining region [13] accumulated higher concentrations of Cr, Co, Cd, Pb, V, and Zn than in the present study. In contrast, AMs exposed to potentially less impacted sites along the Nyl River, Limpopo, South Africa [20] contained either similar or higher element concentrations when compared to this study.

The element concentrations in the water indicated that Bospoort Dam contained higher concentrations than those found in the Olifantsnek Dam. During the first survey, the AMs in the two impoundments had similar concentrations, while, in the second survey, there were some significant differences between the sampling sites. However, these results (AMs) did not indicate a specific site has higher concentrations. Other field studies using AMs also show variations in data.

The AMs contained higher Pb and V concentrations when compared to the transplanted clams. The element accumulation in the AMs indicated site-specific and survey-specific patterns.

4.4. Comparison of Artificial Mussels and Transplanted Clams

For almost all of the elements, in the present study, the transplanted clams contained significantly higher concentrations than the AMs. According to Wu et al. [10], the AM and any bioindicator species bioaccumulate different metal fractions during field exposures. Due to the assembly of the AM, large molecules and particle-bound metals are not able to cross the polyacrylamide gel barrier, which acts as a surrogate on the double lipid layer of a cell membrane. However, metals and metalloids occur in the environment in different complexes or may be attached to particles. Many factors can influence their speciation. In contrast to AMs, the transplanted clams can ingest different metals and metalloids from the food that is available within the environment and living organisms can regulate their concentrations by toxicokinetic processes [10]. The characteristics of the permeability of the AM and the binding capacity of the Chelex beads can also have an influence [10,23]. A combination of all of these factors may explain the differences in the accumulation capacity between AM and TM.

The partitioning (dissolve vs. particulate bound) and speciation of metals are dependent on several chemical and physical properties, such as pH, temperature, and the presence of ligands and particulates [43]. Morquecho and Pitt [43] filtered several stormwater samples to determine the colloidal bound metals, metals bound to ligands, and the ionic/dissolved fraction. After adding Chelex beads to the filtered samples, they found that the ionic form of the metals had bound to the beads, while the particle-bound metals remained in the solution. More than 90%, 80%, 80%, and 30% of the filtered Zn, Cr, Pb, and Cd were in an ionic form, respectively [43]. These percentages reflect the amount of each metal that was bound to the Chelex beads during the filtering process. Therefore, in the present study, the elements being accumulated by the AMs belong to the dissolved fraction, whereas the TMs accumulated the dissolved and particulate bound fraction. This may explain the different accumulation patterns for the AMs and TMs.

Claassens et al. [13] found that the AMs accumulated higher concentrations of Cd, Pb, V, and Zn than the transplanted organisms. This is partly in accordance with the present study showing that the AMs accumulated higher concentrations of Pb, Pt, and V. However, a direct comparison of the concentrations in the AM and TM is problematic since, for the AMs, the concentrations were calculated by dividing the total mass of the element that has bound to the Chelex beads by the mass (200 mg) of the Chelex beads. Assuming that not all binding sites of the Chelex beads were saturated, a lower mass of Chelex beads would have resulted in the same amount of metal uptake but a higher calculated concentration in the AM. Thus, it should be considered in future studies that the total mass is a more reliable measure for the uptake by the AM than the concentration. However, as in all studies on AMs published so far, 200 mg of Chelex beads were used since a comparison between these studies is possible by using the concentrations.

Other studies demonstrated that AMs and transplanted organisms accumulate metals in similar patterns but at different concentrations [13,23]. In contrast, in the present study, the accumulation patterns of the AMs indicated significant accumulation when compared to the initial concentrations, whereas, for the transplanted clams for most elements, no clear bioaccumulation was found during the biomonitoring period. Thus, since the AMs have low initial background levels, they can indicate the bioaccumulation of low element concentrations in biomonitoring studies.

During the second survey, over six weeks, there was a massive water hyacinth (*Eichhornia crassipes*) bloom that occurred in the Bospoort Dam. These water plants covered more than 50% of the impoundment at the end of the exposure period. The water hyacinth can accumulate high levels of metals from solutions such as Cd, Co, Cr, Cu, Ni, Pb, V, and Zn [44,45] and the PGEs Pt, Pd, Os, Ru, Ir, and Rh [42]. When considering the biomass of the water hyacinth in the dam, the plants likely had reduced the metal concentrations in the water. This may explain why the concentrations in the water of Bospoort Dam were not significantly different from those found in the Olifantsnek Dam during the second survey whereas the concentrations during the first survey were much higher in the Bospoort Dam.

During the second survey, transplanted clams and AMs were deployed in the pollution control dam on the mine ground, but none of the transplanted clams survived. It is presumed that the transplanted clams died due to a layer of oil on the water surface and decreasing water levels than due to very high metal concentrations. This shows the practicability of the AMs in ecosystems where adverse environmental conditions affect the survival of living organisms. However, it is possible that the oil layer obstructed the metals from diffusing through the gel layers. The metal and metalloid concentrations within the water from the pollution control dam were significantly higher than the impoundments. However, the AMs did not indicate this difference.

5. Conclusions

Different bioindicator species have different accumulation strategies. They are affected by many factors within the environment and can regulate internal metal concentrations through physiological processes [13,15]. The difference between the results of the AMs and the transplanted clams should not be seen as a shortcoming and was expected based on results from previous studies. The present study demonstrated that the transplanted clams and AMs accumulated different metals at different concentrations, which could be attributed to the accumulation of metals in different forms (dissolved or particle-bound) showing different biological availabilities. The AMs and transplanted clams accumulated similar As concentrations, while the AMs preferentially accumulated Pb, Ni, Pt, and V. The AMs are a promising tool for metals and metalloids occurring at low environmental concentrations close to the analytical detection limit and when organisms for the transplantation during active biomonitoring approaches already show high background levels. Metals can be found in various forms within the aquatic environment. The semi-permeable membrane of the AM permits the bioavailable metal ions to pass through and bind to the Chelex beads. Some metal ions have a stronger binding affinity toward the Chelex beads, where these ions may compete for binding sites within the AMs. For that reason, it is possible that some metals can be more abundant in the environment but the AMs will not be able to indicate these concentrations due to competition by other metals. Nonetheless, in this study, it was shown that the AM successfully accumulated Pt and all the other metals that were studied. Traditional bioindicator bivalves were more successful in indicating exposure to metals such as Cd, Co, Cr, and Zn. Thus, the combination of AMs and living bioaccumulation indicator organisms will provide a holistic assessment of metal and metalloid exposure in aquatic environments.

Water **2020**, *12*, 32

Supplementary Materials: The following are available online at http://www.mdpi.com/2073-4441/12/1/32/s1. Table S1: Metal concentrations measured in the transplanted clams (TM), artificial mussels (AM), and water from two surveys conducted in Bospoort Dam, Olifantsnek Dam, and the pollution control dam. The control refers to concentrations in reference clams that were maintained in the laboratory. Concentrations are presented in μg/g dry mass for transplanted clams, μg/g Chelex for the AMs, and water concentrations in μg/L (mean ± SEM, ND = below detection limit, the * indicates significant difference from the control group in a specific survey). Table S2: Target water quality ranges for domestic use and aquatic ecosystems as set by the Department of Water Affairs and Forestry.

Author Contributions: Conceptualization, B.S. and N.J.S. Formal analysis, M.L., M.N., S.Z., B.S., and N.J.S. Investigation, M.L. Methodology, M.L. and V.W. Resources, B.S. and N.J.S. Supervision, V.W. and S.Z. Validation, V.W., M.N., and S.Z. Writing-original draft, M.L. Writing-review & editing, V.W., S.Z., B.S., and N.J.S. All authors have read and agreed to the published version of the manuscript.

Funding: The financial assistance of the NRF (National Research Foundation, South Africa), [UID108852] towards this research is hereby acknowledged. This work is based on the research and researchers supported by the National Research Foundation (NRF) of South Africa (NRF Project GERM160623173784; grant 105875; NJ Smit, PI) and BMBF/PT-DLR (Federal Ministry of Education and Research, German, grant 01DG17022; B Sures, PI). Opinions, findings, conclusions, and recommendations expressed in this publication are that of the authors, and the NRF and BMBF/PT-DLR accept no liability whatsoever in this regard.

Acknowledgments: This is the contribution number 360 of the North-West University (NWU) Water Research Group. The assistance of Anja Greyling, Hannes Erasmus, and Rian Pienaar is hereby acknowledged for their help during sampling trips and the map of the study site.

Conflicts of Interest: The authors declare no conflict of interest.

References

1. Rauch, S.; Peucker-Ehrenbrink, B. Sources of Platinum Group Elements in the Environment. In *Platinum Metals in the Environmen*; Zereini, F., Wiseman, C.L.S., Eds.; Springer: Heidelberg, Germany, 2015; Volume 1, pp. 3–17.

2. Almécija, C.; Cobelo-García, A.; Wepener, V.; Prego, R. Platinum group elements in stream sediments of mining zones: The Hex River (Bushveld Igneous Complex, South Africa). *J. Afr. Earth Sci.* **2017**, *129*, 934–943. [CrossRef]

3. Erasmus, J.; Malherbe, W.; Zimmermann, S.; Lorenz, A.; Nachev, M.; Wepener, V.; Sures, B.; Smit, N. Metal accumulation in riverine macroinvertebrates from a platinum mining region. *Sci. Total Environ.* **2019**, *703*, 134738. [CrossRef] [PubMed]

4. Zhou, Q.; Zhang, J.; Fu, J.; Shi, J.; Jiang, G. Biomonitoring: An appealing tool for assessment of metal pollution in the aquatic ecosystem. *Anal. Chim. Acta* **2008**, *606*, 135–150. [CrossRef]

5. Gupta, S.K.; Singh, J. Evaluation of mollusc as sensitive indicator of heavy metal pollution in aquatic system: A review. *IIOAB J.* **2011**, *2*, 49–57.

6. Farris, J.L.; Van Hassel, J.H. *Freshwater Bivalve Ecotoxicology*; CRC Press: Boca Raton, FL, USA, 2006. [CrossRef]

7. Wepener, V. Application of active biomonitoring within an integrated water resources management framework in South Africa. *S. Afr. J. Sci.* **2008**, *104*, 367–373.

8. Giarratano, E.; Duarte, C.A.; Amin, O.A. Biomarkers and heavy metal bioaccumulation in mussels transplanted to coastal waters of the Beagle Channel. *Ecotoxicol. Environ. Saf.* **2010**, *73*, 270–279. [CrossRef]

9. Wu, R.S.S.; Lau, T.C. Polymer-ligands: A novel chemical device for monitoring heavy metals in the aquatic environment? *Mar. Pollut. Bull.* **1996**, *32*, 391–396. [CrossRef]

10. Wu, R.S.S.; Lau, T.C.; Fung, W.K.M.; Ko, P.H.; Leung, K.M.Y. An 'artificial mussel' for monitoring heavy metals in marine environments. *Environ. Pollut.* **2007**, *145*, 104–110. [CrossRef]

11. Kibria, G.; Lau, T.; Wu, R. Innovative 'Artificial Mussels' technology for assessing spatial and temporal distribution of metals in Goulburn–Murray catchments waterways, Victoria, Australia: Effects of climate variability (dry vs. wet years). *Environ. Int.* **2012**, *50*, 38–46. [CrossRef]

12. Hossain, M.M.; Kibria, G.; Nugegoda, D.; Lau, T.; Wu, R. *A Training Manual for Assessing Pollution (trace/heavy metals) in Rivers, Estuaries and Coastal waters-Using Innovative "Artificial Mussel (AM) Technology"—Bangladesh Model. Research collaboration between Scientists of the IMSF, University of Chittagong, Bangladesh, RMIT University, Australia, the City University of Hong Kong, and the University of Hong Kong*; FAO: Rome, Italy, 2015; Volume 1, pp. 1–24. [CrossRef]

13. Claassens, L.; Dahms, S.; Van Vuren, J.; Greenfield, R. Artificial mussels as indicators of metal pollution in freshwater systems: A field evaluation in the Koekemoer Spruit, South Africa. *Ecol. Indic.* **2016**, *60*, 940–946. [CrossRef]

14. Leung, K.M.Y.; Furness, R.W.; Svavarsson, J.; Lau, T.C.; Wu, R.S.S. Field validation, in Scotland and Iceland, of the artificial mussel for monitoring trace metals in temperate seas. *Mar. Pollut. Bull.* **2008**, *57*, 790–800. [CrossRef] [PubMed]

15. Degger, N.; Wepener, V.; Richardson, B.J.; Wu, R.S.S. Application of artificial mussels (AMs) under South African marine conditions: A validation study. *Mar. Pollut. Bull.* **2011**, *63*, 108–118. [CrossRef] [PubMed]

16. Rauch, S.; Fatoki, O.S. Impact of Platinum Group Element Emissions from Mining and Production Activities. In *Platinum Metals in the Environment*; Zereini, F., Wiseman, C.L.S., Eds.; Springer: Heidelberg, Germany, 2015; pp. 19–29. [CrossRef]

17. Wittmann, G.; Förstner, U. Heavy metal enrichment in mine drainage: 1. The Rustenburg platinum mining area. *Science* **1976**, *72*, 242–246.

18. Roux, H. Status of the yellowfish populations & River health Programme in NW Province. In Proceedings of the 15th Yellowfish Working Group Conference, Mpumalanga, South Africa, 18–20 February 2011.

19. Mogakabe, D.; Van Ginkel, C. *The Water Quality of the Bospoort Dam*; Internal Report No. N/0000/00/DEQ/0108; Department of Water Affairs and Forestry: Pretoria, South Africa, 2008.

20. Dahms-Verster, S.; Baker, N.; Greenfield, R. A multivariate examination of 'artificial mussels' in conjunction with spot water tests in freshwater ecosystems. *Environ. Monit. Assess.* **2018**, *190*, 427. [CrossRef] [PubMed]

21. Brand, S.J.; Erasmus, J.H.; Labuschagne, M.; Grabner, D.; Nachev, M.; Zimmermann, S.; Wepener, V.; Smit, N.; Sures, B. Bioaccumulation and metal-associated biomarker responses in a freshwater mussel, Dreissena polymorpha, following short-term platinum exposure. *Environ. Pollut.* **2019**, *246*, 69–78. [CrossRef]

22. Zimmermann, S.; Menzel, C.M.; Berner, Z.; Eckhardt, J.-D.; Stüben, D.; Alt, F.; Messerschmidt, J.; Taraschewski, H.; Sures, B. Trace analysis of platinum in biological samples: A comparison between sector field ICP-MS and adsorptive cathodic stripping voltammetry following different digestion procedures. *Anal. Chim. Acta* **2001**, *439*, 203–209. [CrossRef]

23. Degger, N.; Chiu, J.M.Y.; Po, B.H.K.; Tse, A.C.K.; Zheng, G.J.; Zhao, D.-M.; Xu, D.; Cheng, Y.-S.; Wang, X.-H.; Liu, W.-H.; et al. Heavy metal contamination along the China coastline: A comprehensive study using Artificial Mussels and native mussels. *Environ. Manag.* **2016**, *180*, 238–246. [CrossRef]

24. Kushner, E. On determining the statistical parameters for pollution concentration from a truncated data set. *Atmos. Environ.* **1976**, *10*, 975–979. [CrossRef]

25. Cawthorn, R.G. The platinum group element deposits of the Bushveld Complex in South Africa. *Platin. Met. Rev.* **2010**, *54*, 205–215. [CrossRef]

26. Thorsen, S. Past Weather in Rustenburg, South Africa—May 2017. Available online: https://www.timeanddate.com/weather/south-africa/rustenburg/historic?month=5&year=2017 (accessed on 1 October 2018).

27. Prathumratana, L.; Sthiannopkao, S.; Kim, K.W. The relationship of climatic and hydrological parameters to surface water quality in the lower Mekong River. *Environ. Int.* **2008**, *34*, 860–866. [CrossRef]

28. Department of Water Affairs and Forestry (DWAF). *South African Water Quality Guidelines (Second Edition). Volume 1: Domestic Use*; DWAF: Pretoria, South Africa, 1996.

29. Department of Water Affairs and Forestry (DWAF). *South African Water Quality Guidelines. Volume 7: Aquatic Ecosystems*; DWAF: Pretoria, South Africa, 1996.

30. Gerber, R.; Smit, N.J.; van Vuren, J.H.; Wepener, V. Metal concentrations in *Hydrocynus vittatus* (Castelnau 1861) populations from a premier conservation area: Relationships with environmental concentrations. *Ecotoxicol. Environ. Saf.* **2016**, *129*, 91–102. [CrossRef] [PubMed]

31. Kemp, M.; Wepener, V.; de Kock, K.; Wolmarans, C. Metallothionein Induction as Indicator of Low Level Metal Exposure to Aquatic Macroinvertebrates from a Relatively Unimpacted River System in South Africa. *Bull. Environ. Contam. Toxicol.* **2017**, *99*, 662–667. [CrossRef] [PubMed]

32. Hoppstock, K.; Sures, B. Platinum-Group Metals. In *Elements and Their Compounds in the Environment: Occurrence, Analysis and Biological Relevance*, 2nd ed.; Merian, E., Anke, M., Ihnat, M., Stoeppler, M., Eds.; WILEY-VCH Verlag GMBH & Co. KGaA: Weinheim, Germany, 2004; pp. 1047–1086.

33. Zimmermann, S.; Alt, F.; Messerschmidt, J.; von Bohlen, A.; Taraschewski, H.; Sures, B. Biological availability of traffic-related platinum-group elements (palladium, platinum, and rhodium) and other metals to the zebra mussel (*Dreissena polymorpha*) in water containing road dust. *Environ. Toxicol. Chem.* **2002**, *21*, 2713–2718. [CrossRef] [PubMed]

34. Hoppstock, K.; Alt, F. Voltammetric determination of ultratrace platinum and rhodium in biological and environmental samples. In *Anthropogenic Platinum-Group Element Emissions*; Fathi, Z., Friedrich, A., Eds.; Springer: Heidelberg, Germany, 2000; pp. 145–152. [CrossRef]

35. Moldovan, M.; Gomez, M.M.; Palacios, M.A. On-line preconcentration of palladium on alumina microcolumns and determination in urban waters by inductively coupled plasma mass spectrometry. *Anal. Chim. Acta* **2003**, *478*, 209–217. [CrossRef]

36. Monticelli, D.; Carugati, G.; Castelletti, A.; Recchia, S.; Dossi, C. Design and development of a low cost, high performance UV digester prototype: Application to the determination of trace elements by stripping voltammetry. *Microchemical* **2010**, *95*, 158–163. [CrossRef]

37. Cobelo-García, A.; López-Sánchez, D.E.; Almécija, C.; Santos-Echeandía, J. Behavior of platinum during estuarine mixing (Pontevedra Ria, NW Iberian Peninsula). *Mar. Chem.* **2013**, *150*, 11–18. [CrossRef]

38. Ruchter, N.; Sures, B. Distribution of platinum and other traffic related metals in sediments and clams (Corbicula sp.). *Water Res.* **2015**, *70*, 313–324. [CrossRef]

39. van Heerden, D.; van Rensburg, P.J.; Nikinmaa, M.; Vosloo, A. Gill damage, metallothionein gene expression and metal accumulation in *Tilapia sparrmanii* from selected field sites at Rustenburg and Potchefstroom, South Africa. *Afr. J. Aquat. Sci.* **2006**, *31*, 89–98. [CrossRef]

40. Ruchter, N.; Zimmermann, S.; Sures, B. Field studies on PGE in aquatic ecosystems. In *Platinum Metals in the Environment*; Zereini, F., Wiseman, C.L.S., Eds.; Springer: Heidelberg, Germany, 2015; pp. 351–360. [CrossRef]

41. Greenfield, R.; Brink, K.; Degger, N.; Wepener, V. The usefulness of transplantation studies in monitoring of metals in the marine environment: South African experience. *Mar. Pollut. Bull.* **2014**, *85*, 566–573. [CrossRef]

42. Genç, T.O.; Po, B.H.; Yılmaz, F.; Lau, T.C.; Wu, R.S.; Chiu, J.M. Differences in metal profiles revealed by native mussels and artificial mussels in Sarıçay Stream, Turkey: Implications for pollution monitoring. *Mar. Freshw. Res.* **2018**, *69*, 1372–1378. [CrossRef]

43. Morquecho, R.; Pitt, R. Stormwater heavy metal particulate associations. *Proc. Water Environ. Fed.* **2003**, *2003*, 774–803. [CrossRef]

44. Nor, Y.M. The absorption of metal ions by Eichhornia crassipes. *Chem. Speciat. Bioavailab.* **1990**, *2*, 85–91. [CrossRef]

45. Shirinpur-Valadi, A.; Hatamzadeh, A.; Sedaghathoor, S. Study of the accumulation of contaminants by Cyperus alternifolius, Lemna minor, Eichhornia crassipes, and Canna generalis in some contaminated aquatic environments. *Environ. Sci. Pollut. Res. Int.* **2019**, *26*, 21340–21350. [CrossRef] [PubMed]

Article

Use of Larval Morphological Deformities in *Chironomus plumosus* (Chironomidae: Diptera) as an Indicator of Freshwater Environmental Contamination (Lake Trasimeno, Italy)

Enzo Goretti [1,*], Matteo Pallottini [1], Sarah Pagliarini [1], Marianna Catasti [1], Gianandrea La Porta [1], Roberta Selvaggi [1], Elda Gaino [1], Alessandro Maria Di Giulio [2] and Arshad Ali [3]

[1] Dipartimento di Chimica, Biologia e Biotecnologie, Università degli Studi di Perugia, 06123 Perugia, Italy; matteo.pallottini@unipg.it (M.P.); sarah.pagliarini87@gmail.com (S.P.); marianna.catasti@gmail.com (M.C.); gianandrea.laporta@unipg.it (G.L.P.); roberta.selvaggi@unipg.it (R.S.); elda.gaino@unipg.it (E.G.)

[2] Servizio Disinfestazione, USLUmbria1, 06127 Perugia, Italy; alessandro.digiulio@uslumbria1.it

[3] Entomology and Nematology Department, University of Florida, Gainesville, FL 32611, USA; umar@ufl.edu

* Correspondence: enzo.goretti@unipg.it

Received: 31 October 2019; Accepted: 15 December 2019; Published: 18 December 2019

Abstract: The mentum deformity incidence in *Chironomus plumosus* larvae to assess the environmental contamination level in Lake Trasimeno, Central Italy, was investigated. The survey lasted from May 2018 to August 2019. Fifty-one samplings were carried out: 34 in the littoral zone and 17 in the central zone. The deformity assessment was based on 737 and 2767 larval specimens of *C. plumosus* collected from the littoral and central zones, respectively. Comparison of the larval morphometric variables between normal and deformed specimens highlighted that the deformities did not cause alterations of the larval growth. The deformity incidence amounted to 7.22% in the whole Trasimeno's ecosystem, reaching 8.28% in the littoral zone and 6.94% in the central zone. Among the different seasonal cohorts, the spring cohort had overall the highest deformity value (11.41%). The deformity type assessment protocol highlighted that the most common deformity type was "round/filed teeth" (64%). The results of this 2018–2019 survey revealed a low deformity incidence, within the background range of relatively low-impacted freshwaters. Comparison with previous investigations (2000–2010) of the same habitat showed a clear decrease of the deformity incidence. This study further contributes to the evaluation of the mentum deformity in chironomids that represent an indicator endpoint of the anthropogenic contamination level in freshwaters.

Keywords: Chironomidae; *Chironomus plumosus* larvae; mentum deformities; freshwater contamination; Lake Trasimeno

---

1. Introduction

Anthropic activities continuously add urban, industrial, and agricultural wastes into the environment. Common and widespread groups of pollutants, such as heavy metals, pesticides, polycyclic aromatic hydrocarbons, and polychlorinated biphenyls are dispersed into the environment forming a complex mixture whose toxic effects are difficult to quantify [1–3].

Normally, the set of contaminants contributing to the total toxic burden is not well known and, even when the exact composition is known, it is difficult to interpret the estimate of potential effects of simple additivity or synergism or antagonism [4]. In fact, the measurement of concentrations of toxic elements or compounds in the environment does not imply knowledge of the resulting toxicity of their mixture on the biota [5]. Furthermore, the peculiar trophic network of living organisms [6] determines bioaccumulation and biomagnification processes of contaminants [7–9].

For an evaluation of the anthropogenic contamination in a given territory, the biological monitoring of freshwaters seems to be the most appropriate strategy. The rains draining the atmosphere and the soil carry pollutants into the water receptor of the underlying hydrographic basin [10,11].

Among aquatic organisms, benthic macroinvertebrates inhabiting the surficial layer of the bottom sediments in inland waters are valid biological indicators because they are subject to the resulting action of pollutants and thus respond to many ecological stressors [12–16]. Aquatic Diptera typically represent the predominant component of the benthic biocoenosis, and among them chironomids (Chironomidae: Diptera) in particular are characterized by diverse species compositions and have a key role in the trophic network of freshwater ecosystems especially those impaired by substantial organic loads [17]. Therefore, chironomid communities represent an important food source for a wide range of aquatic and terrestrial animals [18], also considering in some cases, the role of pestiferous chironomid species [19].

Some chironomid species are frequently used as endpoints (as test organisms) in ecotoxicological bioassays for the assessment of ecological risks; for example, larval growth rate, adult emergence, and survival [20]. In particular, widely used in bioassays are the larvae of aquatic midge species belonging to the genus *Chironomus*, such as *Chironomus plumosus* (Linnaeus, 1758), *Chironomus riparius* (Meigen, 1804), *Chironomus tentans* (Fabricius, 1805), *Chironomus sancticaroli* (Strixino and Strixino, 1981), and *Chironomus tepperi* (Skuse, 1889) [21–25].

In freshwater ecosystems, these bioindicator chironomid species are common, abundant, and tolerant to polluted sediments. The exposure to potentially contaminated sediments is due to their benthic habits and feeding on fine particulate organic matter (POM: 0.5 µm–1 mm) collected from sediments [18]. Thus, they exhibit suitability in monitoring the pollutant effects in inland waters.

In several laboratory and field studies, an association between environmental contamination level (metals, such as As, Cd, Cr, Cu, Hg, Mn, Ni, Pb, Zn, and organic compounds, such as polychlorinated biphenyls (PCBs) and pesticides) and the onset of sublethal effects as deformities of mouthparts (i.e., mentum, mandibles, and pectin epipharyngis) in chironomid larvae has been observed. In this regard, studies on chironomid mouthpart deformities have been developed since the 1970s [26–28] and a vast amount of literature based on laboratory bioassays [22,23,29–41] and field surveys [21,42–47] endorses that the incidence of these deformities is well associated with the degree of sediment toxicity, whereas no relationship has been detected between deformed phenotypes and organic enrichment; hence, the water quality assessment indices are not always consistent with the use of chironomid deformities [48]. However, some studies have expressed concerns about the association between deformity incidence and toxicants [24,49].

The mouthpart deformities probably are a consequence of the interaction of contaminants, playing as endocrine-disrupting chemicals, with hormones structurally related to estrogen, such as ecdysone, that leads and regulates insect growth and metamorphosis [50]. Therefore, the "altered" ecdysone might interfere during the molting processes in larval development, causing phenotypic anomalies, as the mouthpart deformities in chironomids [51].

Thus, morphological deformities in chironomids are an aspecific response to different contaminants, such as heavy metals and pesticides, and an ecotoxicological endpoint to evaluate the total effects of the exposure to a toxic mixture of peculiar toxicants present in the freshwater environments [4,22]. For this reason, chironomid deformities are a reliable biomarker for assessing environmental toxicity.

The purpose of the present research was: (i) to use the incidence of mentum deformities in *C. plumosus* larvae as an effective tool for biological assessment of the environmental contamination entity in Lake Trasimeno, Central Italy; (ii) to compare the results with previous investigation in Lake Trasimeno (nearly ten years ago [43]) and evaluate the enhancement of environmental contamination over time.

2. Materials and Methods

2.1. Study Area

Lake Trasimeno, the fourth largest lake in Italy, is located within the Tiber River basin in Umbria, Central Italy. It is the most extended lake of peninsular Italy (124.3 km^2 surface). It is a shallow lake of tectonic origin, at hydrometric zero (257.33 meters above sea level (m a.s.l.)) and has an average depth of 4.7 m with a maximum depth of 6.3 m [52,53]. It is characterized by a Mediterranean climate, and its hydrological regime depends on local precipitations, which shows considerable fluctuations in water level over the years [54,55]. These peculiar features and the diffuse anthropogenic sources of pollution (agriculture, livestock farms, industry, and urban settlements) in the lake surrounding land areas make this biotope susceptible to contamination [56]. For its naturalistic and ecological importance, Lake Trasimeno is considered as a Special Area of Conservation (IT5210018; Habitat Directive, 1992/43/EEC) and Special Protection Area (IT5210070; Bird Directive, 2009/147/EC), and it is also a regional park.

2.2. Field Sampling Campaign

The field sampling campaign was carried out in the littoral and central zones of the lake. In the littoral zone, six sampling sites were selected along a 350 m area, representative of the main bottom types (sandy, silty, and macrophyte-covered) present in Lake Trasimeno, situated along the coastline (1.5 m average depth) close to the town of Castiglione del Lago. This area among the littoral zone is most subject to the anthropic impact on the lake. In the central zone of the lake, three sampling sites were selected in the same bathymetric zone (4–5 m depth) (Figure 1). The sampling design in the present study was based on a previous investigation [43] carried out from 2000 to 2010 at Lake Trasimeno.

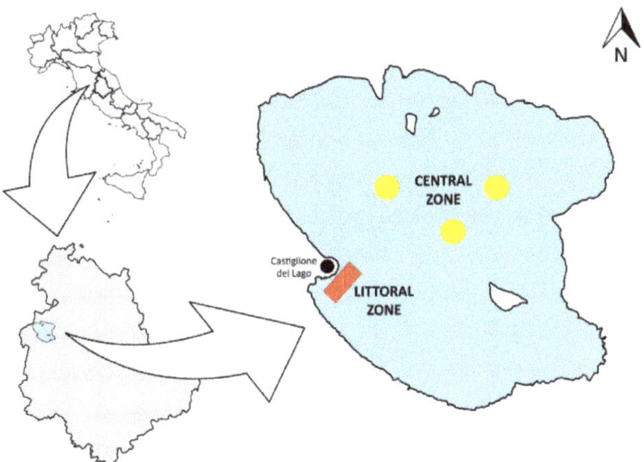

Figure 1. Lake Trasimeno, Umbria, Italy, and location of the field sampling sites.

The survey lasted for 16 months: from May 2018 to August 2019. A total of 51 samplings were carried out: 34 in the littoral zone and 17 in the central zone (Table 1). For each of the two zones, at least one monthly survey was conducted (with the exception of June 2018 for the central zone), with a higher sampling frequency during the spring–summer period.

During each sampling occasion, in the littoral zone 6, quantitative sub-samples of the bottom sediments were collected at each sampling site (a total of 36 sub-samples) with a hand dredge (112 cm^2 sampling surface); in the central zone the sampling technique consisted of a quantitative sampling of the bottom sediments with a compressed air dredge (400 cm^2 sampling surface, 21 mesh per cm nylon net), repeated 5 times per site, with a duration of 1 min each (a total of 15 sub-samples).

Table 1. *Chironomus plumosus* individuals (N), *C. plumosus* relative percentages within the chironomid community (%), Chironomidae densities (individuals m^{-2}), Chironomidae relative percentages within the total macroinvertebrates (%), Naididae densities (individuals m^{-2}), Naididae relative percentages within the total macroinvertebrates (%), other taxa densities (individuals m^{-2}), other taxa relative percentages within the total macroinvertebrates (%), macroinvertebrate densities (individuals m^{-2}); minimum (Min), median, mean, standard deviation (SD), and maximum (Max) of these parameters; (**a**) littoral zone, (**b**) central zone.

				(a) Littoral Zone					
Date	*C. plumosus*	*C. plumosus*	Chironomidae	Chironomidae	Naididae	Naididae	Other Taxa	Other Taxa	Macroinvertebrates
	(N)	(% of Chironomidae)	(ind. m^{-2})	(%)	(ind. m^{-2})	(%)	(ind. m^{-2})	(%)	(ind. m^{-2})
11/05/2018	2	2.94	183	9.24	1793	90.40	7	0.36	1983
18/05/2018	6	4.44	469	28.08	1188	71.17	12	0.74	1669
25/05/2018	4	2.26	646	35.67	1154	63.68	12	0.65	1812
01/06/2018	16	11.85	518	35.13	935	63.36	22	1.51	1476
14/06/2018	7	3.87	599	42.62	578	41.12	228	16.26	1405
22/06/2018	2	1.60	426	39.29	536	49.41	122	11.29	1084
28/06/2018	6	8.33	239	23.76	706	70.23	60	6.01	1006
11/07/2018	16	23.53	242	23.29	670	64.56	126	12.15	1037
19/07/2018	12	29.27	122	9.51	1099	86.02	57	4.47	1277
25/07/2018	8	19.05	152	14.32	814	76.98	92	8.70	1058
01/08/2018	35	51.47	242	18.11	993	74.41	100	7.48	1334
09/08/2018	28	20.90	452	26.55	1182	69.48	68	3.97	1702
22/08/2018	21	24.42	376	20.10	1451	77.51	45	2.39	1872
03/09/2018	72	46.15	490	27.51	1250	70.20	41	2.29	1781
18/09/2018	102	76.69	439	35.76	743	60.60	45	3.64	1226
27/09/2018	12	37.50	179	13.77	1103	85.02	16	1.21	1297
10/10/2018	15	12.93	567	25.65	1641	74.23	3	0.12	2211
17/10/2018	33	13.25	1079	34.09	2066	65.27	20	0.64	3166
24/10/2018	75	35.55	843	47.95	903	51.34	12	0.71	1758
06/11/2018	8	6.11	457	48.29	490	51.71	0	0	947
04/12/2018	9	6.47	472	58.18	339	41.82	0	0	811
16/01/2019	3	1.75	533	53.00	462	45.95	11	1.04	1006
12/02/2019	16	10.53	505	51.70	472	48.30	0	0.00	977
15/03/2019	0	0	258	34.90	469	63.42	12	1.68	739
04/04/2019	6	4.17	463	32.93	944	67.07	0	0	1407
02/05/2019	13	3.75	1008	41.73	1375	56.91	33	1.36	2415
21/05/2019	7	7.95	304	25.70	829	70.19	48	4.10	1181
04/06/2019	16	7.69	722	32.59	1409	63.61	84	3.81	2215
18/06/2019	33	17.19	623	38.20	952	58.45	55	3.35	1629
27/06/2019	11	14.47	252	42.67	328	55.56	11	1.78	591
09/07/2019	68	83.95	262	48.66	262	48.66	14	2.67	539
23/07/2019	24	55.81	130	13.97	760	81.64	41	4.38	931

Table 1. *Cont.*

(a) Littoral Zone

Date	C. plumosus (N)	C. plumosus (% of Chironomidae)	Chironomidae (ind. m^{-2})	Chironomidae (%)	Naididae (ind. m^{-2})	Naididae (%)	Other Taxa (ind. m^{-2})	Other Taxa (%)	Macroinvertebrates (ind. m^{-2})
06/08/2019	72	40.22	656	24.74	1885	71.13	110	4.14	2651
22/08/2019	38	17.33	880	15.13	4655	80.01	283	4.86	5818
Min	0	0.00	122	9.24	262	41.12	0	0.00	539
Median	14	13.09	460	32.76	939	64.91	37	2.34	1369
Mean	23	20.69	464	31.55	1072	64.98	53	3.46	1589
SD	25	21.41	246	13.17	782	12.90	64	3.83	955
Max	102	83.95	1079	58.18	4655	90.40	283	16.26	5818

(b) Central Zone

Date	C. plumosus (N)	C. plumosus (% of Chironomidae)	Chironomidae (ind. m^{-2})	Chironomidae (%)	Naididae (ind. m^{-2})	Naididae (%)	Other Taxa (ind. m^{-2})	Other Taxa (%)	Macroinvertebrates (ind. m^{-2})
29/05/2018	107	81.68	252	16.76	1250	83.24	0	0	1502
18/07/2018	396	98.02	685	75.69	220	24.31	0	0	905
08/08/2018	274	90.43	577	50.96	548	48.45	7	0.59	1132
22/08/2018	274	89.54	542	55.75	430	44.25	0	0	972
21/09/2018	117	100	202	16.40	1028	83.60	0	0	1230
16/10/2018	68	100	120	21.24	445	78.76	0	0	565
28/11/2018	66	100	113	12.88	767	87.12	0	0	880
20/12/2018	107	98.17	208	22.12	732	77.70	2	0.18	942
31/01/2019	82	100	143	22.34	498	77.66	0	0	642
21/02/2019	61	96.83	112	27.57	293	72.43	0	0	405
08/03/2019	86	100	145	27.44	383	72.56	0	0	528
12/04/2019	38	97.44	65	25.49	190	74.51	0	0	255
17/05/2019	69	100	123	28.57	308	71.43	0	0	432
13/06/2019	78	100	145	21.22	538	78.78	0	0	683
11/07/2019	74	98.67	137	37.61	225	61.93	2	0.46	363
31/07/2019	598	99.50	1115	37.54	1853	62.40	2	0	2970
08/08/2019	567	100	1008	43.94	1287	56.06	0	0	2295
Min	38	81.68	65	12.88	190	24.31	0	0.00	255
Median	86	99.50	145	27.44	498	72.56	0	0.00	880
Mean	180	97.07	335	31.97	647	67.95	1	0.08	982
SD	180	5.11	330	16.57	461	16.62	2	0.18	715
Max	598	100.00	1115	75.69	1853	87.12	7	0.59	2970

Benthic macroinvertebrates were separated from the sediment, sorted, and then preserved in 70% ethanol for later examination in the laboratory. The main taxa composing the benthic community of each sample were firstly identified to the family level using suitable taxonomic keys [57]. Chironomid specimens were subsequently identified to genus or species (*Chironomus plumosus*) level through the observation of peculiar features of their head capsules [58–61].

2.3. Mouthpart Deformities of Chironomus plumosus Larvae

Body length and head capsule dimensions (width and length) of chironomid larvae were measured to identify the larval stage (instar), and head capsules were then mounted on slides and fixed with Faure liquid (modified composition [22,43]).

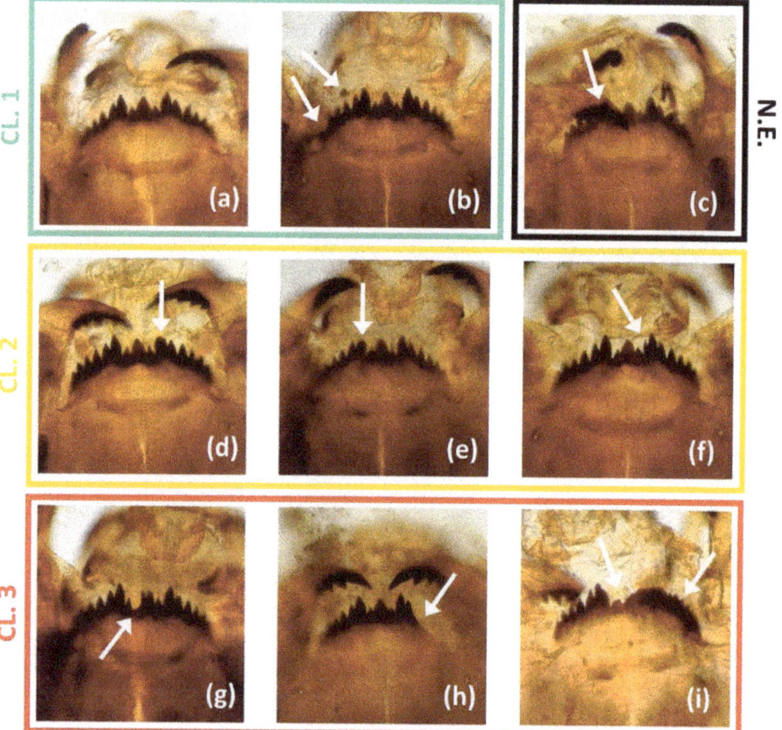

Figure 2. *Chironomus plumosus* images showing Class 1 (CL. 1): (**a**) the normal-shaped mentum, (**b**) the normal-shaped mentum with mechanical damage (arrows), (**c**) the mentum not evaluable (N.E., not completely visible, arrow), and the various mentum deformity types of Class 2 (CL. 2) and Class 3 (CL. 3). (**d**) CL. 2: mentum with a rounded tooth (arrow), (**e**) mentum with a filed tooth (arrow), and (**f**) mentum with an additional tooth (arrow). (**g**) CL. 3: mentum with a Köhn gap (arrow), (**h**) mentum with missing teeth (arrow), and (**i**) mentum with a combination of different deformities (arrows).

Only *Chironomus plumosus* head capsules were considered for mouthpart deformity assessment and examined under a microscope. Specimens that did not have completely visible mentum or those with a cephalic capsule highly damaged during fixation on the microscope slides were excluded from further analyses. *Chironomus plumosus* specimens were classified according to the protocol for morphological response of the mentum, consisting of three classes [22,42–44,62]: Class 1 (CL. 1) specimens without morphological deformity or with mechanical damages (broken or worn menta or teeth) [4,5,63–65]; Class 2 (CL. 2) specimens with weak deformity (one or two round/filed teeth; one missing or additional tooth; one bifid tooth; one serrate tooth; two joined teeth; weak asymmetry); Class 3 (CL. 3) specimens with severe deformity (very round/filed teeth; two or more missing or additional teeth; two or more bifid teeth; serrate teeth; three or more joined teeth; severe asymmetry; Köhn gap; combination of different deformities). Examples of mentum deformity classes of *C. plumosus* larvae are shown in Figure 2.

2.4. Statistical Analysis

Ljung–Box test was used to analyze temporal autocorrelation. Generalized linear models (GLM) were used to analyze differences in head width and length among the different deformity classes, setting larval instar as a covariate and Gaussian family as the link function. The test of equal proportions was carried out to examine differences in the deformity incidences among the seasons [66]. Each deformity class was tested independently and the pairwise comparisons for proportions with the Bonferroni's correction were used as a post-hoc procedure. The results were regarded as significant at a two-tailed *p*-value of 0.05. All calculations were performed using the glm function and the prop.test of the R software [67].

3. Results

3.1. Macroinvertebrates and Chironomids

During this survey at Lake Trasimeno, a total of 31,240 benthic macroinvertebrate specimens were collected: 21,220 in the littoral zone and 10,020 in the central zone. The benthic communities were mainly composed of Oligochaeta (Naididae) and Diptera: (Chironomidae). The littoral zone community was composed of the following means (±SD): Naididae: 64.98% (±12.90%), Chironomidae: 31.55% (±13.17%), and other taxa: 3.46% (±3.83%). A very similar benthic community composition was found in the central zone: Naididae: 67.95% (±16.62%), Chironomidae: 31.97% (±16.57%), and other taxa: 0.08% (±0.18%). The other macroinvertebrate taxa collected were essentially alien invasive species (i.e., *Dreissena polymorpha* (Pallas, 1771) and *Dikerogammarus villosus* (Sowinsky, 1894)) (Table 1).

During the whole sampling period, the mean (±SD) density of Chironomidae was 464 (±246) and 335 (±330) individuals m^{-2}, in the littoral and central zones, respectively; the mean (±SD) density of Naididae was 1072 (±782) and 647 (±461) individuals m^{-2}, in the littoral and central zones, respectively; and the mean (±SD) density of other taxa was 53 (±64) and 1 (±2) individuals m^{-2}, in the littoral and central zones, respectively (Table 1).

A total of 3857 *C. plumosus* larvae was collected, 796 from the littoral zone and 3061 from the central zone. In the littoral zone, *C. plumosus* represented a mean (±SD) of 20.69% (±21.41%) of the total chironomid community (ranging from 0% to 83.95%), while it was always the dominant taxon of the central zone chironomid community (mean 97.07% ± 5.11%), ranging from 81.68% to 100% (Table 1).

The analysis of biometric variables identified the respective instar of each specimen. In the littoral zone most (60.93%) specimens belonged to instar IV, 29.27% to instar III, and 9.80% to instar II. Almost all specimens (99.25%) from the central zone belonged to instar IV, while only 0.72% and 0.03% belonged to instar III and II, respectively.

3.2. Mentum Deformity in Chironomus plumosus

For mentum deformity evaluation of *C. plumosus* larvae, 59 specimens collected from the littoral zone and 294 collected from the central zone of Lake Trasimeno were excluded due to damages to the cephalic capsule or due to incomplete visibility of mentum. Thus, deformity assessment was based on 737 specimens from the littoral zone and on 2767 specimens from the central zone. Ljung–Box test revealed no temporal autocorrelation within deformity classes among samplings. The deformity incidence amounted to 7.22% (1.51% in CL. 3) in the whole Lake Trasimeno's ecosystem, reaching 8.28% (3.12% in CL. 3) in the littoral zone and 6.94% (1.08% in CL. 3) in the central zone of the lake (Table 2).

Table 2. *Chironomus plumosus* incidence of the deformity classes: CL. 1, no deformity; CL. 2, weak deformity; CL. 3, strong deformity and CL. (2 + 3); individuals (N) and percentages (%); (**a**) littoral zone; (**b**) central zone; (**c**) Lake Trasimeno.

(a) Littoral Zone			(b) Central Zone			(c) Lake Trasimeno		
CLASS	N	%	CLASS	N	%	CLASS	N	%
CL. 1	676	91.72	CL. 1	2575	93.1	CL. 1	3251	92.78
CL. 2	38	5.16	CL. 2	162	5.85	CL. 2	200	5.71
CL. 3	23	3.12	CL. 3	30	1.08	CL. 3	53	1.51
CL. (2 + 3)	61	8.28	CL. (2 + 3)	192	6.94	CL. (2 + 3)	253	7.22

The polyvoltine biological cycle of *C. plumosus* at Lake Trasimeno exhibited one winter cohort, one spring cohort, and different summer cohorts during the year. Thus, the mentum deformity classes were analyzed in the different seasonal cohorts as shown in a box-plot of the relative statistical variables in Figure 3. In particular, in the littoral zone (34 samplings), the deformed specimens (CL. 2 + 3) had mean values of 1.0 ± 0, 5.0 ± 3.6, and 10 ± 3.0 in the winter, spring, and summer cohorts, respectively; while in the central zone (17 samplings) the deformed specimens (CL. 2 + 3) had mean values of 4.5 ± 1.5, 9.3 ± 7.5, and 45.7 ± 33.1 in the winter, spring, and summer cohorts, respectively.

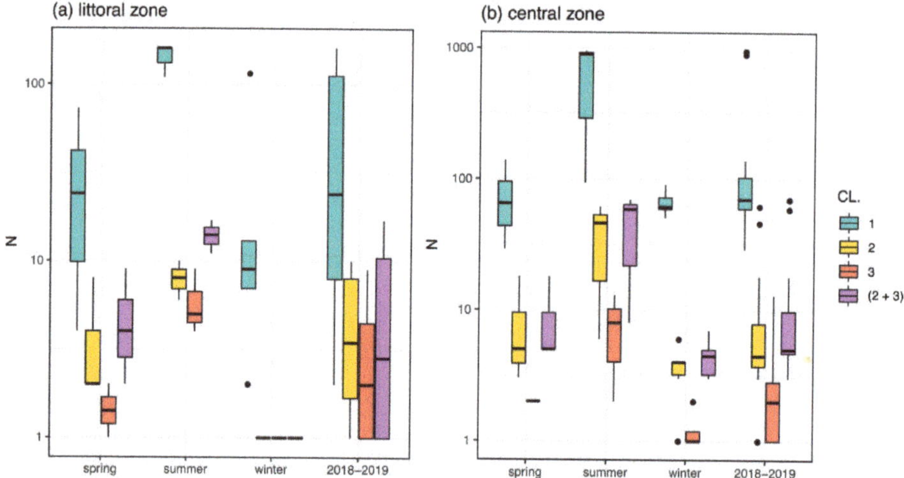

Figure 3. Boxplot of *C. plumosus* specimens (N, log scale) in seasonal cohorts and overall 2018–2019 survey, grouped by deformity classes: CL. 1 (green, no deformity), CL. 2 (yellow, weak deformity), CL. 3 (red, strong deformity), and CL. (2 + 3) (purple). (**a**) Littoral zone; (**b**) central zone. Box represents the interquartile range, thick lines represent the median, whiskers represent the minimum and maximum values, and dots represent the outliers.

In the littoral zone the deformity incidence of the winter cohort was 2.67% (1.33% in CL. 3), the spring cohort was 12.93% (2.59% in CL. 3), and the summer cohort was 8.92% (3.82% in CL. 3); while in the central zone the deformities in the winter cohort amounted to 6.32% (1.17% in CL. 3), 10.73% in the spring cohort (0.77% in CL. 3), and 6.59% in the summer cohort (1.11% in CL. 3) (Figure 4).

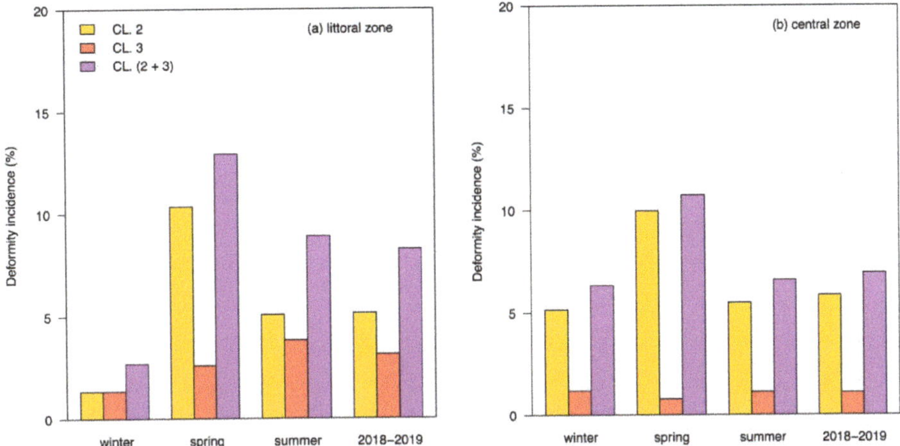

Figure 4. Deformity incidence (%): CL. 2 (yellow, weak deformity); CL. 3 (red, strong deformity); and CL. (2 + 3) (purple) in *C. plumosus* seasonal cohorts and overall 2018–2019 survey; (**a**) littoral zone; (**b**) central zone.

In the littoral zone, the test of equal proportions showed significant differences in the rate of deformity incidence among the seasons for CL. 2 ($p < 0.01$). In contrast, no differences were observed for CL. 3 ($p > 0.6$). Combining CL. 2 and CL. 3, significant differences were detected with an increase in the proportion of specimens with deformities in spring ($p < 0.01$). In the central zone, significant differences were observed for CL. 2, with a deformity incidence in spring significantly higher than that recorded in summer ($p < 0.01$), while no differences were detected for CL. 3 ($p > 0.9$). Combining CL. 2 and CL. 3, no statistical differences emerged.

The deformity assessment in relation to the different larval stages (instars) highlighted that in the littoral zone 7.33% of the deformed specimens (2.85% in CL. 3) belonged to instar IV and only 0.95% (0.27% in CL. 3) to instar III, while in the central zone the instar IV deformed specimens were 6.90% (1.08% in CL. 3) and only 0.04% (none in CL. 3) belonged to instar III. No deformed specimens in both zones belonged to instar II (Table 3).

Figure 5 shows the distribution and relative head size (head length and width) of normal and deformed larvae among instars, in the littoral and central zones of Lake Trasimeno. In the littoral zone, the head length and width means of instar IV specimens (about 61% of total *C. plumosus* larvae) belonging to CL. 1 (normal), CL. 2 (weak deformity), and CL. 3 (strong deformity) were 0.93 and 0.73 mm, 0.93 and 0.73 mm, and 0.95 and 0.75 mm, respectively. Similar trends of head length and width of instar IV specimens (about 99% of total *C. plumosus* larvae) were recorded in the central zone, the mean sizes were 0.98 and 0.78 mm (CL. 1), 0.98 and 0.78 mm (CL. 2), and 1.00 and 0.80 mm (CL. 3) (Table 3). No significant differences were observed in the biometric variables of the head among the three deformity classes ($p > 0.05$).

Table 3. Head length and width (mm) of *C. plumosus* larvae (all instars, instar IV, instar III, and instar II); CL. 1, no deformity; CL. 2, weak deformity; CL. 3, strong deformity; and CL. (2 + 3); number (N), minimum (Min), median, mean, standard deviation (SD), and maximum (Max); (**a**) littoral zone; (**b**) central zone.

		(a) Littoral Zone								(b) Central Zone							
		Head Lenght (mm)				Head Width (mm)				Head Lenght (mm)				Head Width (mm)			
CL		1	2	3	2 + 3	1	2	3	2 + 3	1	2	3	2 + 3	1	2	3	2 + 3
All Instars	N	676	38	23	61	676	38	23	61	2575	162	30	192	2575	162	30	192
	Min	0.18	0.48	0.40	0.40	0.13	0.38	0.38	0.38	0.30	0.53	0.85	0.53	0.25	0.43	0.70	0.43
	Median	0.83	0.91	0.95	0.93	0.65	0.75	0.75	0.75	1.00	1.00	1.00	1.00	0.78	0.78	0.79	0.78
	Mean	0.75	0.88	0.90	0.89	0.58	0.69	0.72	0.70	0.98	0.98	1.00	0.98	0.78	0.78	0.80	0.78
	SD	0.24	0.17	0.16	0.17	0.20	0.13	0.12	0.13	0.09	0.10	0.09	0.10	0.08	0.08	0.08	0.08
	Max	1.20	1.10	1.10	1.10	0.93	0.95	0.90	0.95	1.25	1.25	1.25	1.25	1.00	1.05	1.00	1.05
Instar IV	N	402	33	21	54	402	33	21	54	2555	161	30	191	2555	161	30	191
	Min	0.70	0.75	0.75	0.75	0.50	0.60	0.63	0.60	0.70	0.68	0.85	0.68	0.50	0.55	0.70	0.55
	Median	0.93	0.93	0.95	0.95	0.75	0.75	0.75	0.75	1.00	1.00	1.00	1.00	0.78	0.78	0.79	0.78
	Mean	0.93	0.93	0.95	0.94	0.73	0.73	0.75	0.74	0.98	0.98	1.00	0.98	0.78	0.78	0.80	0.79
	SD	0.10	0.09	0.08	0.08	0.08	0.07	0.07	0.07	0.08	0.09	0.09	0.09	0.07	0.07	0.08	0.07
	Max	1.20	1.10	1.10	1.10	0.93	0.95	0.90	0.95	1.25	1.25	1.25	1.25	1.00	1.05	1.00	1.05
Instar III	N	214	5	2	7	214	5	2	7	19	1	-	1	19	1	-	1
	Min	0.40	0.48	0.40	0.40	0.20	0.38	0.38	0.38	0.50	0.53	-	0.53	0.30	0.43	-	0.43
	Median	0.53	0.50	0.45	0.50	0.40	0.40	0.39	0.40	0.55	0.53	-	0.53	0.40	0.43	-	0.43
	Mean	0.53	0.51	0.45	0.49	0.40	0.40	0.39	0.40	0.55	0.53	-	0.53	0.41	0.43	-	0.43
	SD	0.05	0.03	0.07	0.05	0.05	0.03	0.02	0.03	0.04	-	-	-	0.04	-	-	-
	Max	0.70	0.55	0.50	0.55	0.50	0.45	0.40	0.45	0.63	0.53	-	0.53	0.45	0.43	-	0.43
Instar II	N	60	-	-	-	60	-	-	-	1	-	-	-	1	-	-	-
	Min	0.18	-	-	-	0.13	-	-	-	0.30	-	-	-	0.25	-	-	-
	Median	0.33	-	-	-	0.23	-	-	-	0.30	-	-	-	0.25	-	-	-
	Mean	0.31	-	-	-	0.22	-	-	-	0.30	-	-	-	0.25	-	-	-
	SD	0.05	-	-	-	0.04	-	-	-	-	-	-	-	-	-	-	-
	Max	0.40	-	-	-	0.33	-	-	-	0.30	-	-	-	0.25	-	-	-

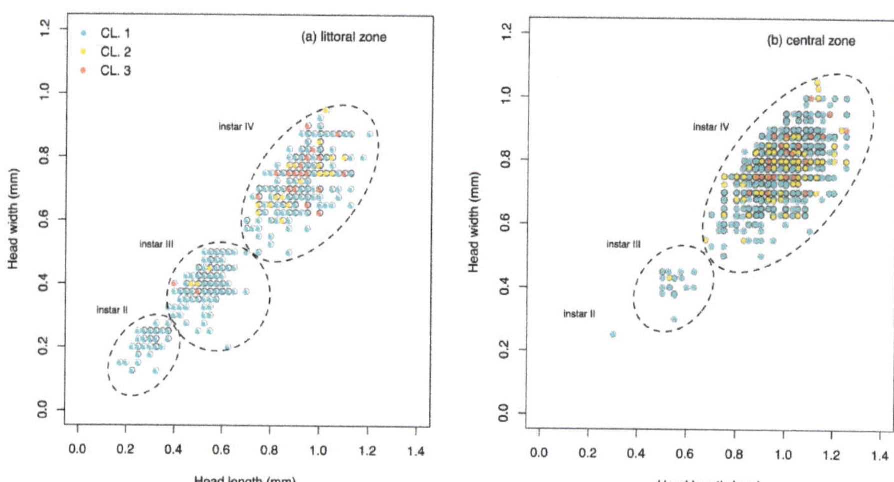

Figure 5. Size (length and width; mm) of *C. plumosus* larvae, different instars (II, III, and IV) are surrounded by dotted circles, mentum in CL. 1 (green, no deformity), CL. 2 (yellow, weak deformity), and CL. 3 (red, strong deformity). (**a**) littoral zone; (**b**) central zone.

Lastly, in Table 4 the different types of *C. plumosus* mentum deformities in Lake Trasimeno are reported. The main CL. 2 deformity types in the littoral zone, central zone, and the whole lake, respectively, were: "one or two round/filed teeth" with 81.58%, 69.75%, and 72.00%; "one missing tooth" with 2.63%, 17.90%, and 15.00%; "one additional tooth" with 5.26%, 3.09%, and 3.50%; "one serrate tooth" with 10.53%, 3.70%, and 5.00%; "one bifid tooth" with 0%, 4.94%, and 4.00%. The main CL. 3 deformity types, listed in the same order, were: "very round/filed teeth" with 39.13%, 33.33%, and 35.85%; "two or more missing teeth" with 21.74%, 16.67%, and 18.87%; "two or more additional teeth" with 8.70%, 0%, and 3.77%; "Köhn gap" with 0%, 20.00%, and 11.32%.

Table 4. *Chironomus plumosus* mentum deformity types: CL. 2 (weak deformity), CL. 3 (strong deformity), and CL. (2 + 3); numbers (N) and percentages (%); (**a**) littoral zone; (**b**) central zone; (**c**) Lake Trasimeno, Central Italy.

Class and Deformity Type	(a) Littoral Zone		(b) Central Zone		(c) Lake Trasimeno	
	(N)	(%)	(N)	(%)	(N)	(%)
CL. 2	**38**		**162**		**200**	
One or Two Round/Filed Teeth	31	81.58	113	69.75	144	72.00
One Missing Tooth	1	2.63	29	17.90	30	15.00
One Additional Tooth	2	5.26	5	3.09	7	3.50
One Bifid Tooth	0	0	8	4.94	8	4.00
One Serrate Tooth	4	10.53	6	3.70	10	5.00
Two Joined Teeth	0	0	1	0.62	1	0.50
Weak Asymmetry	0	0	0	0	0	0
CL. 3	**23**		**30**		**53**	
Very Round/Filed Teeth	9	39.13	10	33.33	19	35.85
Two or More Missing Teeth	5	21.74	5	16.67	10	18.87
Two or More Additional Teeth	2	8.70	0	0	2	3.77
Two or More Bifid Teeth	0	0	0	0	0	0
Serrate Teeth	0	0	0	0	0	0
Three or More Joined Teeth	0	0	0	0	0	0
Severe Asymmetry	0	0	0	0	0	0
Köhn Gap	0	0	6	20.00	6	11.32
Combination of Different Deformities	7	30.43	9	30.00	16	30.19
CL. (2 + 3)	**61**		**192**		**253**	
Round/Filed Teeth	40	65.57	123	64.06	163	64.43
Missing Teeth	6	9.84	34	17.71	40	15.81
Additional Teeth	4	6.56	5	2.60	9	3.56
Bifid Teeth	0	0	8	4.17	8	3.16
Serrate Teeth	4	6.56	6	3.13	10	3.95
Joined Teeth	0	0	1	0.52	1	0.40
Asymmetry	0	0	0	0	0	0
Köhn Gap	0	0	6	3.13	6	2.37
Combination of Different Deformities	7	11.48	9	4.69	16	6.32

4. Discussion

This study analyzed mentum deformities in *C. plumosus* larvae as an endpoint for evaluating the effects of anthropogenic stressors in the shallow Lake Trasimeno. The results of the survey showed a low deformity incidence, amounting to about 7% (1.5% in CL. 3) in the whole Lake Trasimeno's ecosystem, with similar values in the littoral and central zones of the lake, about 8% (3% in CL. 3) and 7% (1% in CL. 3), respectively.

Paleolimnological studies, representing natural background levels during the pre-industrial age, showed deformity incidences between 0% and 0.8% [68,69]. In the contemporary age [70,71] the incidence of background deformities in low-impacted areas is lower than 8% suggesting this threshold as the reference background for deformity incidence in natural inland waters. Therefore, the results of the present research affirm that this biotope has a deformity incidence value comparable to that of inland waters subject to low anthropic impact. Perhaps, this can be associated to a low degree of contamination of the Trasimeno's ecosystem, in particular of sediments that represent a preferential sink for toxic compounds and, at the same time, serve as a habitat for chironomid larvae.

The deformity incidence analysis among the different cohorts of the polyvoltine biological cycle of *C. plumosus* showed that the spring cohort had overall the highest deformity values in Lake Trasimeno, corresponding to 11.41% (1.33% in CL. 3). Perhaps, this peak of deformities in the spring cohort is due to the compromise between metabolic activity and residence time in sediments, compared to the winter cohort (low metabolic activity and long residence time) and to summer cohorts (high metabolic activity and short residence time).

Head size (length and width) of normal and deformed larvae grouped according to their instar showed similar trends, regardless of whether they were found in the littoral or in the central zone of the lake. These results highlight that mentum deformity did not cause alterations of the head size of deformed specimens, meaning a similar larval growth of the latter compared to normal specimens [62].

A comparison of current results with previous investigations (2000–2010) conducted in Lake Trasimeno using the same sampling methods showed a clear decrease in the incidence of mentum deformities in *C. plumosus* larvae, both in the littoral and in the central zone of the lake [43]. In fact, the deformity incidences that were about 20% (9.5% in CL. 3) in the littoral zone of Castiglione del Lago during 2006–2009 and about 13% (3.8% in CL. 3) in the central zone of the lake during 2000–2010, decreased to 8% and 7% in the respective lake areas during the present survey (2018–2019). This 10-year comparison demonstrated a clear decrease in the current deformity index values. Over the years, in effect, there has been a confirmation of this decreasing trend of deformity incidence that, in the littoral zone, decreased from 31% in 2006 to 10% in 2009, until reaching the current level of 8%. A similar phenomenon tended to occur in the central zone of the lake as well, which showed a deformity index of 11% in 2000, 17% in 2001, 10% in 2002, and 13% in October 2010, until reaching the current level of 7%.

In summary, the surveys carried out at Lake Trasimeno allow a space–time biological assessment of lacustrine sediments' contamination showing a qualitative recovery over time of the lake ecosystem both in the central and littoral zone; the latter zone, more easily subject to the introduction of pollutants, showed an incidence of severe deformities (CL. 3) of 3% compared to 1% in the central zone.

However, the relationship between the incidence of strong (CL. 3) and weak (CL. 2) deformities is not easy to interpret, in fact, even if severe deformed phenotypes (CL. 3) were found in specimens subject to sediments contaminated both with organic and inorganic pollutants, their increase is probably influenced not only by the quantity but also by the type of compounds that form the toxic mixture [4,22]. Consequently, as a reliable biomarker, it is more appropriate to consider the two deformity classes together (CL. 2 + 3) for an overall assessment of the toxic effects of sediment contamination [22].

In this regard, some studies based on ecotoxicological laboratory experiments queried the relationship between chironomid deformities and toxicity assessment in freshwater ecosystems [24,49,72–76]. However, the high variability of those experimental results can be influenced by different variables, such as: (i) the endpoint used (mentum, mandibles, pectin epipharyngis); (ii) the deformity types considered; (iii) the larval instars used in the experiment; (iv) the type of cultures used which often cause a high incidence of deformities in the control test; (v) the exposure time of the larvae to the contaminants; and (vi) the type of contaminant mixture used in the test.

On the other hand, in natural environments the association between incidence of chironomid deformities and levels of chemical contamination is very consistent, despite the lack of a standardized protocol for the deformity assignment which can generate cases of inconsistent interpretation. In this respect, in the ring test (based on digital images) [74] involving 25 international experts to assess the status (normal or deformed) of 211 chironomid larvae menta (IV instar) from Lake Saimaa (Finland), the majority of them (at least 2/3) highlighted clearly that the incidence of mentum deformities was high in the polluted site (about 30%) and low in the reference site (<10%), with no significant differences between blind or open assessment.

In addition, the present study contributes to the elaboration of a detailed protocol for the definition of which types of mentum deformities to be considered for an assessment of the deformity incidence in *Chironomus* populations of freshwater biotopes.

The present study data showed that, among deformed menta of *C. plumosus* found in Lake Trasimeno (littoral and central zones), the most common deformity types were "round/filed teeth" with 64% followed by "missing teeth" (16%), a "combination of different deformities" (about 6%), "serrate teeth" (4%), "additional teeth" (4%), "bifid teeth" (3%), "Köhn gap" (2%), and "joined teeth" (0.4%) (Table 4).

In conclusion, the present study on Lake Trasimeno provides a further contribution to the evaluation of mentum deformities in chironomid populations of inland waters. In the biomonitoring programs, the deformity incidence can provide an easy biological tool to assess the effects of sediment-associated toxicants. The mentum deformities assessment is complementary and non-substitutive of the chemical instrumental analyses; in fact, a high incidence of deformities in a freshwater ecosystem is a warning (indicator) for the start of a chemical analytical monitoring program aimed at qualitative and quantitative detection of the toxic compounds causing environmental contamination.

Author Contributions: Conceptualization, E.G. (Enzo Goretti), M.P., S.P., M.C., G.L.P., E.G. (Elda Gaino), and A.A.; Data curation, E.G. (Enzo Goretti), M.P., S.P., M.C., and G.L.P.; Methodology, E.G. (Enzo Goretti), M.P., S.P., M.C., G.L.P., R.S., and Á.M.D.G.; Investigation, E.G. (Enzo Goretti), M.P., S.P., M.C., G.L.P., R.S., and A.M.D.G.; Writing—original draft, E.G. (Enzo Goretti), M.P., G.L.P., E.G. (Elda Gaino), and A.A.; Writing—review and editing, E.G. (Enzo Goretti), M.P., G.L.P., E.G. (Elda Gaino), and A.A. All authors have read and agreed to the published version of the manuscript.

Funding: This work was supported by the "Fondazione Brunello e Federica Cucinelli" within the 2017–2020 research project "Chironomid population control at Lake Trasimeno: evaluation of biological control methods and new systems for the attraction and mechanical-light capture".

Acknowledgments: We thank Andrea Pagnotta (amateur fisherman), Marco Chiappini (Provincia di Perugia), Alberto Fais, and Francesco Giglietti (USL Umbria1), Rosalba Padula and Luca Nicoletti (ARPAUmbria) for their great support in the sampling campaign.

Conflicts of Interest: The authors declare no conflicts of interest.

References

1. Rosenberg, D.M.; Resh, V.H. Introduction to freshwater biomonitoring and benthic macroinvertebrates. In *Freshwater Biomonitoring Benthic Macroinvertebrates*; Chapman and Hall: New York, NY, USA, 1993.

2. Maldonado, A.C.D.; Wendling, B. Manejo de ecossistemas aquáticos contaminados por Metais pesados. *Revista Agropecuária Técnica* **2009**, *30*, 21–32. [CrossRef]

3. Goretti, E.; Pallottini, M.; Ricciarini, M.I.; Selvaggi, R.; Cappelletti, D. Heavy metals bioaccumulation in selected tissues of red swamp crayfish: An easy tool for monitoring environmental contamination levels. *Sci. Total Environ.* **2016**, *559*, 339–346. [CrossRef] [PubMed]

4. Vermeulen, A.C. Elaboration of chironomid deformities as bioindicators of toxic sediment stress: The potential application of mixture toxicity concepts. *Ann. Zool. Fenn.* **1995**, *32*, 265–285.

5. Warwick, W.F. Morphological deformities in Chironomidae (Diptera) larvae as biological indicators of toxic stress. In *Toxic Contaminants and Ecosystem Health; A Great Lakes Focus*; Evans, M.S., Ed.; John Wiley and Sons: New York, NY, USA, 1988; pp. 281–320.

6. Apostolico, F.; Vercillo, F.; La Porta, G.; Ragni, B. Long-term changes in diet and trophic niche of the European wildcat (*Felis silvestris silvestris*) in Italy. *Mamm. Res.* **2016**, *61*, 109–119. [CrossRef]

7. Davutluoglu, O.I.; Seckin, G.; Ersu, C.B.; Ylmaz, T.; Sari, B. Heavy metal content and distribution in surface sediments of the Seyhan River, Turkey. *J. Environ. Manag.* **2011**, *92*, 2250–2259. [CrossRef]

8. Goretti, E.; Pallottini, M.; Cenci Goga, B.T.; Selvaggi, R.; Petroselli, C.; Vercillo, F.; Cappelletti, D. Mustelids as bioindicators of the environmental contamination by heavy metals. *Ecol. Indicat.* **2018**, *94*, 320–327. [CrossRef]

9. Goretti, E.; Pallottini, M.; Rossi, R.; La Porta, G.; Gardi, T.; Cenci Goga, B.T.; Elia, A.C.; Galletti, M.; Moroni, B.; Petroselli, C.; et al. Heavy metal bioaccumulation in honey bee matrix, an indicator to assess the contamination level in terrestrial environments. *Environ. Pollut.* **2019**, *256*, 113388. [CrossRef]

10. Holt, M.S. Sources of chemical contaminants and routes into the freshwater environment. *Food Chem. Toxicol.* **2000**, *38*, 21–27. [CrossRef]

11. Chapman, P.M. The sediment quality triad approach to determining pollution-induced degradation. *Sci. Total Environ.* **1990**, *97*, 815–825. [CrossRef]

12. Cummins, K.W.; Wilzbach, M.A. *Field Procedures for the Analysis of Functional Feeding Groups in Stream Ecosystems*; Contribution 1611; Appalachian Environmental Laboratory, University of Maryland: Frostburg, MD, USA, 1985.

13. Céréghino, R.; Park, Y.S.; Compin, A.; Lek, S. Predicting the species richness of aquatic insects in streams using a limited number of environmental variables. *J. N. Am. Benthol. Soc.* **2003**, *22*, 442–456. [CrossRef]

14. Dolédec, S.; Statzner, B. Reponses of freshwater biota to human disturbances: Contribution of J-NABS to developments in ecological integrity assessments. *J. N. Am. Benthol. Soc.* **2010**, *29*, 286–311. [CrossRef]

15. Pallottini, M.; Goretti, E.; Gaino, E.; Selvaggi, R.; Cappelletti, D.; Cereghino, R. Invertebrate diversity in relation to chemical pollution in an Umbrian stream system (Italy). *C. R. Biol.* **2015**, *338*, 511–520. [CrossRef] [PubMed]

16. Pallottini, M.; Cappelletti, D.; Fabrizi, E.; Gaino, E.; Goretti, E.; Selvaggi, R.; Cereghino, R. Macroinvertebrate functional trait responses to chemical pollution in agricultural-industrial landscapes. *River Res. Appl.* **2017**, *33*, 505–513. [CrossRef]

17. Adler, P.H.; Courtney, G.W. Ecological and Societal Services of Aquatic Diptera. *Insects* **2019**, *10*, 70. [CrossRef]

18. Armitage, P.D.; Pinder, L.C.; Cranston, P. *The Chironomidae: Biology and Ecology of Non-Biting Midges*; Springer Netherlands: Heidelberg, Germany, 1995.

19. Goretti, E.; Coletti, A.; Di Veroli, A.; Di Giulio, A.M.; Gaino, E. Artificial light device for attracting pestiferous chironomids (Diptera): A case study at Lake Trasimeno (Central Italy). *Ital. J. Zool.* **2011**, *78*, 336–342. [CrossRef]

20. OECD. *Test No. 235: Chironomus sp., Acute Immobilisation Test*; OECD Publishing: Paris, France, 2011. [CrossRef]

21. Božanić, M.; Marković, Z.; Živić, M.; Dojčinové, B.; Perić, A.; Stanković, M.; Živić, I. Mouthpart Deformities of *Chironomus plumosus* Larvae Caused by Increased Concentrations of Copper in Sediment from Carp Fish Pond. *Turk. J. Fish. Aquat. Sci.* **2018**, *19*, 251–259. [CrossRef]

22. Di Veroli, A.; Goretti, E.; León Paumen, M.; Kraak, M.H.S.; Admiraal, W. Mouthpart deformities in *Chironomus riparius* larvae exposed to toxicants. *Environ. Pollut.* **2012**, *166*, 212–217. [CrossRef]

23. Martinez, E.A.; Moore, B.C.; Schaumloffel, J.; Dasgupta, N. Morphologic and Growth Responses in *Chironomus tentans* to Arsenic Exposure. *Arch. Environ. Contam. Toxicol.* **2006**, *51*, 529–536. [CrossRef]

24. Gagliardi, B.S.; Long, S.M.; Pettigrove, V.J.; Griffin, P.C.; Hoffmann, A.A. A re-evaluation of chironomid deformities as an environmental stress response: Avoiding survivorship bias and testing noncontaminant biological factors. *Environ. Toxicol. Chem.* **2019**, *38*, 1658–1667. [CrossRef]

25. Corbi, J.J.; Bernegossi, A.C.; Moura, L.; Felipe, M.C.; Issa, C.G.; Silva, M.R.L.; Gorni, G.R. *Chironomus sancticaroli* (Diptera, Chironomidae) as a Sensitive Test Species: Can We Rely on Its Use After Repeated Generations, Under Laboratory Conditions? *Bull. Environ. Contam. Toxicol.* **2019**, *103*, 213–217. [CrossRef] [PubMed]

26. Saether, O.A. A survey of the bottom fauna of the Okanagan valley, British Columbia. *Can. Fish. Mar. Serv. Techn. Rep.* **1970**, *196*, 1–29.

27. Hamilton, A.L.; Saether, O.A. The occurrence of characteristic deformities in the chironomid larvae of several Canadian lakes. *Can. Entomol.* **1971**, *103*, 363–368. [CrossRef]

28. Hare, L.; Carter, J.C.H. The distribution of *Chironomus* (s.s.)? Cucini (*salinarius* group) larvae (Diptera: Chironomidae) in Parry Sound, Georgian Bay, with particular reference to structural deformities. *Can. J. Zool.* **1976**, *54*, 2129–2134. [CrossRef]

29. Janssens de Bisthoven, L.; Vermeulen, A.; Ollevier, F. Experimental Induction of Morphological Deformities in *Chironomus riparius* Larvae by Chronic Exposure to Copper and Lead. *Arch. Environ. Contam. Toxicol.* **1998**, *35*, 249–256. [CrossRef]

30. Reynolds, S.K.; Ferrington, L.C. Differential Morphological Responses of Chironomid Larvae to Severe Heavy Metal Exposure (Diptera: Chironomidae). *J. Kans. Entomol. Soc.* **2002**, *75*, 172–184.

31. Martinez, E.A.; Moore, B.C.; Schaumloffel, J.; Dasgupta, N. Morphological abnormalities in *Chironomus tentans* exposed to cadmium and copper-spiked sediments. *Ecotoxicol. Environ. Saf.* **2003**, *55*, 204–212. [CrossRef]

32. Péry, A.R.R.; Mons, R.; Garric, J. Energy-based modeling to study population growth rate and production for the midge *Chironomus riparius* in ecotoxicological risk assessment. *Ecotoxicology* **2004**, *13*, 647–656. [CrossRef]

33. Ebau, W.; Rawi, C.S.M.; Din, Z.; Al-Shami, S.A. Toxicity of cadmium and lead on tropical midge larvae, *Chironomus kiiensis* Tokunaga and *Chironomus javanus* Kieffer (Diptera: Chironomidae). *Asian Pac. J. Trop. Biomed.* **2012**, *2*, 631–634. [CrossRef]

34. Gagliardi, B.; Pettigrove, V. Removal of intensive agriculture from the landscape improves aquatic ecosystem health. *Agric. Ecosyst. Environ.* **2013**, *176*, 1–8. [CrossRef]

35. Grebenjuk, L.P.; Tomilina, I.I. Morphological Deformations of Hard Chitinized Mouthpart Structures in Larvae of the Genus *Chironomus* (Diptera, Chironomidae) as the Index of Organic Pollution in Freshwater Ecosystems. *Inland Water Biol.* **2014**, *7*, 273–285. [CrossRef]

36. Beneberu, G.; Mengistou, S. Head capsule deformities in *Chironomus* spp. (Diptera: Chironomidae) as indicator of environmental stress in Sebeta River, Ethiopia. *Afr. J. Ecol.* **2014**, *53*, 268–277. [CrossRef]

37. Arimoro, F.O.; Odume, O.N.; Meme, F.K. Environmental drivers of head capsule deformities in *Chironomus* spp. (Diptera: Chironomidae) in a stream in north central Nigeria. *Zool. Ecol.* **2015**, *25*, 70–76. [CrossRef]

38. Odume, O.N.; Palmer, C.G.; Arimoro, F.O.; Mensaha, P.K. Chironomid assemblage structure and morphological response to pollution in an effluent-impacted river, Eastern Cape, South Africa. *Ecol. Indic.* **2016**, *67*, 391–402. [CrossRef]

39. Arimoro, F.O.; Auta, Y.I.; Odume, O.N.; Keke, U.N.; Mohammed, A.Z. Mouthpart deformities in Chironomidae (Diptera) as bioindicators of heavy metals pollution in Shiroro Lake, Niger State, Nigeria. *Ecotoxicol. Environ. Saf.* **2018**, *149*, 96–100. [CrossRef]

40. Akyildiz, G.K.; Bakir, R.; Polat, S.; Duran, M. Mentum Deformities of Chironomid Larvae as an Indicator of Environmental Stress in Büyük Menderes River, Turkey. *Inland Water Biol.* **2018**, *11*, 515–522. [CrossRef]

41. Deliberalli, W.; Cansian, R.L.; Pereira, A.A.M.; Loureiro, R.C.; Hepp, L.U.; Restello, R.M. The effects of heavy metals on the incidence of morphological deformities in Chironomidae (Diptera). *Zoologia* **2018**, *35*. [CrossRef]

42. Di Veroli, A.; Selvaggi, R.; Pellegrino, R.M.; Goretti, E. Sediment toxicity and deformities of chironomid larvae in Lake Piediluco (Central Italy). *Chemosphere* **2010**, *79*, 33–39. [CrossRef]

43. Di Veroli, A.; Selvaggi, R.; Goretti, E. Chironomid mouthpart deformities as indicator of environmental quality: A case study in Lake Trasimeno (Italy). *J. Environ. Monit.* **2012**, *14*, 1473–1478. [CrossRef]

44. Di Veroli, A.; Santoro, F.; Pallottini, M.; Selvaggi, R.; Scardazza, F.; Cappelletti, D.; Goretti, E. Deformities of chironomid larvae and heavy metal pollution: From laboratory to field studies. *Chemosphere* **2014**, *112*, 9–17. [CrossRef]

45. Savić-Zdravković, D.; Jovanović, B.; Đurđević, A.; Stojković-Piperac, M.; Savić, A.; Vidmar, J.; Milošević, D. An environmentally relevant concentration of titanium dioxide (TiO$_2$) nanoparticles induces morphological changes in the mouthparts of *Chironomus tentans*. *Chemosphere* **2018**, *211*, 489–499. [CrossRef] [PubMed]

46. Tomilina, I.I.; Grebenjuk, L.P. Induction of Deformities of the Hard-Chitinized Mouthpart Structures of Larvae *Chironomus riparius* Meigen under Various Contents of Persistent Organic Substances in Bottom Sediments. *Inland Water Biol.* **2019**, *12*, 346–355. [CrossRef]

47. Park, K.; Kwak, I.S. Disrupting effects of antibiotic sulfathiazole on developmental process during sensitive life-cycle stage of *Chironomus riparius*. *Chemosphere* **2018**, *190*, 25–34. [CrossRef] [PubMed]

48. Pallottini, M.; Goretti, E.; Selvaggi, R.; Cappelletti, D.; Dedieu, N.; Cereghino, R. An efficient semi-quantitative macroinvertebrate multimetric index forthe assessment of water and sediment contamination in streams. *Inland Waters* **2017**, *7*, 314–322. [CrossRef]

49. Arambourou, H.; Planelló, R.; Llorente, L.; Fuertes, I.; Barata, C.; Delorme, N.; Noury, P.; Herrero, Ó.; Villeneuve, A.; Bonnineau, C. *Chironomus riparius* exposure to field-collected contaminated sediments: From subcellular effect to whole-organism response. *Sci. Total Environ.* **2019**, *671*, 874–882. [CrossRef]

50. He, H.; Xi, G.; Lu, X. Molecular cloning, characterization, and expression analysis of an ecdysone receptor homolog in *Teleogryllus emma* (Orthoptera: Gryllidae). *J. Insect Sci.* **2015**, *15*. [CrossRef]

51. Vermeulen, A.C.; Liberloo, G.; Dumont, P.; Ollevier, F.; Goddeeris, B.R. Exposure of *Chironomus riparius* larvae (Diptera) to lead, mercury, and β-sitosterol: Effects on mouthpart deformation and moulting. *Chemosphere* **2000**, *41*, 1581–1591. [CrossRef]

52. Marchegiano, M.; Francke, A.; Gliozzi, E.; Wagner, B.; Arizteguia, D. High-resolution palaeohydrological reconstruction of central Italy during the Holocene. *Holocene* **2018**, *29*, 481–492. [CrossRef]

53. Frondini, F.; Dragoni, W.; Morgantini, N.; Donnini, M.; Cardellini, C.; Caliro, S.; Melillo, M.; Chiodini, G. An Endorheic Lake in a Changing Climate: Geochemical Investigations at Lake Trasimeno (Italy). *Water* **2019**, *11*, 1319. [CrossRef]

54. Ludovisi, A.; Gaino, E. Meteorological and water quality changes in Lake Trasimeno (Umbria, Italy) during the last fifty years. *J. Limnol.* **2010**, *69*, 174–188. [CrossRef]

55. Marchegiano, M.; Francke, A.; Gliozzi, E.; Arizteguia, D. Arid and humid phases in central Italy during the Late Pleistocene revealed by the Lake Trasimeno ostracod record. *Palaeogeogr. Palaeoclimatol. Palaeoecol.* **2018**, *490*, 55–69. [CrossRef]

56. Mancinelli, G.; Papadia, P.; Ludovisi, A.; Migoni, D.; Bardelli, R.; Fanizzi, F.P.; Vizzini, S. Beyond the mean: A comparison of trace- and macroelement correlation profiles of two lacustrine populations of the crayfish *Procambarus clarkii*. *Sci. Total Environ.* **2018**, *624*, 1455–1466. [CrossRef] [PubMed]

57. Tachet, H.; Richoux, P.; Bournaud, M.; Usseglio-Polatera, P. *Invertébrés d'eau douce. Systématique, biologie, écologie*; CNRS Éditions: Paris, France, 2010.

58. Ferrarese, U.; Rossaro, B. Chironomidi, 1. In *Guida Per Il Riconoscimento Delle Specie Animali Delle Acque Interne Italiane*; CNR AQ/1/129; CNR: Roma, Italy, 1981.

59. Rossaro, B. Chironomidi, 2. In *Guida Per Il Riconoscimento Delle Specie Animali Delle Acque Interne Italiane*; CNR AQ/1/171; CNR: Roma, Italy, 1982.

60. Ferrarese, U. Chironomidi, 3. In *Guida Per Il Riconoscimento Delle Specie Animali Delle Acque Interne Italiane*; CNR AQ/1/204; CNR: Roma, Italy, 1983.

61. Nocentini, A.M. Chironomidi, 4. In *Guida Per Il Riconoscimento Delle Specie Animali Delle Acque Interne Italiane*; CNR AQ/1/233; CNR: Roma, Italy, 1985.

62. Janssens De Bisthoven, L.; Yts, P.; Goddeeris, B.; Ollevier, F. Sublethal parameters in morphologically deformed *Chironomus* larvae: Clues to understand their bioindicator value. *Freshw. Biol.* **1998**, *39*, 179–191. [CrossRef]

63. Madden, C.P.; Austin, A.D.; Sutter, P.J. Pollution monitoring using chironomid larvae: What is a deformity? In *Chironomids from Genes to Ecosystems*; Craston, P., Ed.; CSRIO Publication: East Melbourne, Australia, 1995; pp. 89–94.

64. Nazarova, L.B.; Riss, H.W.; Kahlheber, A.; Werding, B. Some observations of buccal deformities in chironomid larvae (Diptera: Chironomidae) from the Ciénaga Grande de Santa Marta, Colombia. *Caldasia* **2004**, *26*, 275–290.

65. Cao, Y.; Zhang, E.; Tang, H.; Langdon, P.; Ning, D.; Zheng, W. Combined effects of nutrients and trace metals on chironomid composition and morphology in a heavily polluted lake in central China since the early 20th century. *Hydrobiologia* **2016**, *779*, 147–159. [CrossRef]

66. Newcombe, R.G. Interval estimation for the difference between independent proportions: Comparison of eleven methods. *Stat. Med.* **1998**, *17*, 873–890. [CrossRef]

67. R Core Team. *R: A Language and Environment for Statistical Computing*; R Foundation for Statistical Computing: Vienna, Austria, 2018; Available online: https://www.R-project.org/ (accessed on 13 December 2019).

68. Warwick, W.F. Paleolimnology of the Bay of Quinte, Lake Ontario: 2800 years of cultural influence. *Can. Bull. Fish. Aquat. Sci.* **1980**, *206*, 1–118.

69. Wiederholm, T. Incidence of deformed chironomid larvae (Diptera Chironomidae) in Swedish Lakes. *Hydrobiologia* **1984**, *109*, 243–249. [CrossRef]

70. Warwick, W.F. Morphological abnormalities in Chironomidae (Diptera) larvae as measures of toxic stress in freshwater ecosystems: Indexing antennal deformities in *Chironomus* Meigen. *Can. J. Fish. Aquat. Sci.* **1985**, *42*, 1881–1914. [CrossRef]

71. Bird, G.A. Use of chironomid deformities to assess environmental degradation in the Yamaska River, Quebec. *Environ. Monit. Assess.* **1994**, *30*, 163–175. [CrossRef]

72. Vogt, C.; Langer-Jaesrich, M.; Elsässer, O.; Schmitt, C.; Van Dongen, S.; Köhler, H.R.; Oehlmann, J.; Nowak. Effects of inbreeding on mouthpart deformities of *Chironomus riparius* under sublethal pesticide exposure. *Environ. Toxicol. Chem.* **2013**, *32*, 423–425. [CrossRef] [PubMed]

73. Arambourou, H.; Beisel, J.N.; Branchu, P.; Debat, V. Exposure to sediments from polluted rivers has limited phenotypic effects on larvae and adults of *Chironomus riparius*. *Sci. Total Environ.* **2014**, *484*, 92–101. [CrossRef] [PubMed]

74. Salmelin, J.; Vuori, K.M.; Hämäläinen, H. Inconsistency in the analysis of morphological deformities in Chironomidae (Insecta: Diptera) larvae. *Environ. Toxicol. Chem.* **2015**, *34*, 1891–1898. [CrossRef]
75. Salmelin, J.; Karjalainen, A.K.; Hämäläinen, H.; Leppänen, M.T.; Kiviranta, H.; Kukkonen, J.V.K.; Vuori, K.M. Biological responses of midge (*Chironomus riparius*) and lamprey (*Lampetra fluviatilis*) larvae in ecotoxicity assessment of PCDD/F-, PCB- and Hg-contaminated river sediments. *Environ. Sci. Pollut. Res.* **2016**, *23*, 18379–18393. [CrossRef] [PubMed]
76. Gagliardi, B.S.; Pettigrove, V.J.; Long, S.M.; Hoffmann, A.A. A Meta-Analysis Evaluating the Relationship between Aquatic Contaminants and Chironomid Larval Deformities in Laboratory Studies. *Environ. Sci. Technol.* **2016**, *50*, 12903–12911. [CrossRef] [PubMed]

Article

Preliminary Analysis of the Diet of *Triturus carnifex* and Pollution in Mountain Karst Ponds in Central Apennines

Mattia Iannella [1], Giulia Console [1,*], Paola D'Alessandro [1], Francesco Cerasoli [1], Cristina Mantoni [1], Fabrizio Ruggieri [2], Francesca Di Donato [2] and Maurizio Biondi [1]

[1]   Department of Life, Health and Environmental Sciences, Environmental Sciences Sect., University of L'Aquila, Via Vetoio, Coppito, 67100 L'Aquila, Italy; mattia.iannella@univaq.it (M.I.); paola.dalessandro@univaq.it (P.D.); francesco.cerasoli@univaq.it (F.C.); cristina.mantoni@graduate.univaq.it (C.M.); maurizio.biondi@univaq.it (M.B.)

[2]   Department of Physical and Chemical Sciences (DSFC), University of L'Aquila, Via Vetoio, Coppito, 67100 L'Aquila, Italy; fabrizio.ruggieri@univaq.it (F.R.); francesca.didonato3@graduate.univaq.it (F.D.D.)

*   Correspondence: giulia.console@graduate.univaq.it

Received: 29 October 2019; Accepted: 18 December 2019; Published: 20 December 2019

Abstract: Mountain karst ponds are sensitive environments, hosting complex trophic networks where amphibians play a major role, often as top predators. The diet of the Italian crested newt (*Triturus carnifex*) is still poorly known for populations occupying mountain karst ponds. These are traditionally used as livestock's watering points, leading to water pollution due to excreta and wading behavior. The aim of this paper is to understand the relationship between *T. carnifex* diet composition, assessed through the stomach flushing technique, and physical and chemical characteristics in mountain ponds, focusing on parameters altered by livestock pressure, such as ammonium concentration and dissolved oxygen. The high diversity of prey items found within the newts' gut contents confirms the generalist diet even in mountain ponds. The number of prey taxa, their relative abundance and Shannon–Wiener diversity index show variations among the sampled sites, related to livestock organic pollution. Moreover, we report the very first European records of microplastic items in amphibians' stomach content, which also represent the first evidence for Caudata worldwide. Our findings suggest that livestock pressure directly influences *T. carnifex* diet and highlight that the emerging issue of plastics is a threat even in remote high-altitude environments.

Keywords: diet; *Triturus carnifex*; mountain karst ponds; pollution; microplastics

---

1. Introduction

High-altitude ecosystems are nowadays more and more studied because of their susceptibility to ongoing climate change and habitat loss [1–3]. Many species have been proven to shift their ranges to higher altitudes [4–6] or, when the habitat does not permit these migrations, to possibly go extinct [7–9].

Karst mountain ponds are sensitive water bodies because of their exposure to several highly variable environmental factors. They are subject to seasonal snow coverage, noticeable daily and annual thermal excursion, UV-radiation and water-level changes [10–13]. As a result, the biotic component is often unique as well, showing adaptations to temperature fluctuation, long diapause or hibernation and short breeding phenology [14–16]. The species belonging to these animal communities are often of conservation concern, because of their specialization to mountain environments and to low connectivity among ponds [10,17].

In the Apennines, ponds are usually located within grasslands of karstic plateaus, and are traditionally used as watering points for livestock [18–20]. Therefore, the occurrence of cattle around and inside the ponds determine the eutrophication of the water through the direct and indirect introduction of their excreta [21,22], as we also frequently observed during field campaigns. Even though most of these ponds fall within National and Regional Parks, they are still vulnerable and sensitive to many occurring environmental changes, because of their limited size [23,24]. Considering the alternation of turbid and clear water due to the alternative stable state of ponds [25], it is evident that the increase of non-managed livestock in ponds (especially horses and cows) determines nutrient enrichment, shifting towards turbid water conditions through sediment resuspension and alteration of chemicals, such as ammonium and oxygen concentrations [26,27]. Such physical and chemical modification affects both the macroinvertebrates and the vertebrates in ponds. In particular, the macroinvertebrate community structure changes and negatively influences the amphibians' community; this cascade effect depends on the environmental conditions of the ponds, because of a complex interaction between abiotic and biotic components [22,28,29]. Further, amphibians themselves show high sensitivity to climate [30,31], contaminants [32–34] and habitat changes [35,36].

Despite their highly threatened status [37,38] and their importance in food webs [39,40], amphibians' dietary ecology in high altitude lakes or ponds has been poorly investigated so far [41,42].

In this paper, we analyze how the ponds' alteration due to dense livestock presence could affect amphibians' diet, through the diet analysis of the Italian crested newt, *Triturus carnifex*, at mountain sampling sites. We study this aspect, which is still little known, focusing on both livestock-used ponds of different extents and artificial waters (rainwater collection cistern and troughs), characterized by different physical and chemical water conditions, in protected areas.

We explored, through a correspondence analysis, the association between physical and chemical variables and diversity in prey taxa in ponds with different degrees of livestock pollution. We also performed multiple statistical comparisons to test possible significant differences in the variability of newts' diet between natural and artificial sites and among the single sampling sites. We further assessed the presence of human-derived pollutants, such as plastics, which possibly affect freshwater ecosystems even in high altitudes.

2. Materials and Methods

2.1. Target Species

The Italian crested newt *Triturus carnifex* (Laurenti, 1768) was the target species of this study; it shows a disjunct distribution between the Italian and Balkan peninsulas [43]. In Italy, the species occurs from the sea level up to 1980 m [44]; in the Central Apennines, it reaches 1818 m a.s.l. in Lake Pantaniello [45,46].

The terrestrial environment preferred by the Italian crested newt ranges from deciduous beech forests to typically Mediterranean xeric meadows; the shelters must be fresh and moist, often found under downed woods, in forest litter and in the burrows of small mammals. The aquatic phase occurs in lakes, ponds, springs and puddles, preferably rich in aquatic vegetation for egg deposition.

Triturus carnifex is classified as "Near threatened" by the Italian IUCN red list, with a declining population trend [47]. The main threats are the destruction or alteration of its reproductive sites at different levels, the shift to intensive agricultural practices, and climate change, which often leads to the desiccation of their preferred habitats [48–50].

2.2. Study Area

The study area included the Central Apennines' mountain karst plains (900–1700 m a.s.l.) (Figure 1), characterized by a scarcity of surface water. In this area, we selected 9 natural and artificial sampling sites, classified as ponds, troughs and a cistern; all of them fell within two protected areas, the Gran Sasso and Monti della Laga National Park and the Sirente Velino Regional Park. In the former, the ponds named Fossetta (two neighboring ponds, coded as GSA and GSB), Fossa di Paganica (GSC) and Racollo (GSD) were studied. In the latter, the Tempra pond (SVA), the Piane di Iano pond (SVB), the

Pagliare di Tione no-roof rainwater collection cistern, hereafter named only "cistern" (SVC), the Piano Canale (SVD) and Fonte dell'Acqua (SVE) troughs were sampled. Relevant pieces of information about the sampling sites are reported in Table 1.

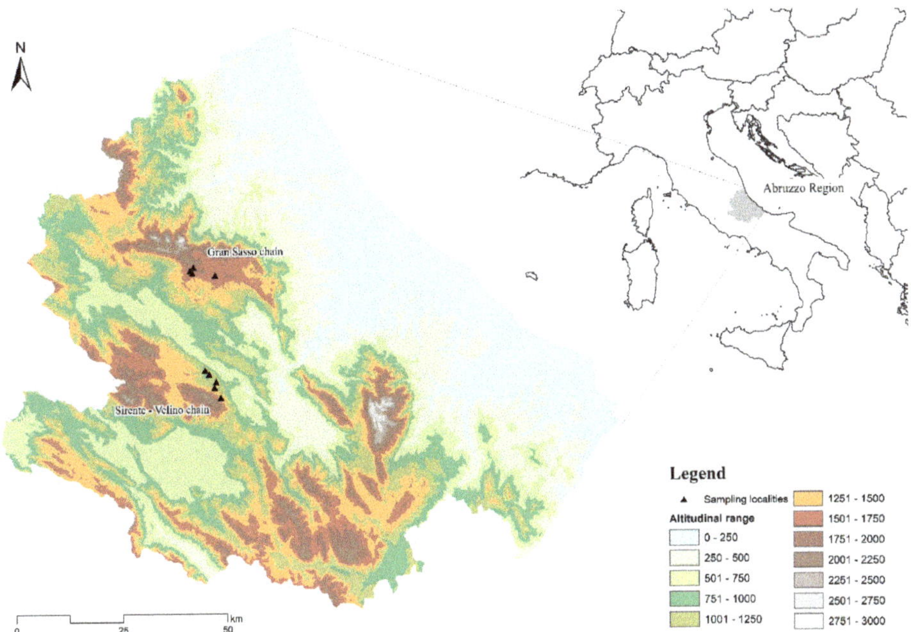

Figure 1. Study area with altitude classes and sampling localities (black triangles).

Table 1. Sampling sites considered in the analyses and the corresponding geographic information.

Name	Code	Type	Area (m^2)	Max. Depth (m)	Latitude	Longitude	Elevation (m a.s.l.)
Fossetta A	GSA	Pond	1870	1.2	42°24′17″ N	13°35′14″ E	1658
Fossetta B	GSB	Pond	358	1.2	42°24′19″ N	13°35′13″ E	1658
Fossa di Paganica	GSC	Pond	4500	1.5	42°24′55″ N	13°39′27″ E	1680
Racollo	GSD	Pond	7000	0.8	42°23′37″ N	13°39′27″ E	1570
Tempra	SVA	Pond	870	1.5	42°10′22″ N	13°38′30″ E	1183
Piane di Iano	SVB	Pond	780	1.2	42°11′21″ N	13°37′22″ E	960
Pagliare di Tione	SVC	Cistern	125	2.5	42°11′55″ N	13°36′41″ E	1080
Piano Canale	SVD	Trough	15	0.5	42°08′25″ N	13°39′17″ E	1342
Fonte dell'Acqua	SVE	Trough	26	0.5	42°08′21″ N	13°39′10″ E	1167

The GSA, GSB, GSD and SVB ponds show remarkable water level variations during summer; they are characterized by turbid waters (because of the silty bottom) and scarce vegetation (mainly composed by *Potamogeton*, *Ceratophyllum* and *Myriophyllum*) due to the constant presence of livestock (sheep, cows and horses), which enter the ponds for some meters from the banks (Figure 2). On the contrary, GSC and SVA show clearer waters, probably due to the wide extent (and the gravel bottom) of the former and due to the low livestock frequentation of the latter. In all these ponds, water is supplied only from snow melting and rainfall. SVC is not used by livestock as a watering point because of its limited accessibility due to perimeter walls; it has clear water and abundant aquatic vegetation. Finally, SVD and SVE show clear water and aquatic vegetation (including algae) as well; they are used as watering points, but livestock does not enter them.

Figure 2. Livestock near Fossa di Paganica pond, GSC (**a**); horses grazing aquatic plants (**b**) and cow cooling off during summer (**c**) in the Fossetta pond, GSA.

2.3. Sampling Methods

Newts (excluding larvae) were sampled starting from 2015 to 2019 (June–September, following the aquatic phase of the target species), employing both active and passive sampling methods. Dip netting sessions were performed, following a circular transect to cover the entire pond's perimeter. Funnel traps were also used as passive capture method, leaving them in the sampling sites for a 12-h period. The captured newts were stomach flushed using all the necessary equipment as described in [51]; the stomach contents were stored in Eppendorf safe-lock tubes with 70% ethanol.

A multi-parametric probe Eutech PD650 (EUTECH Instruments, Singapore, Singapore) was used to measure pH, dissolved oxygen (DO in mg/L), redox potential (mV) and the possible presence of ammonium (NH_4^+ concentration, mg/L), resulting from livestock excreta directly released within the ponds (or nearby the corresponding banks). All of the parameters were measured once for each sampling session.

2.4. Diet Analysis

The prey items were identified at the lowest taxon allowed by the level of digestion, generally as orders.

The sampled newts for each sampling site, the number of different prey groups and the total number of prey items found within the stomach contents of each newt are reported in Supplementary Material Table S1.

Diversification of predated taxa was calculated for each sampled newt within each sampling site through the Shannon–Wiener diversity index (H'), by means of the 'diversity' function of the 'vegan' [52] package in R [53]. The resulting H' values are reported in Supplementary Material Table S1.

Appropriate statistical tests were conducted in R to check for potential discrepancies in the diversification of prey groups among the sampling sites, as well as among the types of sampling sites (i.e., ponds, cistern and troughs) and their "class" (i.e., natural, comprising the ponds, and artificial, including the cistern and the troughs).

A Shapiro–Wilk test and QQ plots were used to check the normality of the residuals from a linear model built considering the H′ index as the response variable and the three above-mentioned grouping factors as covariates. Levene tests were performed to check homoscedasticity across the levels of the three grouping arrangements. Kruskall–Wallis tests were conducted to assess the significance of the considered grouping factors.

Subsequently, multiple pairwise Wilcoxon Rank Sum tests were conducted, using the 'pairwise.wilcox.test' function in R [53] with the Benjamini and Hochberg p-value adjustment method, to investigate differences among the single levels of the grouping factor(s) resulting as significant.

Moreover, we considered for each site the relative percentages of the different prey taxa per "average newt" (i.e., the ratio between the number of prey items, for each taxon, and the number of captured newts).

Correspondence analysis was performed through the 'ca' package [54] in RStudio [53], using both chemical parameters and relative prey abundance for each sampling sites as variables.

Together with prey items and other materials (e.g., parts of plants and sediments), we unexpectedly found microplastics in some stomach contents. All microplastics were observed and measured through a Leica M205C (Leica Microsystems, Wetzlar, Germany) binocular microscope. Photomicrographs were taken using a Leica DFC500 camera and the Zerene Stacker software (Zerene Systems LLC, Richland, WA, USA) version 1.04.

3. Results

The stomach contents of 239 Italian crested newts were collected from the nine sampling sites. The number of sampled newts and the total number of prey items for the different prey groups (for each site) are reported in Supplementary Material Tables S1 and S2.

Prey items were counted and classified in 19 different taxa, namely Crustacea (4750 prey items), Diptera (1800), Haplotaxida (328), Odonata (220), Hemiptera (191), Mollusca (135), Ephemeroptera (75), Coleoptera (53), Araneae (35), Acarina (17), Hymenoptera (13), Dermaptera (five), Caudata (four), Trichoptera (three), Tricladida (three), Lepidoptera (two), Julida (two), Neuroptera (one) and Mermithida (one). A total of 641 items could not be determined, due to the high level of deterioration; 161 fragments of plants, 93 particles of inorganic sediments probably deriving from the bottom, 59 microplastic items in the form of fibres, splinters and other undeterminable fragments were also found.

Crustacea (among which 47% were Ostracoda, 33% Branchiopoda and 19% Maxillopoda) and Diptera (among which 81% were Chironomidae and 9% Culicidae) showed the highest frequencies per average newt in almost all the sampling sites, with Crustacea exceeding 85% of the total prey items in GSD and SVB; differently, the highest Diptera frequency is found in SVA (Figure 3). GSC showed the highest number of taxa (14) per average newt, followed by SVD (13) and an equal number for SVA and SVE (11); the remaining five sampling sites showed a total number of prey taxa ≤ 10 (GSA = 10, SVC = 10, GSD = 8, GSB = 7, SVB = 6).

With respect to the analyses performed to assess the level of diversification within the diet of the single newts among the sampling sites, the performed Shapiro–Wilk test (W = 0.95, $p = 4.49 \times 10^{-7}$) and the inspected QQ plot led us to reject the hypothesis of the normality of the residuals from the linear model built upon the H′ index, while the Levene tests did not reject homoscedasticity for the sampling site (F = 0.97, $p = 0.46$), site type (F = 1.35, $p = 0.26$) and site class (F = 0.06, $p = 0.80$) grouping factors.

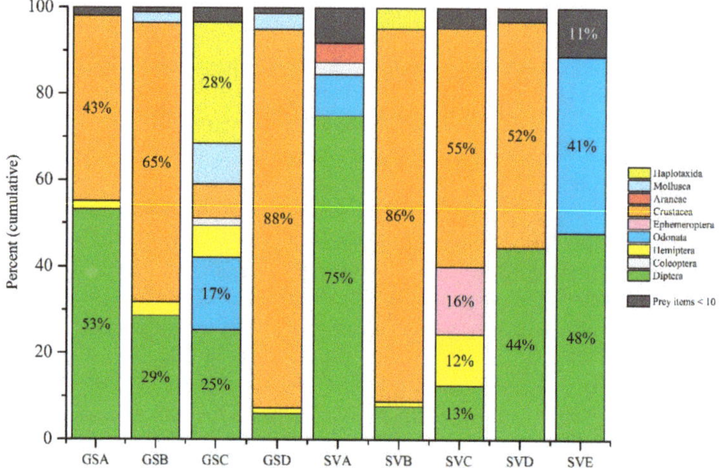

Figure 3. Frequency of prey items (taxa) for each sampled site. Taxa < 10 prey items were aggregated into a single class (grey); percentages < 10% are not reported within the figure for graphical purposes.

Given the normality assumption was not met, a Kruskall–Wallis test was used to check the significance of the three grouping factors—significant differences in the means of the single levels resulted in only considering the sampling sites as a grouping factor ($\chi^2 = 18.01$, df = 8, $p = 0.02$).

From the Wilcoxon Rank Sum tests successively performed on this grouping factor, GSB presented as significantly more diversified in terms of taxa predated by the single newts with respect to SVE ($p = 0.04$); it is also worth noticing the low median H′ value of GSA, GSD, SVB and SVE with respect to the remaining sampling sites (Figure 4a).

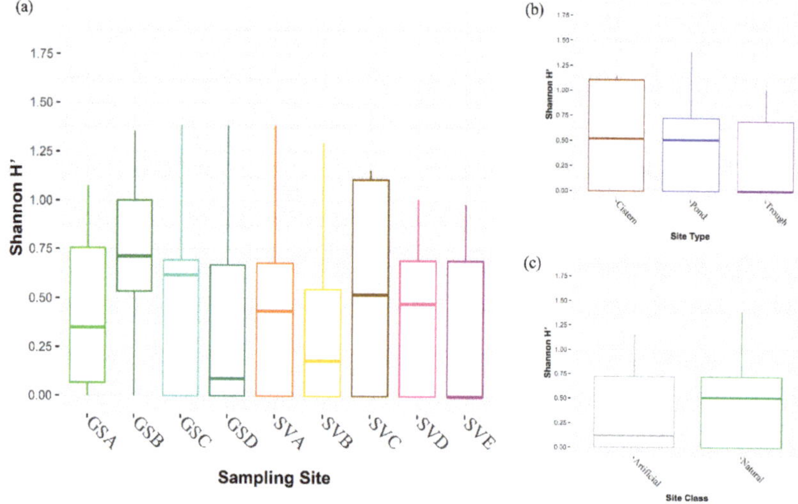

Figure 4. Boxplots of Shannon–Wiener H′ index values across sampling sites (**a**), site types (**b**), and site classes (**c**).

Notwithstanding, the site type ($\chi^2 = 4.29$, df = 2, $p = 0.12$) and site class ($\chi^2 = 2.32$, df = 8, $p = 0.13$) were shown to be non-significant grouping factors by the Kruskall–Wallis tests, considering the differences among the levels of these two factors emerging from boxplots in Figure 4b,c, single Mann–Whitney U tests were performed between natural and artificial sites, and among the three site types.

With respect to site class, the diets of the single newts within natural sites could be considered to be more diversified than in artificial ones (U = 5798.5, p = 0.06, alternative = 'greater') under a less constraining 0.1 threshold for the p-value; differently, considering site types, diets of the single newts in ponds were seen to be significantly richer than in (averages across the replicates) troughs (U = 4781, p = 0.02, alternative = 'greater'), while no significant pattern emerged for the other two pairs (i.e., ponds–cistern and cistern–troughs).

The chemical parameters measured for each sampling site (averages across the replicates) are reported in Table 2. GSA, GSB, GSD and SVB show very low values of dissolved oxygen, along with relatively high values of ammonium (up to 0.26 mg/L in SVB), leading to classify them as "low water quality sites".

Table 2. Chemical parameters measured for each sampling site.

Name	Code	pH	Redox Potential (mV)	DO (mg/L) (DO (%sat), Temp. (°C))	NH_4^+ (mg/L)
Fossetta A	GSA	7.3	−6.9	3.0 (38%, 20.8 °C)	0.12
Fossetta B	GSB	7.6	−17.7	3.6 (47%, 21.1 °C)	0.17
Fossa di Paganica	GSC	8.3	38.1	4.5 (62%, 24.0 °C)	0.00
Racollo	GSD	7.6	−7.9	2.3 (32%, 23.2 °C)	0.03
Tempra	SVA	9.5	−5.1	7.0 (90%, 24.2 °C)	0.06
Piane di Iano	SVB	7.5	−28.7	3.7 (45%, 17.9 °C)	0.26
Pagliare di Tione	SVC	8.5	16.2	7.2 (87%, 21.4 °C)	0.00
Piano Canale	SVD	8.1	27.2	7.8 (96%, 20.3 °C)	0.00
Fonte dell'Acqua	SVE	7.9	24.7	7.4 (93%, 23.4 °C)	0.00

SVA and GSC can be classified as "halfway sites", while SVC, SVD and SVE, which represent the artificial sites, could be grouped as "high water quality sites" (Table 2).

The correspondence analysis (Figure 5) resulted in a neat prevalence of Dim 1 (80.4%) along which natural sites (GSA, GSB, GSD, SVA and SVB) and artificial ones (SVC, SVD and SVE) are clearly separated, except for the GSC pond, which groups with the latter. Ammonium is distant from DO and redox potential and placed near to the ponds group; differently, pH does not seem to form any relevant group with the other variables. Mollusca, Crustacea, Hemiptera, Coleoptera and Diptera are placed in the negative portion of Dim 1, near the ponds group; on the contrary, Odonata, Trichoptera and Caudata are the main taxa placed close to the artificial sites. Tricladida, Mermithida, Haplotaxida, Julida, Dermaptera, Ephemeroptera and Hymenoptera are located between the two groups of sampling sites, in proximity to the artificial sites. Acarina, Araneae and Lepidoptera are distant from both the ponds and artificial sites groups, while Neuroptera is completely separated along the positive portion of the Dim 2 (not shown), probably due to their little contribution in diet diversity.

Microplastics were found in the stomach contents of newts inhabiting four sampling sites, namely GSC, GSD, SVD and SVE. A sample for each type of microplastic found is represented in Figure 6. Two net-like plastic fragments were found in the stomach flush of a newt sampled in GSD, measuring 3.7 mm and 3.1 mm, respectively (Figure 6a). Two textile-like fibres (out of many others found in all the sampling sites) are shown in Figure 6b, measuring 1.8 mm (blue) and 1.2 mm (red). Finally, a blue microplastic fragment (1.44 mm in size), was found in the stomach contents of a newt sampled in the SVE trough along with a little (0.1 mm) yellow fragment (Figure 6c).

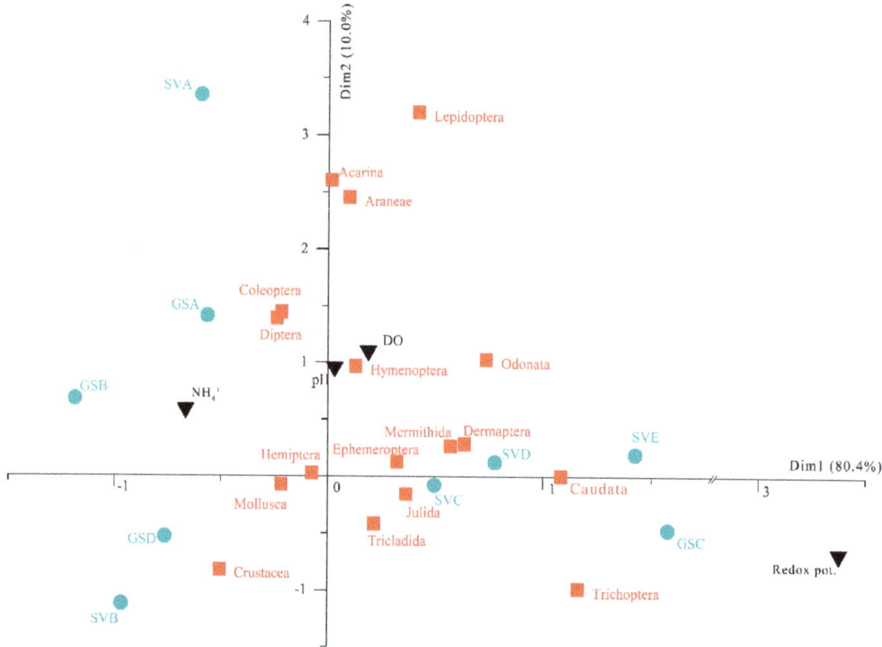

Figure 5. Correspondence analysis for both prey items and chemical parameters measured in all sampling sites. Blue circles: sampling sites, red squares: prey taxa and black triangles: physical and chemical parameters.

To characterize these items, ATR-FT-IR analyses were carried out using a Perkin Elmer (USA) Spectrum Two™ FTIR (PerkinElmer, Waltham, MA, USA) instrument equipped with the Attenuated Total Reflectance (ATR) accessory with a single bounce diamond crystal. The microplastics were analyzed at a 1 cm^{-1} resolution within the 4000–400 cm^{-1} range, with air used for the background spectrum. A consistent force was applied using the pressure monitoring system integrated with the instrument to maximize the spectrum intensity; the optimal number of scans was determined by collecting the spectra of the plastic fragments. An increase in the quality of the signal-to-noise ratio between 10 and 50 scans at a spectral resolution of 1 cm^{-1} was observed. After 50 scans, the improvement in spectral quality was less evident, thus, all samples were collected at 50 scans.

Analyses by ATR-FT-IR spectroscopy enabled the identification of microplastics by detecting absorbance bands. In Figure 7, the IR spectra of a fragment of polyethylene resulting from the analysis of the blue item pictured in Figure 6c are reported, corresponding to the bands of absorbance intrinsic of the stretching of C–H (3000–2770 cm^{-1}) and the bending of C–H (1500–1450 cm^{-1}) bonds. Additionally, ATR-FT-IR measurements enabled the detection of bands attributed to the scissoring of C–H bonds (750–700 cm^{-1}).

The IR analysis of the fragments pictured in Figure 6a shows the characteristic bands of a medical pressure-sensitive adhesive (spectrum in Supplementary Material Figure S1a). The fibres reported in Figure 6b are probably part of mountain technical clothes. The analysis of the corresponding IR spectrum (Supplementary Material Figure S1b) shows the presence of the typical bands of polyacrylic polymers, although the signal-to-noise ratio is moderately low.

Finally, the chop of Figure 6c is likely to be a part of a plastic bottle cap, considering its polyethylene composition revealed by the IR spectra.

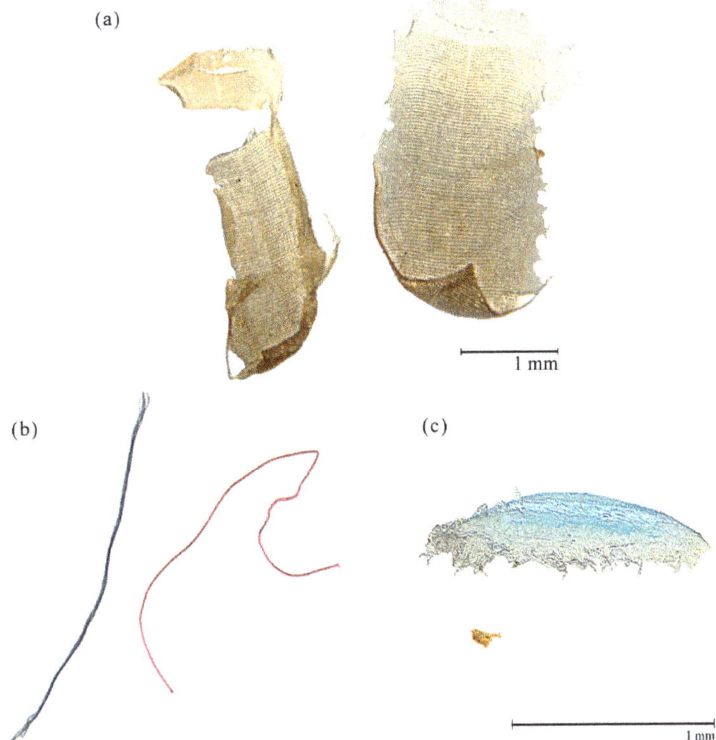

Figure 6. Microplastics found within *Triturus carnifex* stomach contents. (**a**) the biggest fragments found, probably deriving from a medical patch; (**b**) blue and red fibres, probably deriving from mountain clothes; (**c**) chop, probably deriving from a plastic bottle and very small yellow fragment (bottom) from an undeterminable origin.

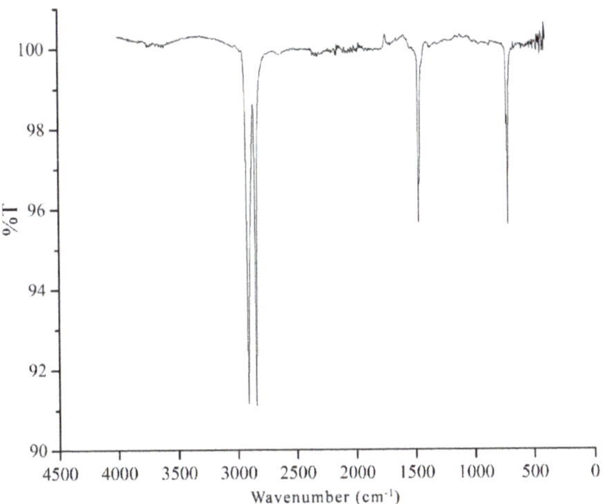

Figure 7. IR spectrum of the polyethylene chop (displayed in Figure 6c, above) found within an Italian crested newt's stomach contents.

4. Discussion

The Italian crested newt exhibits a generalist diet [44,55,56], consuming a considerable variety of preys [57]. In our samples, aquatic invertebrates presented the great majority of the collected preys; the most predated taxa were Crustacea, followed by Diptera, with a neat prevalence of small-sized prey items for both, while the others 17 taxa found in the stomach contents were less represented.

The study area has been used for centuries for grazing, and its management has been subject to national regulation for at least 100 years [58]; while in the past the studied territory was mainly pastured by sheep, nowadays cattle and horses represent the main grazers. This determines a change in grazing management, since usually only sheep are looked after by the shepherd during the whole day (personal field observations). The differences in the chemical and physical conditions observed in the sampled sites may be the direct consequence of such kind of pasture management: in fact, while sheep do not usually wade the ponds, cows and horses are frequently observed in them. Thus, ponds are subject to livestock wading and defecation, determining an alteration of water quality, while drinking troughs or tanks does not usually undergo these pressures [29,59]. This could explain the notable differences between the natural sampled sites and the artificial ones, in terms of abiotic conditions.

The correspondence analysis permitted us to segregate two main groups. Sites showing good physical and chemical parameters (i.e., the three artificial sites and the GSC pond) are associated along the positive half of Dim 1 with taxa with narrow ecological tolerances to disturbance and organic pollution (e.g., Odonata, Ephemeroptera and Trichoptera). Differently, GSD and SVB are placed on the negative portion of Dim 1 and Dim 2 and are mainly associated with Crustacea, which showed predominance in the diet of newts from these sites, leading to low prey diversity (Figures 3 and 4a). These two sites are part of a larger group comprising also GSA and GSB (as shown by the correspondence plot), in which all sites are associated with ammonium, probably reflecting the ease of livestock wading in these ponds. SVA, which shows low ammonium and high DO concentrations when compared to the aforementioned sites, is distant from this group along Dim 2; this can be explained by the low livestock density around this site (personal observation).

The controversial impact of grazing on amphibians' communities is still debated, as reviewed by Howell, Mothes, Clements, Catania, Rothermel and Searcy [29]; some studies highlighted the negative effects of grazing on amphibians, e.g., [60,61], while others documented a positive or intermediate response, due to the species-specific requirements of the target amphibians [62–64].

Based on our research, ponds, which represent typical habitats for the newts in grasslands of karst plains and show high biodiversity in uncontaminated conditions [13], are negatively influenced by livestock pressure, especially when they can be easily waded. Indeed, we found livestock contamination altering the chemical and physical water parameters in ponds (e.g., low redox potential, high ammonium and low DO); this can be related to the absence of highly sensitive taxa and the consequent low number of different predated taxa per site.

Another important aspect is that the only water supply for these ponds derives from snow melting and rain, resulting in a low rate of water replacement, especially during late summer, when the pasture pressure increases.

On the other hand, the artificial sampling sites, notwithstanding their small extent, present a high number of prey taxa, probably due to the constant water flow, which positively contributes to the high quality of water in these sites. This is further confirmed by the presence of some taxa which exhibit a reduced tolerance to the disturbances found in ponds, as also shown by the correspondence analysis.

Fossa di Paganica (GSC) is the only pond which does not follow this trend, showing conditions similar to the artificial sampling sites; in fact, it presents the highest number of prey taxa among all sites, and good chemical parameters. This could be explained by the large dimensions of the pond, which allow for the buffering of the negative effects of cattle and horses wading and defecating. Furthermore, the diffused presence of macrophytes found in GSC could favour macroinvertebrate diversity, as already observed in other karst ponds [65]. This is reflected by the high number of prey taxa found in this site, which could anyway be an underestimation of the total macroinvertebrate diversity.

Notwithstanding the evidence of livestock contamination of ponds and the corresponding influence on the number of prey taxa, the analysis of the Shannon H′ variations among the sites showed that prey diversification for each newt is occasionally high if compared to the number of prey taxa per sampling site (e.g., GSB and SVA, see Figure 4a).

Previous studies showed that the low diversity of available prey may hamper larval development, due to low intake of dietary antioxidants [66,67] leading to reduced immune response capability and mass growth [68]. Thus, we speculate that newts, also considering their generalist diet, may be led to diversify their preys as much as possible, to integrate a possible lack of nutrients and avoid the negative effects mentioned before.

Furthermore, amphibian feeding is also conditioned by interspecific competition [69], which can be changed by the noticeable disequilibria found in most of the sampled sites, considering the absence of invertebrate predators (sometimes also acting as preys), such as Coleoptera, Hemiptera and Odonata.

In this work, we present the first European record of microplastics in amphibians' stomach contents and the first one ever recorded for Caudata.

While several studies about microplastics in marine ecosystems have been published in recent decades [70], field studies examining their presence in freshwater ecosystems started only recently [71]. Despite both Caudata and Anuran being among the most threatened animal groups worldwide, the only papers assessing the presence of microplastics in gut content referred to Anuran species [71,72]. Moreover, the previous records of microplastics in Anuran stomach contents were collected in human-altered freshwater habitats, such as farmlands surrounding river deltas [72] and plantations [71]; on the contrary, we found evidences of microplastic ingestion in scarcely populated, high-elevation territories, even covered by protected areas.

The chemical composition of the collected microplastics, emerging from the performed ATR-FT-IR spectroscopy analyses, shows that they are associated to human activities carried out in mountain environments, such as pastoralism and tourism; indeed, the majority of microplastic items are textile fibres (Figure 6b). The presence of these fibres in sampling sites may be due to waste displacement by the wind from the neighbouring roads and mountain retreats, or to livestock-related items, such as horse bridles or cow tags.

While the effects of microplastic ingestion by amphibians have been already assessed in laboratory studies with contrasting results [73,74], there is no information about the potential negative impacts on amphibians' communities in freshwater habitats.

5. Conclusions

High-altitude environments are nowadays proven to be very sensitive to ongoing environmental changes; the alteration of their freshwater habitats leads to rapid and dramatic loss of biodiversity. While, on one hand, pasture has shaped some peculiar environments for centuries, on the other hand it may represent a serious threat if not correctly managed. Human presence has, in this case, a dual role; it may cause a loss of diversity (e.g., unmanaged pasture) but it may also lead to the persistence of amphibian populations, because of the construction of artificial watering points, which currently host highly diverse communities. This is the very first assessment of the livestock-related detrimental effects on newts' diets in mountain karstic ponds; ongoing targeted sampling campaigns will enrich the knowledge about these threats, which widely affect mountain ecosystems.

The unexpected discovery of microplastics in many stomach contents reinforces the alarm about the ubiquitous plastic pollution of natural habitats, even in high-altitude protected territories; it is thus worth conducting future targeted research about this emerging issue to properly adapt conservation strategies.

Supplementary Materials: The following are available online at http://www.mdpi.com/2073-4441/12/1/44/s1, Figure S1: IR spectra of (a) medical pressure-sensitive adhesive and (b) fibres pictured in Figure 6a,b, respectively, found within Italian crested newts' stomach contents. Table S1: Sampled newts for each sampling site, number of different prey groups, Shannon H', and total number of prey items found within the stomach content of each newt. Table S2: Macroinvertebrates found within stomach contents for all the sampled newts within each sampling site.

Author Contributions: Conceptualization, M.I. and M.B.; methodology, M.I., G.C. and F.C.; software, M.I., G.C. and P.D.; formal analysis, G.C., C.M. and P.D.; data curation, G.C. and C.M.; chemical analyses, F.R. and F.D.D.; writing—original draft preparation, all authors; writing—review and editing, all authors; funding acquisition, M.I. and M.B. All authors have read and agreed to the published version of the manuscript.

Funding: This work was supported by the A.I.M. Project—PON R & I 2014–2020 No. 1870582.

Acknowledgments: We are particularly thankful to the Scientific Department of Gran Sasso—Monti della Laga. National Park and to the Scientific and Naturalistic Office of the Sirente-Velino Regional Park for the technical support provided during the collection of some field data (permissions n. 0000219/15 UT-RAU-SCNZ448 and N. VII-02-02-P6213).

Conflicts of Interest: The authors declare no conflict of interest.

References

1. Rogora, M.; Frate, L.; Carranza, M.; Freppaz, M.; Stanisci, A.; Bertani, I.; Bottarin, R.; Brambilla, A.; Canullo, R.; Carbognani, M.; et al. Assessment of climate change effects on mountain ecosystems through a cross-site analysis in the Alps and Apennines. *Sci. Total Environ.* **2018**, *624*, 1429–1442. [CrossRef]

2. Dirnböck, T.; Essl, F.; Rabitsch, W. Disproportional risk for habitat loss of high-altitude endemic species under climate change. *Glob. Chang. Biol.* **2011**, *17*, 990–996. [CrossRef]

3. Brambilla, M.; Pedrini, P.; Rolando, A.; Chamberlain, D.E. Climate change will increase the potential conflict between skiing and high-elevation bird species in the Alps. *J. Biogeogr.* **2016**, *43*, 2299–2309. [CrossRef]

4. Di Musciano, M.; Carranza, M.; Frate, L.; Di Cecco, V.; Di Martino, L.; Frattaroli, A.; Stanisci, A. Distribution of plant species and dispersal traits along environmental gradients in central Mediterranean summits. *Diversity* **2018**, *10*, 58. [CrossRef]

5. Cerasoli, F.; Iannella, M.; Biondi, M. Between the hammer and the anvil: How the combined effect of global warming and the non-native common slider could threaten the European pond turtle. *Manag. Biol. Invasions* **2019**, *10*, 428. [CrossRef]

6. Chamberlain, D.; Brambilla, M.; Caprio, E.; Pedrini, P.; Rolando, A. Alpine bird distributions along elevation gradients: The consistency of climate and habitat effects across geographic regions. *Oecologia* **2016**, *181*, 1139–1150. [CrossRef] [PubMed]

7. Cerasoli, F.; Thuiller, W.; Guéguen, M.; Renaud, J.; D'Alessandro, P.; Biondi, M. The role of climate and biotic factors in shaping current distributions and potential future shifts of European Neocrepidodera (Coleoptera, Chrysomelidae). *Insect Divers. Conserv.* **2019**. [CrossRef]

8. Brunetti, M.; Magoga, G.; Iannella, M.; Biondi, M.; Montagna, M. Phylogeography and species distribution modelling of *Cryptocephalus barii* (Coleoptera: Chrysomelidae): Is this alpine endemic species close to extinction? *ZooKeys* **2019**, *856*, 3. [CrossRef] [PubMed]

9. Wiens, J.J. Climate-Related local extinctions are already widespread among plant and animal species. *PLoS Biol.* **2016**, *14*. [CrossRef] [PubMed]

10. Oertli, B.; Biggs, J.; Céréghino, R.; Grillas, P.; Joly, P.; Lachavanne, J.B. Conservation and monitoring of pond biodiversity: Introduction. *Aquat. Conserv. Mar. Freshw. Ecosyst.* **2005**, *15*, 535–540. [CrossRef]

11. Niedrist, G.; Psenner, R.; Sommaruga, R. Climate warming increases vertical and seasonal water temperature differences and inter-annual variability in a mountain lake. *Clim. Chang.* **2018**, *151*, 473–490. [CrossRef]

12. Sommaruga, R. The role of solar UV radiation in the ecology of alpine lakes. *J. Photochem. Photobiol. B Biol.* **2001**, *62*, 35–42. [CrossRef]

13. Hinden, H.; Oertli, B.; Menetrey, N.; Sager, L.; Lachavanne, J.B. Alpine pond biodiversity: What are the related environmental variables? *Aquat. Conserv. Mar. Freshw. Ecosyst.* **2005**, *15*, 613–624. [CrossRef]

14. Miaud, C.; Guyetant, R.; Faber, H. Age, size, and growth of the alpine newt, *Triturus alpestris* (Urodela: Salamandridae), at high altitude and a review of life-history trait variation throughout its range. *Herpetologica* **2000**, *56*, 135–144.

15. Alonso, M.; Garcia-De-Lomas, J. Systematics and ecology of *Linderiella baetica* n. sp. (Crustacea, Branchiopoda, Anostraca, Chirocephalidae), a new species from southern Spain. *Zoosystema* **2009**, *31*, 807–828. [CrossRef]

16. Mushet, D.M.; Euliss Jr, N.H.; Chen, Y.; Stockwell, C.A. Complex spatial dynamics maintain northern leopard frog (*Lithobates pipiens*) genetic diversity in a temporally varying landscape. *Herpetol. Conserv. Biol.* **2013**, *8*, 163–175.

17. Lund, J.O.; Wissinger, S.A.; Peckarsky, B.L. Caddisfly behavioral responses to drying cues in temporary ponds: Implications for effects of climate change. *Freshw. Sci.* **2016**, *35*, 619–630. [CrossRef]

18. Catorci, A.; Gatti, R.; Vitanzi, A. Relationship between phenology and above-ground phytomass in a grassland community in central Italy. In *Nature Conservation. Environmental Science and Engineering (Environmental Science)*; Gafta, D., Akeroyd, J., Eds.; Springer: Heidelberg, Germany, 2006; pp. 309–327.

19. Filibeck, G.; Cancellieri, L.; Sperandii, M.G.; Belonovskaya, E.; Sobolev, N.; Tsarevskaya, N.; Becker, T.; Berastegi, A.; Bückle, C.; Che, R. Biodiversity patterns of dry grasslands in the Central Apennines (Italy) along a precipitation gradient: Experiences from the 10th EDGG Field Workshop. *Bull. Eurasian Dry Grassl. Group* **2018**, *36*, 25–41.

20. Primi, R.; Filibeck, G.; Amici, A.; Bückle, C.; Cancellieri, L.; Di Filippo, A.; Gentile, C.; Guglielmino, A.; Latini, R.; Mancini, L.D. From Landsat to leafhoppers: A multidisciplinary approach for sustainable stocking assessment and ecological monitoring in mountain grasslands. *Agric. Ecosyst. Environ.* **2016**, *234*, 118–133. [CrossRef]

21. Ruggiero, A.; Solimini, A.G.; Carchini, G. Limnological aspects of an Apennine shallow lake. *Ann. Limnol. Int. J. Limnol.* **2004**, *40*, 89–99. [CrossRef]

22. Solimini, A.G.; Bazzanti, M.; Ruggiero, A.; Carchini, G. Developing a multimetric index of ecological integrity based on macroinvertebrates of mountain ponds in central Italy. In *Pond Conservation in Europe*; Oertli, B., Céréghino, R., Biggs, J., Declerck, S., Hull, A., Miracle, M.R., Eds.; Springer: Dordrecht, The Netherlands, 2007; pp. 109–123.

23. Calhoun, A.J.; Mushet, D.M.; Bell, K.P.; Boix, D.; Fitzsimons, J.A.; Isselin-Nondedeu, F. Temporary wetlands: Challenges and solutions to conserving a 'disappearing' ecosystem. *Biol. Conserv.* **2017**, *211*, 3–11. [CrossRef]

24. Hunter, M.L., Jr.; Acuña, V.; Bauer, D.M.; Bell, K.P.; Calhoun, A.J.; Felipe-Lucia, M.R.; Fitzsimons, J.A.; González, E.; Kinnison, M.; Lindenmayer, D. Conserving small natural features with large ecological roles: A synthetic overview. *Biol. Conserv.* **2017**, *211*, 88–95. [CrossRef]

25. Ruggiero, A.; Solimini, A.; Carchini, G. The alternative stable state concept and the management of Apennine mountain ponds. *Aquat. Conserv. Mar. Freshw. Ecosyst.* **2005**, *15*, 625–634. [CrossRef]

26. Solimini, A.G.; Ruggiero, A.; Anello, M.; Mutschlechner, A.; Carchini, G. The benthic community structure in mountain ponds affected by livestock watering in nature reserves of Central Italy. *Int. Ver. Theor. Angew. Limnol. Verh.* **2000**, *27*, 501–505. [CrossRef]

27. Ruggiero, A.; Solimini, A.; Carchini, G. Pratica dell'alpeggio e conservazione ambientale negli stagni montani dell'Appennino Centrale. In Proceedings of the Atti del Convegno Nazionale Conservazione Dell'ambiente e Rischio Idrogeologico, Perugia, Italy, 11–12 December 2002; pp. 687–694.

28. Wissinger, S.A.; Oertli, B.; Rosset, V. Invertebrate communities of alpine ponds. In *Invertebrates in Freshwater Wetlands*; Batzer, D., Boix, D., Eds.; Springer: Cham, Switzerland, 2016; pp. 55–103.

29. Howell, H.J.; Mothes, C.C.; Clements, S.L.; Catania, S.V.; Rothermel, B.B.; Searcy, C.A. Amphibian responses to livestock use of wetlands: New empirical data and a global review. *Ecol. Appl.* **2019**. [CrossRef]

30. Iannella, M.; Cerasoli, F.; Biondi, M. Unraveling climate influences on the distribution of the parapatric newts *Lissotriton vulgaris meridionalis* and *L. italicus*. *Front. Zool.* **2017**, *14*, 55. [CrossRef]

31. Iannella, M.; D'Alessandro, P.; Biondi, M. Evidences for a shared history for spectacled salamanders, haplotypes and climate. *Sci. Rep.* **2018**, *8*, 16507. [CrossRef]

32. Hayes, T.B.; Stuart, A.A.; Mendoza, M.; Collins, A.; Noriega, N.; Vonk, A.; Johnston, G.; Liu, R.; Kpodzo, D. Characterization of atrazine-induced gonadal malformations in African clawed frogs (*Xenopus laevis*) and comparisons with effects of an androgen antagonist (cyproterone acetate) and exogenous estrogen (17β-estradiol): Support for the demasculinization/feminization hypothesis. *Environ. Health Perspect* **2006**, *114*, 134–141.

33. Buck, J.C.; Scheessele, E.A.; Relyea, R.A.; Blaustein, A.R. The effects of multiple stressors on wetland communities: Pesticides, pathogens and competing amphibians. *Freshw. Biol.* **2012**, *57*, 61–73. [CrossRef]

34. Patar, A.; Giri, A.; Boro, F.; Bhuyan, K.; Singha, U.; Giri, S. Cadmium pollution and amphibians–Studies in tadpoles of *Rana limnocharis*. *Chemosphere* **2016**, *144*, 1043–1049. [CrossRef]

35. Ferreira, M.; Beja, P. Mediterranean amphibians and the loss of temporary ponds: Are there alternative breeding habitats? *Biol. Conserv.* **2013**, *165*, 179–186. [CrossRef]

36. Ficetola, G.F.; Rondinini, C.; Bonardi, A.; Baisero, D.; Padoa-Schioppa, E. Habitat availability for amphibians and extinction threat: A global analysis. *Divers. Distrib.* **2015**, *21*, 302–311. [CrossRef]

37. Stuart, S.N.; Chanson, J.S.; Cox, N.A.; Young, B.E.; Rodrigues, A.S.; Fischman, D.L.; Waller, R.W. Status and trends of amphibian declines and extinctions worldwide. *Science* **2004**, *306*, 1783–1786. [CrossRef] [PubMed]

38. May, R.M. Ecological science and tomorrow's world. *Philos. Trans. R. Soc. B Biol. Sci.* **2010**, *365*, 41–47. [CrossRef]

39. Best, M.L.; Welsh, J.; Hartwell, H. The trophic role of a forest salamander: Impacts on invertebrates, leaf litter retention, and the humification process. *Ecosphere* **2014**, *5*, 1–19. [CrossRef]

40. Davic, R.D.; Welsh, H.H. On the ecological roles of salamanders. *Annu. Rev. Ecol. Evol. Syst.* **2004**, *35*, 405–434. [CrossRef]

41. Salvidio, S.; Costa, A.; Crovetto, F. Individual trophic specialization in the Alpine newt increases with increasing resource diversity. *Ann. Zool. Fenn.* **2019**, *56*, 17–24. [CrossRef]

42. Vignoli, L.; Bombi, P.; D'Amen, M.; Bologna, M.A. Seasonal variation in the trophic niche of a heterochronic population of *Triturus alpestris apuanus* from the south-western Alps. *Herpetol. J.* **2007**, *17*, 183–191.

43. Schabetsberger, R.; Jehle, R.; Maletzky, A.; Pesta, J.; Sztatecsny, M. Delineation of terrestrial reserves for amphibians: Post-breeding migrations of Italian crested newts (*Triturus c. carnifex*) at high altitude. *Biol. Conserv.* **2004**, *117*, 95–104. [CrossRef]

44. Vanni, S.; Andreone, F.; Tripepi, S. *Istituto Superiore per La Protezione e La Ricerca Ambientale*; Anfibi d'Italia. Lanza, B., Nistri, A., Vanni, S., Eds.; Ministero dell'Ambiente e della Tutela del Territorio e del Mare: Rome, Italy, 2007; Volume 29.

45. Naviglio, L. Aspetti naturalistici del Lago Pantaniello. *Nat. Montagna.* **1984**, *31*, 49–57.

46. Ferri, V.; Di Tizio, L.; Pellegrini, M. *Atlante Degli Anfibi d'Abruzzo*; Ianieri–Talea: Pescara, Italy, 2007; p. 199.

47. Andreone, F.; Corti, C.; Ficetola, F.; Razzetti, E.; Romano, A.; Sindaco, R. *Triturus carnifex*. The IUCN Red List of Threatened Species. 2009: e.T59474A11947714. Available online: https://www.iucnredlist.org/species/59474/11947714 (accessed on 5 December 2018).

48. Beebee, T.J.C.; Griffiths, R.A. The amphibian decline crisis: A watershed for conservation biology? *Biol. Conserv.* **2005**, *125*, 271–285. [CrossRef]

49. D'Amen, M.; Bombi, P. Global warming and biodiversity: Evidence of climate-linked amphibian declines in Italy. *Biol. Conserv.* **2009**, *142*, 3060–3067. [CrossRef]

50. Collins, J.P.; Crump, M.L.; Lovejoy Iii, T.E. *Extinction in Our Times: Global Amphibian Decline*; Oxford University Press: New York, NY, USA, 2009.

51. Solé, M.; Rödder, D. Dietary assessments of adult amphibians. In *Amphibian Ecology and Conservation: A Handbook of Techniques*; Oxford University Press: New York, NY, USA, 2010; pp. 167–184.

52. Oksanen, J.; Blanchet, F.G.; Friendly, M.; Kindt, R.; Legendre, P.; McGlinn, D.; Minchin, P.R.; O'Hara, R.; Simpson, G.L.; Solymos, P. 2019. Vegan: Community Ecology Package. R package version 2.5-4. Available online: https://cran.r-project.org/web/packages/vegan (accessed on 1 September 2019).

53. R Core Team. *R: A Language and Environment for Statistical Computing*; R Foundation for Statistical Computing: Vienna, Austria, 2017; Available online: http://www.R-project.org (accessed on 15 June 2017).

54. Nenadic, O.; Greenacre, M. Correspondence Analysis in R, with two-and three-dimensional graphics: The ca package. *J. Stat. Softw.* **2007**, *20*, 1–13.

55. Costa, A.; Salvidio, S.; Posillico, M.; Matteucci, G.; De Cinti, B.; Romano, A. Generalisation within specialization: Inter-individual diet variation in the only specialized salamander in the world. *Sci. Rep.* **2015**, *5*, 13260. [CrossRef]

56. Vignoli, L.; Bologna, M.A. Dietary patterns and overlap in an amphibian assemblage at a pond in Mediterranean central Italy. *Vie Milieu-Life Environ.* **2009**, *59*, 47–57.

57. Fasola, M.; Canova, L. Feeding habits of *Triturus vulgaris, T. cristatus* and *T. alpestris* (Amphibia, Urodela) in the northern Apennines (Italy). *Ital. J. Zool.* **1992**, *59*, 273–280.

58. Regio Decreto Legislativo. *Riordinamento E Riforma Della Legislazione in Materia di Boschi E di Terreni Montani*; Italy, Ed.; NORMATTIVA: Rome, Italy, 1923.

59. Buono, V.; Bissattini, A.M.; Vignoli, L. Can a cow save a newt? The role of cattle drinking troughs in amphibian conservation. *Aquat. Conserv. Mar. Freshw. Ecosyst.* **2019**, *29*, 964–975. [CrossRef]

60. Schmutzer, A.C.; Gray, M.J.; Burton, E.C.; Miller, D.L. Impacts of cattle on amphibian larvae and the aquatic environment. *Freshw. Biol.* **2008**, *53*, 2613–2625. [CrossRef]

61. Hoverman, J.T.; Gray, M.J.; Miller, D.L.; Haislip, N.A. Widespread occurrence of ranavirus in pond-breeding amphibian populations. *Eco. Health* **2012**, *9*, 36–48. [CrossRef]

62. Badillo-Saldaña, L.M.; Ramírez-Bautista, A.; Wilson, L.D. Effects of establishment of grazing areas on diversity of amphibian communities in tropical evergreen forests and mountain cloud forests of the Sierra Madre Oriental. *Rev. Mex. Biodivers.* **2016**, *87*, 133–139. [CrossRef]

63. Moreira, L.F.; Moura, R.G.; Maltchik, L. Stop and ask for directions: Factors affecting anuran detection and occupancy in Pampa farmland ponds. *Ecol. Res.* **2016**, *31*, 65–74. [CrossRef]

64. Howard, A.; Munger, J.C. *Effects of Livestock Grazing on the Invertebrate Prey Base and on the Survival and Growth of Larvae of the Columbia Spotted Frog, Rana Luteiventris*; Bureau of Land Management, Idaho State Office: Boise, ID, USA, 2003.

65. Zelnik, I.; Gregorič, N.; Tratnik, A. Diversity of macroinvertebrates positively correlates with diversity of macrophytes in karst ponds. *Ecol. Eng.* **2018**, *117*, 96–103. [CrossRef]

66. Byrne, P.G.; Silla, A.J. Testing the effect of dietary carotenoids on larval survival, growth and development in the critically endangered southern corroboree frog. *Zoo. Biol.* **2017**, *36*, 161–169. [CrossRef] [PubMed]

67. Clugston, R.D.; Blaner, W.S. Vitamin A (retinoid) metabolism and actions: What we know and what we need to know about amphibians. *Zoo. Biol.* **2014**, *33*, 527–535. [CrossRef] [PubMed]

68. Szuroczki, D.; Koprivnikar, J.; Baker, R.L. Dietary antioxidants enhance immunocompetence in larval amphibians. *Comp. Biochem. Physiol. Part A Mol. Integr. Physiol.* **2016**, *201*, 182–188. [CrossRef]

69. Arribas, R.; Díaz-Paniagua, C.; Caut, S.; Gomez-Mestre, I. Stable isotopes reveal trophic partitioning and trophic plasticity of a larval amphibian guild. *PLoS ONE* **2015**, *10*. [CrossRef]

70. Hidalgo-Ruz, V.; Gutow, L.; Thompson, R.C.; Thiel, M. Microplastics in the marine environment: A review of the methods used for identification and quantification. *Environ. Sci. Technol.* **2012**, *46*, 3060–3075. [CrossRef]

71. Döring, B.; Mecke, S.; Kieckbusch, M.; O'Shea, M.; Kaiser, H. Food spectrum analysis of the Asian toad, *Duttaphrynus melanostictus* (Schneider, 1799) (Anura: Bufonidae), from Timor Island, Wallacea. *J. Nat. Hist.* **2017**, *51*, 607–623. [CrossRef]

72. Hu, L.; Chernick, M.; Hinton, D.E.; Shi, H. Microplastics in small waterbodies and tadpoles from Yangtze River Delta, China. *Environ. Sci. Technol.* **2018**, *52*, 8885–8893. [CrossRef]

73. De Felice, B.; Bacchetta, R.; Santo, N.; Tremolada, P.; Parolini, M. Polystyrene microplastics did not affect body growth and swimming activity in *Xenopus laevis* tadpoles. *Environ. Sci. Pollut. Res.* **2018**, *25*, 34644–34651. [CrossRef] [PubMed]

74. Da Costa Araújo, A.P.; de Melo, N.F.S.; de Oliveira Junior, A.G.; Rodrigues, F.P.; Fernandes, T.; de Andrade Vieira, J.E.; Rocha, T.L.; Malafaia, G. How much are microplastics harmful to the health of amphibians? A study with pristine polyethylene microplastics and *Physalaemus cuvieri*. *J. Hazard. Mater.* **2020**, *382*, 121066. [CrossRef] [PubMed]

Article

AQUALIFE Software: A New Tool for a Standardized Ecological Assessment of Groundwater Dependent Ecosystems

Giovanni Strona [1], Simone Fattorini [2], Barbara Fiasca [2], Tiziana Di Lorenzo [3], Mattia Di Cicco [2], Walter Lorenzetti [4], Francesco Boccacci [4] and Diana M. P. Galassi [2,*]

1   Research Centre for Ecological Change, University of Helsinki, P.O. Box 4 FI-00014 Helsinki, Finland; giovanni.strona@helsinki.fi

2   Department of Life, Health and Environmental Sciences, University of L'Aquila, 67100 L'Aquila, Italy; simone.fattorini@univaq.it (S.F.); barbara.fiasca@univaq.it (B.F.); diciccomattiauni@yahoo.it (M.D.C.)

3   Italian National Research Council, Research Institute on Terrestrial Ecosystems, Via Madonna del Piano 10, Sesto Fiorentino, 50019 Florence, Italy; tiziana.dilorenzo@cnr.it

4   Gis3W S.r.l, Viale Verdi 24, Montecatini Terme, 51016 Pistoia, Italy; lorenzetti@gis3w.it (W.L.); boccacci.francesco@gmail.com (F.B.)

*   Correspondence: dianamariapaola.galassi@univaq.it

Received: 30 October 2019; Accepted: 2 December 2019; Published: 6 December 2019

Abstract: We introduce a suite of software tools aimed at investigating multiple bio-ecological facets of aquatic Groundwater Dependent Ecosystems (GDEs). The suite focuses on: (1) threats posed by pollutants to GDE invertebrates (Ecological Risk, ER); (2) threats posed by hydrological and hydromorphological alterations on the subsurface zone of lotic systems and groundwater-fed springs (Hydrological-Hydromorphological Risk, HHR); and (3) the conservation priority of GDE communities (Groundwater Biodiversity Concern index, GBC). The ER is assessed by comparing tolerance limits of invertebrate species to specific pollutants with the maximum observed concentration of the same pollutants at the target site(s). Comparison is based on an original, comprehensive dataset including the most updated information on tolerance to 116 pollutants for 474 freshwater invertebrate species. The HHR is assessed by accounting for the main direct and indirect effects on both the hyporheic zone of lotic systems and groundwater-fed springs, and by scoring each impact according to the potential effect on subsurface invertebrates. Finally, the GBC index is computed on the basis of the taxonomical composition of a target community, and allows the evaluation of its conservation priority in comparison to others.

Keywords: AQUALIFE software; groundwater dependent ecosystems; threats; biodiversity

1. Introduction

Groundwater dependent ecosystems (GDEs) are ecosystems whose biological structure and ecological processes are directly or indirectly influenced by groundwater [1,2]. They may include springs, wetlands, streams and rivers, hyporheic zones, lakes, coastal lagoons, and saturated aquifers (also indicated as Subsurface Groundwater Dependent Ecosystems, SGDEs). Aquatic GDEs are known to host a high biodiversity, including highly specialized, rare, and often narrow endemic species [3–5]. Preserving their biodiversity is therefore identified as a conservation priority [5–11]. Many GDEs, and hence their biodiversity, are known to be at risk of impairment worldwide [4,7,10,12,13].

Some regulations for the protection of GDEs have recently entered in force in the Australian subcontinent [14,15] and, to a lesser extent, in Europe [10]. Although these ecosystems are cited in Annex II, Section 2, "Groundwater", pt. 2.2. of the EU Directive 2000/60/EC [16], and later in the Directive 2006/118/EC, also called Groundwater Daughter Directive [17], only a few GDEs have been included in Nature 2000 network or designated as Ramsar areas [18,19]. In the United States, a mosaic of laws acts directly or indirectly towards the protection of groundwater and GDEs by controlling contamination, managing conflicts among users, and quantifying abstraction but a comprehensive governance is still missing [20,21]. Proposals for management and recovery of GDEs have been advanced for particular situations, such as mining [12] and groundwater withdrawal [22]. However, current proposals [10] for management actions remain quite vague and protocols to assess the impacts of chemical and hydrological-hydromorphological threats on aquatic GDEs and their conservation priorities are still lacking.

To cope with one of the major challenges of integrating scientific knowledge and environmental management policy, the European Union Life + project "LIFE12 BIO/IT/000231—AQUALIFE" developed a set of procedures and indicators for the assessment of aquatic GDE biodiversity and its threats. The project implemented this set of procedures and indicators into the software suite called "AQUALIFE Software" (hereafter, AS), which we introduce here.

The AS consists of a user-friendly online interface, and it is free to use (after registration) at: http://app.aqualifeproject.eu.

2. Software Capabilities

AS allows users to compute a set of different metrics to evaluate the ecological risk, the hydrological-hydromorphological risk, and the conservation priority of aquatic GDEs. The software has been mainly developed using the Python programming language [23] in combination with the Django web framework [24].

The target for the application of the software is, in principle, a set of monitoring sites and their associated biological communities. The software was generated by assembling a rich biodiversity dataset for selected GDEs in the Abruzzo region (Italy), which was chosen as a pilot area for AS development. Users can use their own biodiversity datasets, together with the concentration of the pollutants measured in their sites and, if working on the hyporheic habitat of streams and rivers or groundwater-fed springs, the hydrological-hydromorphological impacts detected in their study sites. All the AS outputs can be saved online, accessed, and downloaded at any later stage.

AS comprises three sections, namely: Ecological Risk (ER), Hydrological-Hydromorphological Risk (HHR), and Conservation Priority Index (GBC). The first two focus on a particular aspect of the GDEs' vulnerability while the last one on the conservation value of GDE biodiversity. Each section includes three subsections. The first subsection, I—Info, provides general information about the rationale of the main section, as well as technical details regarding the associated computation procedures. The second subsection, R—Risk for the first two sections and I—Index for the third section, allows users to perform the analyses. Finally, the third subsection, H—History, allows users to explore and access saved results from previous analyses.

In the following, we provide a detailed description of the different software sections. In doing that, we mainly focus on the technical aspects of computation procedures and on practical aspects of the user experience. Our aim is to provide stakeholders with enough information to start actively using the software in a fully-conscious manner.

2.1. Section 1: Ecological Risk

The Ecological Risk (ER) quantifies the probability that chemicals produced by human activity (hereafter referred as "pollutants") harm the health status of a biological community inhabiting an aquatic GDE [5,25–28]. Since chemical pollutants usually occur in mixtures in GDEs, ER is quantified as in Equation (1) [29,30]:

$$ER = RQ_{mix} = \Sigma RQ_i = \Sigma \frac{MEC_i}{PNEC_i} \tag{1}$$

where RQ_i is the risk posed by the ith pollutant in the mixture, MEC_i stands for the Measured Environmental Concentration and $PNEC_i$ for the Predicted No-Effect Concentration of the ith pollutant in the mixture. If more environmental concentrations are available for the ith pollutant, MEC_i of Equation (1) is equal to the highest concentration measured by the user, to account for the worst-case scenario [31]. $PNEC_i$ is the concentration at which the ith pollutant is predicted to have no significant detrimental effect on the GDE community [32]. $PNEC_i$ is computed as follows [31]:

$$PNEC_i = \sqrt[n]{PNEC_1 \times PNEC_2 \times \ldots PNEC_n} \tag{2}$$

where

$$PNEC_n = \frac{LC50(EC50)}{AFa} \tag{3}$$

or

$$PNEC_n = \frac{NOEC}{AFb} \tag{4}$$

$PNEC_i$ is the geometric mean of $PNEC_n$ where $PNEC_n$ is the predicted no-effect concentration of the ith pollutant on the nth taxon of a GDE community represented by n taxa [31,33,34]. LC_{50} (Lethal Concentration 50%) and EC_{50} (Effect Concentration 50%) are the concentrations of the ith pollutant producing, respectively, a lethal effect or a maximal effect, on 50% of individuals tested in short-term trials ($\leq$96 h) and NOEC (No-Observed Effect Concentration) is the concentration producing no significant effects under long-term exposures (>96 h). AF is an assessment factor that reflects the uncertainty in extrapolating $PNEC_n$ values from laboratory toxicity tests [27], being AFa equal to 1000 and AFb equal to 100 [35,36]. If more than one LC_{50} and/or EC_{50} and/or NOEC values are available for the nth taxon, the toxicity data are aggregated by computing their geometric mean [31].

AS provides an internal database of PNECs for 116 pollutants known to be potential threats to GDEs; they were derived from 8806 toxicity values related to 474 taxa either found in GDEs or in other freshwater habitats according to information availability (Tables S1 and S2). Toxicity data were retrieved from the US EPA database ECOTOX "Aquatic Data" [37]. When unavailable, LC_{50} values were estimated by US EPA ECOSAR v.2.0 [38]. We used the most updated literature to amend errors (for example, wrong habitat attribution of the records, wrong attribution of the species to the freshwater environment, and wrong species names) as well as to expand the AS database with additional information (e.g., new toxicity data for freshwater species).

In freshwater studies, it is common to focus on tolerance limits of algae, macrophytes, invertebrates, and small fish; however, to assemble the dataset used by AS, we focused only on invertebrates. The rationale behind this choice stems from the circumstantial evidence that in aquatic GDEs there is virtually no primary productivity (except for rare cases of chemoautotrophy [39]) and, as for Europe, no underground fish are known.

AS assigns a risk magnitude and a color to ER based on five classes (Table 1), following the assumption that, if $RQ_{mix} > 1$, an ER is expected [27,28].

Table 1. ER rating, magnitude and color per rank.

ER Rating	ER Magnitude	ER Color Ranking
$0 \leq RQ_{mix} < 1$	no risk	
$1 \leq RQ_{mix} < 10$	low risk	
$10 \leq RQ_{mix} < 100$	medium risk	
$100 \leq RQ_{mix} < 1000$	high risk	
$RQ_{mix} \geq 1000$	very high risk	

Because of the dominance of the crustaceans in certain GDEs (e.g., SGDEs), AS allows users to select one out of two subsets of the toxicity database: (i) *"crustacean species"* (choice suggested for SGDEs such as the saturated aquifers where crustaceans are the dominant group); or (ii) *"freshwater invertebrates"* (choice suggested for springs and the hyporheic zone of streams and rivers, where also non-crustacean species are present, and often with high abundances).

The computation of RQ_{mix} by AS is straightforward. After loading the Ecological Risk page, the user can select the pollutants from a drop-down list, enter the MECs for the pollutants detected in the GDE sites under study, select one out the two subsets of toxicity database, and start the calculation of RQ_{mix} by pressing the "RUN" button. After that, AS will provide the user with both the individual PNEC values of the selected pollutants, and the RQ_{mix} value along with the ER magnitude and color.

AS allows the user to save the analyses and access them at a later moment. To do this, the user will have to associate the analysis to a geographical location (either a location indicated in previous analyses or a new one; in the latter case, the user will have to assign an arbitrary name to the locality and will have the possibility to provide its geographical coordinates, in form of latitudinal/longitudinal decimal degrees). An interactive map will show the position of the target site(s). After that, a "SAVE" button will become available. The last step consists in attributing an identifier name to the analysis, allowing its later retrieval. Saved analyses are accessible by the user in the History Page of the Ecological Risk section. Clearly, users have personalized histories, and do not have access to other user pages. Data from the History subsection can be directly exported to spreadsheet documents for being processed outside the software environment.

2.2. Section 2: Hydrological-Hydromorphological Risk

The second section of AS aims at identifying and scoring potential risks posed by man-induced alterations of the hydrological connectivity (Hydrological-Hydromorphological Risk, HHR) of specific GDE types such as hyporheic zones of streams and rivers and groundwater-fed springs. We designed our procedure taking as a model the Index of Morphological Quality and the Morphological Alteration Index, two operational tools of the Italian "Hydromorphological evaluation system, analysis and monitoring of rivers" named IDRAIM [40].

Several factors can potentially alter the hydrological connectivity of the subsurface habitats from surface waters as well as from the underlying aquifer [41,42], critically influencing hyporheic invertebrate communities [43–45]. Among them, important ones are alteration of streamflow dynamics, changes in streambed and river banks structure and morphology, and reduction of streambed sediments heterogeneity (i.e., colmation or clogging [46]), mainly arising from the presence of dams [42,47–49], barriers, water derivation works, and groundwater overexploitation [21] (HHYC = altered habitat due to hydrological changes [50]).

The presence of solid walls and/or embankments limiting permeability with the alluvial plain of streams and rivers, concreted riverbed, presence of knick points in rivers, artificial riverbed aggradation, channelization [51], clogging or colmation [52–54], artificial elevation of the riverbed, and removal of perifluvial vegetation [55,56] may lead to an alteration of the vertical connectivity river/aquifer [48,57] (HMOC = altered habitats due to morphological changes [50]).

Concerning groundwater-fed springs, alteration of the discharge in the eucrenal-hypocrenal of the spring environment, and in the epirhytron (mountain streams) and alteration of the spring morphology affect severely the crenocenosis and may also cause a disconnection between the spring and the stream [58,59]. If the aquifer feeding the spring is overexploited, an alteration of the spring discharge is expected [58,60–62]. Spring casing and/or concretion (piping, impoundment, and diversion of spring flows), removal of riparian vegetation [63] for maintenance works are reflected in HMOC impacts, such as alteration of the spring morphology with interruption of the spring/aquifer and spring/stream connections, habitat losses, and changes in the nutrient dynamics after vegetation removal [64].

The computation of the HHR is based on the pressures generating effective impacts on GDE invertebrate faunas, with a distinction between those insisting on the hyporheic zones (Table 2) and those affecting springs (Table 3). The list of pressures has been compiled based on expert judgment and is available in the INFO subsection of the main menu under the heading "Hydrological-Hydromorphological Risk" (Tables 2 and 3), which includes a description of the pressures and the indication of the impacts, along with the relative category (i.e., impacts altering habitats due to hydrological changes (HHYC) and impacts altering habitats due to morphological changes

(HMOC)) [50]. A cutoff of 10 ("at risk" vs. "no risk") was identified as the threshold value of the overall impact, which corresponds to the maximum partial impact score attributed to the alteration of the river/aquifer connection exerted by different pressures, considering this alteration sufficient to produce the disappearance of a GDE. All the scores assigned to each impact are stored into the AS.

Table 2. Relevant pressures and related impacts for hyporheic habitats. Pressures and impacts are as so in the Guidance No. 35 of the Water Framework Directive Reporting Guidance 2016, Annexes 1b and 8n [50].

Relevant Pressures: Names and Descriptions		Impacts: Names and Categories	
Hydrological alteration Abstraction or flow diversion Dams, barriers, and locks	Presence of transversal works with reduction of discharge flow rates (detention and/or diversion, dams, weirs)	Alteration of discharge flow rates	HHYC-Altered habitats due to hydrological changes
Hydromorphological alteration	Presence of transversal works with reduction of discharge flow rates (detention and/or diversion, dams, weirs) limiting sediment and woody debris transport on riverbed	Alteration of the longitudinal connectivity in sediment and woody debris in streams/rivers benthic and hyporheic zones	
Abstraction or flow diversion	Presence of solid walls and/or embankments limiting permeability with the alluvial plain of streams and rivers	Alteration of the lateral connectivity in streams/rivers benthic and hyporheic zones	
Dams, barriers, and locks	Concreted riverbed, presence of knick points in rivers or channels where the longitudinal (lengthways) slope of the bed suddenly steepens (incision), artificial riverbed aggradation, clogging, overexploitation of aquifers	Alteration of the vertical connectivity river/aquifer	HMOC-Altered habitats due to morphological changes
Physical alteration of channel/bed/riparian area/shore	Artificially straightened channel	Morphological alteration of the river stretch and riverbed sediment composition	
Groundwater alteration of water level or volume	Removal of perifluvial or riparian vegetation	Increase of riverbank erosion	
	Large woody debris removal	Large woody debris removal	

For hyporheic habitats, most of the pressures are to be evaluated in a river stretch of at least 100 m upstream of the target site. As the effects of dams or river derivations can occur several tens of kilometers downstream, the user will assess their effectiveness at the target site.

A similar set of criteria was adopted to assess the HHR for groundwater-fed springs, where the main pressures are represented by hydrological alteration due to abstraction or flow diversion, dams, or barriers downstream from the spring, which may determine HHYC and HMOC impacts (Table 3 and Table 5).

A score was assigned to each impact based on expert judgment, considering the deviation from reference conditions of unmodified rivers or spring systems. Therefore, the score reflects the degree of alteration of a GDE site related to a given pressure (Tables 4 and 5).

Table 3. Relevant pressures and related impacts for groundwater-fed springs are as so in the Guidance No. 35 of the Water Framework Directive Reporting Guidance 2016, Annexes 1b and 8n [50].

Relevant Pressures		Impacts	
Hydrological alteration Abstraction or flow diversion Dams, barriers, and locks	Presence of transversal works with reduction of discharge flow rates (detention and/or diversion, dams, weirs) downstream the spring emergence	Alteration of discharge flow rate	HHYC–Altered habitats due to hydrological changes
	Overexploitation of aquifers	Alteration of the vertical connectivity	
Physical alteration of channel/bed/riparian area/shore	Spring used for different purposes or concreted at its emergence	Morphological alteration of spring habitats with loss of the ecotonal nature of GW-fed springs	HMOC–Altered habitats due to morphological changes
	Riparian vegetation removal	Habitat and microhabitat losses, nutrient dynamics alteration, current velocity alteration	

Table 4. Impact scores for hyporheic habitats.

Impacts	Impact Score
Alteration of discharge flow rates	6
Alteration of the longitudinal connectivity limiting sediment and woody debris transport on riverbed	4
Alteration of the lateral connectivity	3
Alteration of the vertical connectivity river/aquifer	10
Morphological alteration of the river stretch and riverbed sediment composition	6
Alteration of riparian vegetation	2
Lowering of woody debris storage in streams/rivers	1

Table 5. Impact scores for groundwater-fed springs.

Impacts	Impact Score
Alteration of discharge flow rates	6
Alteration of the ecological features of the spring (e.g., change from rheocrenic to limnocrenic spring as effect of damming)	4
Alteration of the vertical connectivity	10
Morphological alteration of spring habitats with loss of the ecotonal nature of GW-fed springs	5
Alteration of the riparian vegetation, habitat and microhabitat losses, nutrient dynamics alteration, current velocity alteration	3

The computation of HHR is performed by AS in two simple steps. In the first step, users are asked to select the target habitat (either hyporheic or spring habitats).

Depending on this first selection, a pre-defined list of pressures and impacts becomes available, that is either Table 2 or Table 3. The second step consists in choosing from this list the impact(s) recorded at the target site (Table 4 or Table 5). After that, the "RUN" button leads the user to the results window, which provides a risk score (Table 6), and a summary of the user selections.

Table 6. HHR rating, risk ranking, and color per rank.

HHR Rating	Risk Ranking	HHR Color Ranking
HHR < 10	no risk	
HHR ≥ 10	at risk	

As for ER, results can be associated with a geographic location and stored in an internal database that the user can access at any moment through the "History" subsection.

2.3. Section 3: Conservation Priority Index

The third section of AS allows the computation of three conservation indices: Sum of Species Scores (SSS), Biodiversity Conservation Concern (BCC), and Groundwater Biodiversity Concern (GBC). These indices aim at prioritizing GDE communities on the basis of their conservation importance. Although species richness is an obvious criterion to select conservation priority areas, not all species have the same conservation importance. For example, areas that host more imperiled species should be considered more important [65,66]. However, estimating species extinction risk may be extremely difficult, especially for less known invertebrates [67]. For this reason, there is increasing interest in the use of rarity measures as a proxy for the extinction risk of invertebrates [68–70]. Rarity is a relative and multidimensional concept, because a species can be more or less rare compared to other species for various characteristics [71–73]. Once the rarity of a species has been assessed for the various dimensions that are considered, this information can be used to obtain scores that express the overall species' conservation priority. Under these assumptions, AS uses a two-step protocol to assess the conservation priority of GDEs based on invertebrate species. The first step consists in computing the conservation priority of each species occurring in GDE sites based on selected species traits. The second step is to compute site conservation indices using species scores.

The computation of the indices requires users to fill and upload: (1) a dataset where each species (but other taxonomic units can be used) is associated with a set of traits that expresses its conservation importance (the "Dataset" file must contain species in rows, traits in columns); and (2) a matrix of each species abundance across the compared sites (if abundance data are not available, presence/absence can be used) (the "Sampling" file must contain species in rows and sites in columns). Both need to be provided to AS as spreadsheets (.xls). By processing these two input files, AS will compute both a conservation score for each species in the dataset, and three conservation indices for each site (with different scores of the GBC index being associated with different conservation priority categories).

Trait scores are attributed by the user to the collected species for five dimensions (geography, ecology, biology, population, and evolutionary history), each divided into different traits (Table 7). For each trait, the higher is the score, the higher is the species' conservation importance. Different traits have different score ranges. For example, in the "Geography" dimension, the trait "Degree of Endemicity" can obtain a score from 1 to 4, while the trait "Extent of occurrence" can obtain a score from 1 to 7 [74].

Table 7. Traits and their scores used to calculate the Conservation Priority Index.

Dimension	Trait	Score Range
1. GEOGRAPHY	Extent of occurrence	1–7
	Degree of endemicity	1–4
2. ECOLOGY	Degree of GW dependence	1–5
	Ecological specialization to GW	1–3
	Size of ecological niche	1–5
	Tolerance to variation of climatic parameters	1–5
	Microhabitat preferences	1–5
3. BIOLOGY	Trophic role	1–5
	Life time span	1–3
4. POPULATION	Frequency	1–4
	Abundance	1–4
5. EVOLUTIONARY HISTORY	Evolutionary origin	1–3
	Phylogenetic rarity	1–5

Except for "Frequency" and "Abundance", which are calculated by AS, scores for the other traits are assigned to each species present in the dataset by the user on the basis of the best available information. Thus, these scores are based on expert judgment. Frequency and Abundance scores are calculated by AS using site data because these traits are not species-specific, but site-specific (the same species may show different population characteristics at different sites). "Frequency" is given by AS calculating first the frequency of occurrence of each species across the total number of studied sites, and then by attributing to each species a score (from 1 to 4) based on quartiles (with score 1 assigned to species falling above the third quartile, which are the most widely distributed species, and score 4 assigned to species with frequency below the first quartile, which are the species with the narrowest distribution). The "Abundance" trait is computed by AS in a similar way: relative abundance is computed for all species, and then to each species is attributed a score (from 1 to 4) based on quartiles. If abundance data are not available, and only presence/absence data are used, this trait is automatically excluded from calculations.

Trait scores (assigned to each species by the user) belonging to the same dimension are multiplied by AS to obtain a synthetic score for each dimension. Once this operation has been made for all the species in the dataset, a distribution of synthetic scores becomes available for each dimension. These synthetic scores have different ranges. For example, the score for Dimension 1 (Geography) may vary between 1 and 28 (for a species that ranked highest for the two traits included in this dimension), while the score for Dimension 2 (Ecology) may vary between 1 and 1875 (for a species that obtained the maximum score for each of the five traits included in this dimension). For this reason, scores resulting from multiplications are divided using quartiles, and values falling within the same interquartile range receive the same score, ranging from 1 (values below the first quartile) to 4 (values above the third quartile). The use of quartiles to rescale original products into four classes with integer values (i.e., 1, 2, 3, and 4) was adopted in AS to make dimensions comparable. After the quartilization of products, the new scores are summed up to obtain the overall species score. This sum of scores (which can theoretically range from 5, for a species which scored 1 for all five dimensions, to 20, for a species which scored 4 for all dimensions) is rescaled to vary between 1 and 4 by simply dividing by 5. In this way, each species receives a score represented by a value varying between 1 and 4 (SS, Species Scores).

AS generates and uses the SS scores to compute three distinct site-specific indices, and particularly:

(1) Sum of Species Scores (SSS) normalized between 0 and 1 using the formula (Equation (5)):

$$SSSrescaled = (SSS - SSS_{min})/(SSS_{max} - SSS_{min}) \qquad (5)$$

with SSS_{min} and SSS_{max} being, respectively, the minimum and maximum SSS scores recorded across all studied sites.

(2) Biodiversity Conservation Concern (also varying between 0 and 1) (Equation (6)):

$$BCC = (SSS/N - 1)/(SS_{max} - 1) \qquad (6)$$

where N is the total number of species present in a site and SS_{max} is the maximum value of the SS values.

(3) Groundwater Biodiversity Concern (GBC) computed as the mean of SSSrescaled and BCC. GBC values are finally used to identify different priority classes (Table 8).

Table 8. GBC rating, priority ranking and color per rank.

GBC Rating	Priority Ranking	GBC Color Ranking
$0 \leq GBC \leq 0.2$	very low priority	
$0.2 < GBC \leq 0.4$	low priority	
$0.4 < GBC \leq 0.6$	medium priority	
$0.6 < GBC \leq 0.8$	high priority	
$0.8 < GBC \leq 1$	very high priority	

After uploading the files including the table of species scores for each trait, and the matrix of species occurrences/abundances, by pressing the "RUN" button users will be prompted to the results page, which will include both the individual species scores of all species included in the analysis, as well as the site-specific scores, including the raw and rescaled SSS, the BCC, and the GBC (together with the corresponding conservation priority class).

As for the previous sections, users can save all their results, access them at any moment, and export them for additional analysis outside the AS.

3. Example of Application: The Hyporheic Zone of the River Sagittario

To understand the main outputs of AS, namely ER, HHR, and GBC, six hyporheic sites (ABRO86/3S, ABRO87/3S, ABRO88/3S, ABRO89/3S, ABRO90/3S, and ABRO91/3S; Figure 1) of the River Sagittario (Italy) were selected from the AQUALIFE network of sites (Tables S3–S5). The River Sagittario in the Sulmona plain is among the most altered rivers in central Italy [75], thus representing a model environmental system for assessing the effects of different types of pressures on hyporheic biodiversity. At each site, we performed two temporal surveys, namely in December 2014 (winter) and in June 2015 (summer). Hyporheic samples were collected pumping 10 L of interstitial water by a membrane pump connected to steel pipes screened at −40 cm deep and filtered through a 60 μm hand net. Faunal samples were preserved in a 70% alcohol solution. In the laboratory, specimens were sorted under a stereomicroscope at 16× magnification and identified to the lowest taxonomic level possible using a LEICA M205C stereomicroscope. Based on the human activities carried out in the catchment of the River Sagittario, 108 chemicals (including N-compounds, phosphates, metals, volatile organic compounds, and pesticides) were analyzed per each sample taking a 2 L volume. The concentrations of most chemicals (101 out of 108) were below the limit of detection. The MECs of the seven chemical compounds out of 108 (namely, nitrites, nitrates, sulfates, chloride, phosphates, ammonium, and αHCH) exceeding the limits of detection are reported in Table S3. The hydrological-hydromorphological impacts affecting the GDE sites are reported in Table S4. Overall, 58 invertebrate taxa were identified (Table S5) and the trait scores, used by AS to calculate the GBC index, were assigned (Table S5). The invertebrate assemblages were mostly composed by stygobionts or species highly dependent on groundwater (Table S5; see the column "Ecological specialization to GW"). AS was run for each site and the results are summarized in Table 9 and Figure 1. The ER outputs revealed that all sites were at medium risk along the whole river sector (Table 9). No hydrological-hydromorphological risk was reported for the upstream sites ABRO86/3S and ABRO87/3S (HHR < 10) while the four remaining downstream sites showed an HHR > 10 (Table 9). Finally, the highest GBC value (GBC = 0.667) was reached in ABRO86/3S where medium ER (RQ_{mix} = 58.85) and HHR = 6 (no risk) were calculated by AS. In the remaining sites the GBC was in the range very low–medium. The lowest GBC (GBC = 0.172) was assigned by AS to the site ABRO89/3S with low species richness, medium ER (RQ_{mix} = 72.10), and HHR = 13 (at risk) (Table 9 and Figure 1). The application of AS to the River Sagittario has highlighted that the hyporheic zone has only one site with a significant conservation importance. The GDE assemblages experience a medium risk due to pollution of a limited number of substances in all sites, while the hydrological-hydromorphological alteration of the GDE sites under study occurs only in the most downstream sites. This scenario requires management choices that are discussed in detail in the next paragraph.

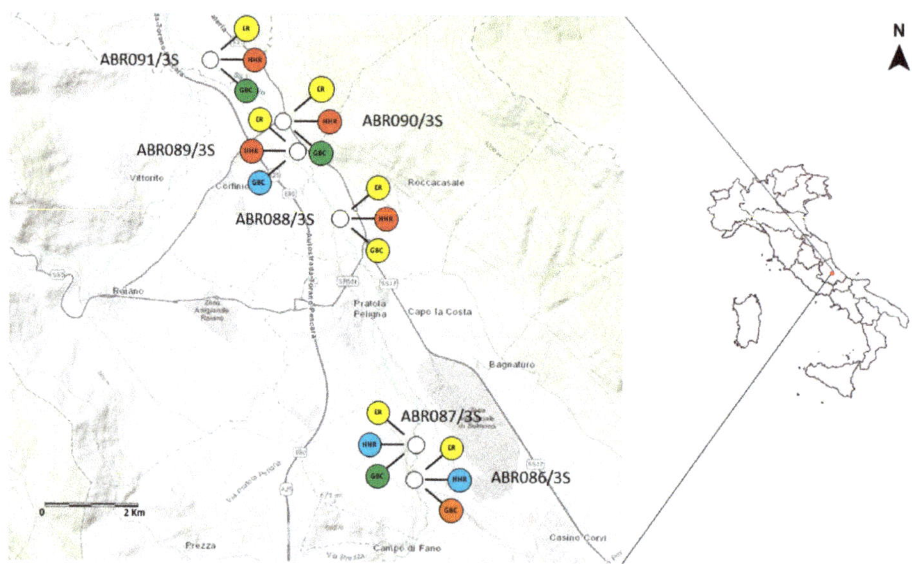

Figure 1. Sampling sites (white circles) of the hyporheic zone of the River Sagittario. Color-coded circles are used to indicate the ranks of the ecological risk (ER), the hydrological-hydromorphological risk (HHR), and the groundwater biodiversity concern index (GBC).

Table 9. AS outputs for ER (Environmental Risk), HHR (Hydrological-Hydromorphological Risk) and the conservation priority indices (SSS, Sum of Species Scores; SSS rescaled, Sum of Species Scores rescaled; BCC, Biodiversity Conservation Concern; GBC, Groundwater Biodiversity Concern) for the example of application in the hyporheic zone of the River Sagittario (Italy).

Sampling Site	ER (RQ_{mix})	ER Ranking	HHR	HHR Ranking	SSS	SSS Rescaled	BCC	GBC	Priority Ranking
ABRO86/3S	58.85	medium	6	no risk	16	1	0.333	0.667	high
ABRO87/3S	83.75	medium	6	no risk	13.8	0.421	0.324	0.372	low
ABRO88/3S	32.11	medium	28	at risk	15	0.737	0.381	0.559	medium
ABRO89/3S	72.10	medium	20	at risk	12.2	0	0.344	0.172	very low
ABRO90/3S	51.22	medium	16	at risk	13.6	0.368	0.314	0.341	low
ABRO91/3S	49.05	medium	15	at risk	13.8	0.421	0.324	0.372	low

4. Discussion

The introduction of the concept of GDE in the context of the Water Framework Directive and the subsequent issuing of specific guidelines have certainly represented a considerable advancement in the protection of aquatic and, in particular, groundwater dependent ecosystems [10]. Due to their communicative effectiveness, scores and color-coded maps have been strongly recommended by legislations worldwide to illustrate the ecological status, the risk, and the monitoring network of water bodies [16]. In fact, scores and colors are intuitive and of immediate understanding even for non-experts and have the advantage of making people aware of information that has been long confined to technicians.

AS provides a user-friendly toolkit to assess (i) the risk effectively or potentially posed to the GDE communities by chemical pollutants; (ii) the hydrological-hydromorphological risk (available for ecotonal GDEs, such as hyporheic zones and groundwater-fed springs); and (iii) the conservation priority of GDE sites based on the community composition. Due to its structure and, above all, thanks to the outputs that are given in scores and color-coded maps, AS fully meets the requirements of intuitive communication of the legislations.

In AS, each index may work either alone or in conjunction with the others, without the limitations of single-function applications. AS creatively addresses also a risk scale for the ecological risk, based on the RQ_{mix} rating, thus allowing the user to measure how far each site is from the condition of "no risk". This graded scale of risk, even if not provided for by any international regulation, allows the user to make an informed choice on which sites require priority recovery and which do not. ER has more classes (five) than HHR (two). This is due to the higher level of knowledge that is available for the ecological risk compared to the hydrological-hydromorphological risk for which scoring is available only for surface water bodies, thus preventing a fine-tuning of the HHR scoring for the GDEs. On a regulatory basis, water bodies are still divided in at risk vs. not at risk, but a further division in classes of the ecological risk is possible ([27] and references therein).

The major advantage of the software is that of providing integrated information. The integration of the most commonly assessed ecological risk with the hydrological-hydromorphological risk is certainly a key-point of AS. In fact, through the HHR index, it is immediately evident to the user that chemical pollution is not always the main factor of GDE degradation. In fact, the greatest impact on GDEs is often due to human activities such as river bed reshaping, grinding, bank defenses, embankments, coatings, burial, vegetation cutting, etc. The attention paid by AS to the physical components of GDEs makes it possible to broaden the horizon of the management measures of protection and recovery [56]. We recommend computing ER along with HHR, because hydrological and hydromorphological impacts occurring in the river channels or groundwater-fed springs can lead to an increasing pollutant concentration both in the hyporheic zone or in spring water and in the feeding aquifers. For instance, areas with intensive agriculture, where a massive riparian vegetation removal has been practiced along river banks, are characterized by higher concentrations of nitrates and ammonium, mainly in the hyporheic zone [26,76] and in the SGDEs, i.e., the saturated aquifers [31,76,77]. The third tool of AS allows the assessment of the conservation priority of sites on the basis of the uniqueness of the species collected in aquatic GDE (and SGDE) sites. This set of indicators (SSS, BBC, and GBC) may work in combination with ER and HHR, based on the assumption that the higher the ER and HHR scores, the higher the risk of biodiversity loss, especially for the most sensitive species. This assumption is not necessarily true, since many variables come into play to define whether a given site is permanently disturbed or not. Furthermore, individual sites are often interconnected, for example if they belong to the same aquifer, or to the hyporheic zone of the same river, thus ensuring a recolonization of perturbed sites from unpolluted aquifer sectors or hyporheic stretches [78].

AS can support European Member States in the preparation of the GDE management plans under the EU Directive 2000/60/EC [16]. Every six years (duration of a management plan) all Member States must indicate the GDEs at risk or not at risk and the related management measures. The actions concern: (i) a new census of anthropic pressures, (ii) a monitoring program, which can be distinguished in surveillance monitoring (which applies exclusively to not-at-risk GDEs) and operational monitoring (which applies to at-risk GDEs); and (iii) protection measures involving interdictions of impactful human activities in the entire GDE or part of it. The list of pressures must be updated every six years, regardless of the risk conditions of the GDE. The choice between the types of monitoring programs has profound economic impact. Surveillance monitoring is largely sustainable because it can be performed only in one point of a given GDE, on basic physico-chemical (e.g., pH, electrical conductivity, and nitrates) and hydrological-hydromorphological parameters (e.g., flow rate, water level, damming, etc.) and with a frequency of one time every six years. On the other hand, the operational monitoring is much more expensive because it involves a significant number of sites and parameters and a frequency of at least once a year for the entire duration of the management plan. In addition, protection measures are often required in at-risk GDEs. To facilitate decision-makers in the choice of the type of monitoring program, we propose to refer to the individual values of the indicators, as shown in Table 10, where ER and HHR provide indications for the type of monitoring and GBC for the protection measures. Based on this rationale, if ER and HHR indicate that the GDE site is not at risk, the monitoring must be based on surveillance pursuant to the EU Directive 2000/60/EC [16] and no protective action is required regardless of the conservation value of the GDE. If, instead, ER and HHR indicate that the GDE is at risk, the monitoring must always be operational, both for the chemical and hydrological-hydromorphological parameters, but the protection actions may be necessary only at certain risk values. For example, protection measures are not necessary if GBC = 0.7 (High Priority) and ER = 9 and HHR < 10. The cases in which AS shows its full utility is when ER does not indicate risk but HHR does (or vice versa). In this case, the monitoring must be operational only for the chemical (or hydrological-hydrogeological) parameters, with a relative cost containment. The study case of the hyporheic zone of the River Sagittario allowed the application of the ER and the HHR together with the GBC (Figure 1). Based on the obtained results, the River Sagittario basin management plan should take into account primarily the restoration of the connectivity between river flowing waters and the underlying aquifer via the hyporheic zone to preserve benthic and hyporheic fauna and the ecosystems services they provide. Four sites out of six were at hydrological-hydromorphological risk for the presence of a dam, water diversion for agriculture, presence of solid walls and/or embankments limiting permeability with the alluvial plain, channelization, artificially straightened channel, and artificial elevation of the riverbed. According to Table 10, the monitoring of the chemical-physical parameters should be operational in all sampling sites while that of the hydrological-hydromorphological variables should be operational in four sites out of six, being based on surveillance in the remaining two sites. Finally, the GBC index indicated that protection measures are required for just one site out of six, namely ABRO86/3S.

Table 10. Decision-making table. Surveillance (SMPs) and operational monitoring program (OMPs) and protection measures (PMs) are indicated according to the values of ER, HHR and GBC indices.

ER Values	ER Monitoring Measures	HHR Values	HHR Monitoring Measures	GBC Values	Measures
$0 \leq RQ_{mix} < 1$	SMPs	HHR < 10	SMPs	$0 \leq GBC \leq 0.2$	No PMs
$1 \leq RQ_{mix} < 10$	OMPs	HHR ≥ 10	OMPs	$0.2 < GBC \leq 0.4$	PMs if RQ ≥ 1000 and HHR ≥ 10
$10 \leq RQ_{mix} < 100$	OMPs			$0.4 < GBC \leq 0.6$	PMs if RQ > 100 and HHR ≥ 10
$100 \leq RQ_{mix} < 1000$	OMPs			$0.6 < GBC \leq 0.8$	PMs if RQ > 10 and HHR ≥ 10
$RQ_{mix} \geq 1000$	OMPs			$0.8 < GBC \leq 1$	PMs if RQ > 1 and/or HHR ≥ 10

From the 2000s until today, other valid tools for assessing the ecological and environmental health in GDEs (but not their conservation values) were presented. For instance, the GW-Fauna-Index [79] was the first attempt to classify GDEs (hyporheic zones and alluvial aquifers) in terms of ecologically relevant conditions using factors which best correlated with stygofauna, namely detritus, temperature, and dissolved oxygen. In 2019, an approach based on three microbiological parameters (prokaryotic cell density, prokaryotic intracellular ATP concentrations, and assimilable organic carbon) was proposed as an integration of the existing routine of GDE monitoring [80]. A single index value obtained from multivariate analyses of the microbial parameters proved to best discriminate between disturbed (organic contamination with hydrocarbons, surface water intrusion, and agricultural activities) and undisturbed alluvial aquifers [80]. A more structured and comprehensive framework for measuring GDE health is the Groundwater Health Index (GHI) [81,82] that was presented in 2011 and successively refined in 2017. Relying upon a combination of microbial, stygofaunal, water chemistry, and environmental indicators, GHI can discriminate three distinct ecosystem health classifications, namely "similar to reference", "mild deviation from reference", and "major deviation from reference".

Other tools were presented to preserve biodiversity of GDEs based on their conservation value (but not on their quality status). A conservation strategy to minimize the risk of groundwater biodiversity loss due to human activities in European GDEs was proposed by designing a network of reserve areas that collectively include most groundwater species [8]. Three area selection methods (species richness hotspots, endemism hotspots, and complementarity) were proposed to preserve 1059 groundwater species of hyporheic zones, springs, and alluvial and karst aquifers in six European regions. Compared to the aforementioned tools, AS offers high flexibility, and can be used, depending on the user needs, in different GDEs and at different spatial scales. In addition, AS is the only available tool that calculates the chemical and hydrological-hydromorphological risks faced by GDEs along with their conservation priority.

However, along with the merits, AS faces also some methodological limitations such as those related, for example, to the number of pollutants for which AS provides the PNECs. Emerging contaminants, such as pharmaceutical compounds, are not represented in the internal database. The HHR index is currently applicable only to ecotonal GDEs, such as hyporheic zones and springs. A HHR for the aquifers (that takes into account quarries, for example, waste materials from stone processing, and the widespread phenomenon of fracking) is still missing. Finally, the scores assigned to each impact are based on expert judgment, which will be potentially tuned with the advancement of knowledge on the GDE functioning and the relative impacts. For this reason, AS will likely be useful for the authorities in charge of GDE management; however, the version presented here should be considered a preliminary workbench toward an appealing refinement of the software itself.

5. Conclusions

The purpose of this paper is to present the AQUALIFE software, which is the main product of the AQUALIFE project. AS consists of a user-friendly online interface, and is free to use (after registration) at: http://app.aqualifeproject.eu. AS is a relatively fast, economical tool of investigation that can be used by operators with basic knowledge of aquatic ecosystems dependent on groundwater and that, through training courses, can complete their knowledge and acquire the necessary experience. AS allows the user to assess the conservation priority of GDEs based on species conservation importance along with chemical and hydrological-hydromorphological risks. Thanks to its novelty, AS has the potential to become a user-friendly standard tool not only for researchers, but also for practitioners and decision makers. The scientific awareness of the existence of a unique biodiversity in the GDEs (and SGDEs) and the role it plays in terms of ecosystem services has not had any impact on the perception and cognition of this reality at the regulatory level; in this regard, AS acts as a tool for ensuring communication between scientists and end-users of environmental information. Being effective both for the calculation of conservation priorities and in terms of potential impact assessment, AS represents a profitable tool to set up appropriate policies to achieve concrete GDE improvement results.

Supplementary Materials: The following materials are available online at http://www.mdpi.com/2073-4441/11/12/2574/s1, Table S1: PNECs calculated for 116 pollutants; Table S2: PNECs calculated for 474 species and 116 pollutants; Table S3: Input data for the calculation of ER and HHR in six sites of the River Sagittario (example of application); Table S4: Species abundances for the calculation of the GBC in six sites of the River Sagittario (example of application); Table S5: Species scores for the calculation of the GBC in six sites of the River Sagittario (example of application).

Author Contributions: Conceptualization, D.M.P.G., G.S., S.F., B.F., and T.D.L.; methodology, D.M.P.G., S.F., and B.F.; software, G.S., W.L., and F.B.; validation, D.M.P.G., S.F., B.F., and M.D.C.; analysis, D.M.P.G. and S.F.; data curation, D.M.P.G. and B.F.; writing—original draft preparation, D.M.P.G., G.S., S.F., T.D.L., and B.F.; writing—review and editing, D.M.P.G., T.D.L., B.F., S.F., and M.D.C.; visualization, M.D.C.; supervision, D.M.P.G.; project administration, B.F.; and funding acquisition, D.M.P.G.

Funding: This research was funded by the European Community (LIFE12 BIO/IT/000231 AQUALIFE).

Acknowledgments: The Stygobiology Laboratory (University of L'Aquila, Department of Life, Health and Environmental Sciences) is greatly acknowledged for the field activity and sorting of the invertebrates during the AQUALIFE project.

Conflicts of Interest: The authors declare no conflict of interest.

References

1. Hatton, T.J.; Evans, R. *Dependence of Ecosystems on Groundwater and Its Significance to Australia*; CSIRO (Land and Water): Canberra, Australia, 1998.
2. Eamus, D.; Froend, R. Groundwater-dependent ecosystems: The where, what and why of GDEs. *Aust. J. Bot.* **2006**, *54*, 91–96. [CrossRef]
3. Gibert, J.; Deharveng, L. Subterranean ecosystems: A truncated functional biodiversity. *Bioscience* **2002**, *52*, 473–481. [CrossRef]
4. Humphreys, W.F. Aquifers: The ultimate groundwater-dependent ecosystems. *Aust. J. Bot.* **2006**, *54*, 115–132. [CrossRef]
5. Mammola, S.; Cardoso, P.; Culver, D.C.; Deharveng, L.; Ferreira, R.L.; Fiŝer, C.; Galassi, D.M.P.; Griebler, C.; Halse, S.; Humphreys, W.F.; et al. Scientists' Warning on the Conservation of Subterranean Ecosystems. *Bioscience* **2019**, *69*, 641–650. [CrossRef]
6. Ferreira, D.; Malard, F.; Dole-Olivier, M.-J.; Gibert, J. Obligate groundwater fauna of France: Diversity patterns and conservation implications. *Biodivers. Conserv.* **2007**, *16*, 567. [CrossRef]
7. Deharveng, L.; Stoch, F.; Gibert, J.; Bedos, A.; Galassi, D.; Zagmajster, M.; Brancelj, A.; Camacho, A.; Fiers, F.; Martin, P.; et al. Groundwater biodiversity in Europe. *Freshw. Biol.* **2009**, *54*, 709–726. [CrossRef]
8. Michel, G.; Malard, F.; Deharveng, L.; Di Lorenzo, T.; Sket, B.; De Broyer, C. Reserve selection for conserving groundwater biodiversity. *Freshw. Biol.* **2009**, *54*, 861–876. [CrossRef]
9. Galassi, D.M.P.; Fiasca, B.; Tosto, D. Patterns of Copepod diversity (Copepoda: Cyclopoida, Harpacticoida) in springs of central Italy: Implications for conservation issues. In *Studies on Freshwater Copepoda: A Volume in Honour of Bernard Dussart*; Brill: Buckinghamshire, UK, 2011; Volume 16, pp. 199–226.
10. European Commission. *Technical Report on Groundwater Associated Aquatic Ecosystems*; Technical Report No. 9; European Commission: Luxembourg, 2015.
11. Marmonier, P.; Maazouzi, C.; Baran, N.; Blanchet, S.; Ritter, A.; Saplairoles, M.; Dole-Olivier, M.J.; Galassi, D.M.P.; Eme, D.; Doledec, S.; et al. Ecology-based evaluation of groundwater ecosystems under intensive agriculture: A combination of community analysis and sentinel exposure. *Sci. Total Environ.* **2018**, *613*, 1353–1366. [CrossRef]
12. Jaffé, R.; Prous, X.; Zampaulo, R.; Giannini, T.C.; Imperatriz-Fonseca, V.L.; Maurity, C.; Oliveira, G.; Brandi, I.V.; Siqueira, J.O. Reconciling Mining with the Conservation of Cave Biodiversity: A Quantitative Baseline to Help Establish Conservation Priorities. *PLoS ONE* **2016**, *11*. [CrossRef]
13. EPA. *EPA's Draft Report on the Environment Technical Document*; EPA: Washington, DC, USA, 2003.
14. Tomlinson, M.; Boulton, A.J.; Hancock, P.J.; Cook, P.G. Deliberate omission or unfortunate oversight: Should stygofaunal surveys be included in routine groundwater monitoring programs? *Hydrogeol. J.* **2007**, *15*, 1317. [CrossRef]
15. Serov, P.; Kuginis, L.; Williams, J.P. *Risk Assessment Guidelines for Groundwater Dependent Ecosystems, Volume 1—The Conceptual Framework*; NSW Department of Primary Industries, Office of Water: Sydney, Australia, 2012.
16. Council of the European Communities. Directive 2000/60/EC of the European Parliament and the Council of 23 October 2000 establishing a framework for Community action in the field of water policy. *Off. J. Eur. Commun.* **2000**, *327*, 1–73.
17. Council of the European Communities. Directive 2006/118/EC of the European Parliament and of the Council of 12 December 2006 on the protection of groundwater against pollution and deterioration. *Off. J. Eur. Union* **2006**, *372*, 19–31.
18. *Millennium Ecosystem Assessment, Ecosystems and Human Well-being: Synthesis*; Island Press: Washington, DC, USA, 2005.
19. Kløve, B.; Ala-Aho, P.; Bertrand, G.; Boukalova, Z.; Erturk, A.; Goldscheider, N.; Ilmonen, J.; Karakaya, N.; Kupfersberger, H.; Kvaerner, J.; et al. Groundwater dependent ecosystems. Part I: Hydroecological status and trends. *Environ. Sci. Policy* **2011**, *14*, 770–781. [CrossRef]
20. Megdal, S.B.; Gerlak, A.K.; Varady, R.G.; Huang, L.Y. Groundwater Governance in the United States: Common Priorities and Challenges. *Groundwater* **2015**, *53*, 677–684. [CrossRef] [PubMed]

21. Saccò, M.; Blyth, A.; Bateman, P.W.; Hua, Q.; Mazumder, D.; Nicole, W.; Humphreys, W.F.; Laini, A.; Griebler, C.; Grice, K. New light in the dark—A proposed multidisciplinary framework for studying functional ecology of groundwater fauna. *Sci. Total Environ.* **2019**, *662*, 963–977. [CrossRef]

22. Taylor, R.G.; Scanlon, B.; Doll, P.; Rodell, M.; van Beek, R.; Wada, Y.; Longuevergne, L.; Leblanc, M.; Famiglietti, J.S.; Edmunds, M.; et al. Ground water and climate change. *Nat. Clim. Chang.* **2013**, *3*, 322–329. [CrossRef]

23. Phyton Software Foundation (US). Available online: https://www.python.org/ (accessed on 3 October 2019).

24. Django Project. Available online: https://www.djangoproject.com (accessed on 3 October 2019).

25. Avramov, M.; Schmidt, S.I.; Griebler, C. A new bioassay for the ecotoxicological testing of VOCs on groundwater invertebrates and the effects of toluene on *Niphargus inopinatus*. *Aquat. Toxicol.* **2013**, *130*, 1–8. [CrossRef]

26. Di Lorenzo, T.; Borgoni, R.; Ambrosini, R.; Cifoni, M.; Galassi, D.M.P.; Petitta, M. Occurrence of volatile organic compounds in shallow alluvial aquifers of a Mediterranean region: Baseline scenario and ecological implications. *Sci. Total Environ.* **2015**, *538*, 712–723. [CrossRef]

27. Di Lorenzo, T.; Castano-Sanchez, A.; Di Marzio, W.D.; Garcia-Doncel, P.; Nozal Martinez, L.; Galassi, D.M.P.; Iepure, S. The role of freshwater copepods in the environmental risk assessment of caffeine and propranolol mixtures in the surface water bodies of Spain. *Chemosphere* **2019**, *220*, 227–236. [CrossRef]

28. Di Lorenzo, T.; Di Marzio, W.D.; Fiasca, B.; Galassi, D.M.P.; Korbel, K.; Iepure, S.; Pereira, J.L.; Reboleira, A.; Schmidt, S.I.; Hose, G.C. Recommendations for ecotoxicity testing with stygobiotic species in the framework of groundwater environmental risk assessment. *Sci. Total Environ.* **2019**, *681*, 292–304. [CrossRef]

29. Calamari, D.; Vighi, M. A proposal to define quality objectives for aquatic life for mixtures of chemical-substances. *Chemosphere* **1992**, *25*, 531–542. [CrossRef]

30. Vighi, M.; Altenburger, R.; Arrhenius, Å.; Backhaus, T.; Bödeker, W.; Blanck, H.; Consolaro, F.; Faust, M.; Finizio, A.; Froehner, K.; et al. Water quality objectives for mixtures of toxic chemicals: Problems and perspectives. *Ecotoxicol. Environ. Saf.* **2003**, *54*, 139–150. [CrossRef]

31. Di Lorenzo, T.; Cifoni, M.; Fiasca, B.; Di Cioccio, A.; Galassi, D.M.P. Ecological risk assessment of pesticide mixtures in the alluvial aquifers of central Italy: Toward more realistic scenarios for risk mitigation. *Sci. Total Environ.* **2018**, *644*, 161–172. [CrossRef] [PubMed]

32. Pennington, D. Extrapolating ecotoxicological measures from small data sets. *Ecotoxicol. Environ. Saf.* **2003**, *56*, 238–250. [CrossRef]

33. Backhaus, T.; Faust, M. Predictive Environmental Risk Assessment of Chemical Mixtures: A Conceptual Framework. *Environ. Sci. Technol.* **2012**, *46*, 2564–2573. [CrossRef]

34. Gustavsson, M.; Kreuger, J.; Bundschuh, M.; Backhaus, T. Pesticide mixtures in the Swedish streams: Environmental risks, contributions of individual compounds and consequences of single-substance oriented risk mitigation. *Sci. Total Environ.* **2017**, *598*, 973–983. [CrossRef]

35. ECHA. *Chapter, R.10: Characterisation of Dose [Concentration]-Response for Environment. Guidance for the Implementation of REACH—Guidance on Information Requirements and Chemical Safety Assessment*; European Chemical Agency: Helskinki, Finland, 2008.

36. European Commission. *Technical Guidance for Deriving Environmental Quality Standard*; European Commission: Brussels, Belgium, 2018.

37. ECOTOX Database. Available online: https://cfpub.epa.gov/ecotox/ (accessed on 3 October 2019).

38. ECOSAR Software. Available online: http://www.epa.gov/oppt/newchems/tools/21ecosar.htm (accessed on 3 October 2019).

39. Galassi, D.M.P.; Fiasca, B.; Di Lorenzo, T.; Montanari, A.; Porfirio, S.; Fattorini, S. Groundwater biodiversity in a chemoautotrophic cave ecosystem: How geochemistry regulates microcrustacean community structure. *Aquat. Ecol.* **2017**, *51*, 75–90. [CrossRef]

40. Rinaldi, M.; Surian, N.; Comiti, F.; Bussettini, M. *IDRAIM—Sistema di Valutazione Idromorfologica, Analisi E Monitoraggio dei Corsi d'acqua—Versione Aggiornata 2016*; ISPRA: Rome, Italy, 2016.

41. Brunner, P.; Cook, P.G.; Simmons, C.T. Disconnected Surface Water and Groundwater: From Theory to Practice. *Ground Water* **2011**, *49*, 460–467. [CrossRef]

42. Siergieiev, D.; Ehlert, L.; Reimann, T.; Lundberg, A.; Liedl, R. Modelling hyporheic processes for regulated rivers under transient hydrological and hydrogeological conditions. *Hydrol. Earth Syst. Sci.* **2015**, *19*, 329–340. [CrossRef]

43. Hancock, P.J. The response of hyporheic invertebrate communities to a large flood in the Hunter River, New South Wales. *Hydrobiologia* **2006**, *568*, 255–262. [CrossRef]

44. Arroita, M.; Flores, L.; Larranaga, A.; Martinez, A.; Martinez-Santos, M.; Pereda, O.; Ruiz-Romera, E.; Solagaistua, L.; Elosegi, A. Water abstraction impacts stream ecosystem functioning via wetted-channel contraction. *Freshw. Biol.* **2017**, *62*, 243–257. [CrossRef]

45. Liu, D.S.; Zhao, J.; Chen, X.B.; Li, Y.Y.; Weiyan, S.P.; Feng, M.M. Dynamic processes of hyporheic exchange and temperature distribution in the riparian zone in response to dam-induced water fluctuations. *Geosci. J.* **2018**, *22*, 465–475. [CrossRef]

46. Mathers, K.L.; Hill, M.J.; Wood, P.J. Benthic and hyporheic macroinvertebrate distribution within the heads and tails of riffles during baseflow conditions. *Hydrobiologia* **2017**, *794*, 17–30. [CrossRef]

47. Sawyer, A.H.; Cardenas, M.B.; Bomar, A.; Mackey, M. Impact of dam operations on hyporheic exchange in the riparian zone of a regulated river. *Hydrol. Process.* **2009**, *23*, 2129–2137. [CrossRef]

48. Boano, F.J.; Harvey, W.; Marion, A.; Packman, A.I.; Revelli, R.; Ridolfi, L.; Wörman, A. Hyporheic flow and transport processes: Mechanisms, models, and biogeochemical implications. *Rev. Geophys.* **2014**, *52*, 603–679. [CrossRef]

49. Magliozzi, C.; Grabowski, R.; Packman, A.; Krause, S. Scaling down hyporheic exchange flows: From catchments to reaches. *Hydrol. Earth Syst. Sci. Discuss.* **2017**. [CrossRef]

50. European Commission. *WFD Reporting Guidance 2016. Final Draft 6.0.6*; European Commission: Brussels, Belgium, 2016.

51. Loheide, S.P.; Booth, E.G. Effects of changing channel morphology on vegetation, groundwater, and soil moisture regimes in groundwater-dependent ecosystems. *Geomorphology* **2011**, *126*, 364–376. [CrossRef]

52. Pacioglu, O.; Shaw, P.; Robertson, A. Patch scale response of hyporheic invertebrates to fine sediment removal in two chalk rivers. *Fundam. Appl. Limnol.* **2012**, *181*, 283–288. [CrossRef]

53. Jones, I.; Growns, I.; Arnold, A.; McCall, S.; Bowes, M. The effects of increased flow and fine sediment on hyporheic invertebrates and nutrients in stream mesocosms. *Freshw. Biol.* **2015**, *60*, 813–826. [CrossRef]

54. Pacioglu, O.; Robertson, A. The invertebrate community of the chalk stream hyporheic zone: Spatio-temporal distribution patterns. *Knowl. Manag. Aquat. Ecosyst.* **2017**, *418*, 10. [CrossRef]

55. Hancock, P.J. Human impacts on the stream-groundwater exchange zone. *Environ. Manag.* **2002**, *29*, 763–781. [CrossRef] [PubMed]

56. Merill, L.; Tonjes, D.J. A Review of the Hyporheic Zone, Stream Restoration, and Means to Enhance Denitrification. *Crit. Rev. Sci. Technol.* **2014**, *44*, 2337–2379. [CrossRef]

57. Wondzell, S.M. Effect of morphology and discharge on hyporheic exchange flows in two small streams in the Cascade Mountains of Oregon, USA. *Hydrol. Process.* **2006**, *20*, 267–287. [CrossRef]

58. Fumetti von, S.; Dmitrovic, D.; Pesic, V. The influence of flooding and river connectivity on macroinvertebrate assemblages in rheocrene springs along a third-order river. *Fundam. Appl. Limnol.* **2017**, *190*, 251–263. [CrossRef]

59. Pesic, V.; Dmitrovic, D.; Savic, A.; Milosevic, D.; Zawal, A.; Vukasinovic-Pesic, V.; Fumetti von, S. Application of macroinvertebrate multimetrics as a measure of the impact of anthropogenic modification of spring habitats. *Aquat. Conserv. Mar. Freshw. Ecosyst.* **2019**, *29*, 341–352. [CrossRef]

60. Salameh, E. Over-exploitation of groundwater resources and their environmental and socio-economic implications: The case of Jordan. *Water Int.* **2008**, *33*, 55–68. [CrossRef]

61. Tal, A.; Katz, D. Rehabilitating Israel's streams and rivers. *Int. J. River Basin Manag.* **2012**, *10*, 317–330. [CrossRef]

62. Siwek, J.; Pociask-Karteczka, J. Springs in South-Central Poland—Changes and threats. *Episodes* **2017**, *40*, 38–46. [CrossRef]

63. Dwire, K.A.; Mellmann-Brown, S.; Gurrieri, J.T. Potential effects of climate change on riparian areas, wetlands, and groundwater-dependent ecosystems in the Blue Mountains, Oregon, USA. *Clim. Serv.* **2018**, *10*, 44–52. [CrossRef]

64. Sabater, F.; Butturini, A.; Mart, E.; Oz, I.; Romaní, A.; Wray, J.; Sabater, S. Effect of Riparian Vegetation Removal on Nutrient Retention in a Mediterranean Stream. *J. N. Am. Benthol. Soc.* **2000**, *19*, 609–620. [CrossRef]

65. Possingham, H.P.; Wilson, K.A. Biodiversity—Turning up the heat on hotspots. *Nature* **2005**, *436*, 919–920. [CrossRef] [PubMed]

66. Funk, S.M.; Fa, J.E. Ecoregion Prioritization Suggests an Armoury Not a Silver Bullet for Conservation Planning. *PLoS ONE* **2010**, *5*, e8923. [CrossRef] [PubMed]

67. Cardoso, P.; Erwin, T.L.; Borges, P.A.V.; New, T.R. The seven impediments in invertebrate conservation and how to overcome them. *Biol. Conserv.* **2011**, *144*, 2647–2655. [CrossRef]

68. Fattorini, S. Use of insect rarity for biotope prioritisation: The tenebrionid beetles of the Central Apennines (Italy). *J. Insect Conserv.* **2010**, *14*, 367–378. [CrossRef]

69. Fattorini, S. Biotope prioritisation in the Central Apennines (Italy): Species rarity and cross-taxon congruence. *Biodivers. Conserv.* **2010**, *19*, 3413–3429. [CrossRef]

70. Fattorini, S.; Cardoso, P.; Rigal, F.; Borges, P.A.V. Use of Arthropod Rarity for Area Prioritisation: Insights from the Azorean Islands. *PLoS ONE* **2012**, *7*. [CrossRef] [PubMed]

71. Kattan, G.H. Rarity and vulnerability: The birds of the Cordillera Central of Colombia. *Conserv. Biol.* **1992**, *6*, 64–70. [CrossRef]

72. Rabinowitz, D. Seven forms of rarity. In *The Biological Aspects of Rare Plant Conservation*; Synge, H., Ed.; Sinauer Associates: Chichester, UK, 1981.

73. Rabinowitz, D.; Cairns, S.; Dillon, T. Seven forms of rarity and their frequency in the flora of the British Isles. In *Conservation Biology: The Science of Scarcity and Diversity*; Soulè, M.E., Ed.; British Ecological Society: London, UK, 1986; pp. 182–204.

74. Fattorini, S.; Fiasca, B.; Di Lorenzo, T.; Di Cicco, M.; Galassi, D.M.P. A new protocol for assessing the conservation priority of groundwater dependent ecosystems. *Aquat. Conserv. Mar. Freshw. Ecosyst.* submitted for publication.

75. Caschetto, M.; Galassi, D.M.P.; Petitta, M.; Aravena, R. Evaluation of the sources of nitrogen compounds and their influence on the biological communities in the hyporheic zone of the Sagittario River, Italy: An isotopic and biological approach. *Ital. J. Geosci.* **2017**, *136*, 145–156. [CrossRef]

76. Di Lorenzo, T.; Cannicci, S.; Spigoli, D.; Cifoni, M.; Baratti, M.; Galassi, D.M.P. Bioenergetic cost of living in polluted freshwater bodies: Respiration rates of the cyclopoid *Eucyclops serrulatus* under ammonia-N exposures. *Fundam. Appl. Limnol.* **2016**, *188*, 147–156. [CrossRef]

77. Di Lorenzo, T.; Cifoni, M.; Lombardo, P.; Fiasca, B.; Galassi, D.M.P. Ammonium threshold values for groundwater quality in the EU may not protect groundwater fauna: Evidence from an alluvial aquifer in Italy. *Hydrobiologia* **2015**, *743*, 139–150. [CrossRef]

78. Scharf, B.; Brunke, M. The recolonization of the river Elbe with benthic and hyporheic Ostracoda (Crustacea) after the reunion of Germany in 1989. *Int. Rev. Hydrobiol.* **2013**, *98*. [CrossRef]

79. Hahn, H.J. The GW-Fauna-Index: A first approach to a quantitative ecological assessment of groundwater habitats. *Limnologica* **2006**, *36*, 119–137. [CrossRef]

80. Fillinger, L.; Hug, K.; Trimbach, A.M.; Wang, H.; Kellermann, C.; Meyer, A.; Bendinger, B.; Griebler, C. The D-A-(C) index: A practical approach towards the microbiological-ecological monitoring of groundwater ecosystems. *Water Res.* **2019**, *163*, 114902. [CrossRef] [PubMed]

81. Korbel, K.; Hose, G.C. The weighted groundwater health index: Improving the monitoring and management of groundwater resources. *Ecol. Indic.* **2017**, *75*, 164–181. [CrossRef]

82. Korbel, K.L.; Hose, G.C. A tiered framework for assessing groundwater ecosystem health. *Hydrobiologia* **2011**, *661*, 329–349. [CrossRef]

MDPI
St. Alban-Anlage 66
4052 Basel
Switzerland
Tel. +41 61 683 77 34
Fax +41 61 302 89 18
www.mdpi.com

Water Editorial Office
E-mail: water@mdpi.com
www.mdpi.com/journal/water